CLIMATE CHANGE AND MICROBIAL DIVERSITY

Advances and Challenges

CLIMATE CHANGE AND MICROBIAL DIVERSITY

Advances and Challenges

Edited by
Suhaib A. Bandh, PhD
Javid A. Parray, PhD
Nowsheen Shameem, PhD

AAP APPLE ACADEMIC PRESS

First edition published 2023

Apple Academic Press Inc.
1265 Goldenrod Circle, NE,
Palm Bay, FL 32905 USA

4164 Lakeshore Road, Burlington,
ON, L7L 1A4 Canada

CRC Press
6000 Broken Sound Parkway NW,
Suite 300, Boca Raton, FL 33487-2742 USA

4 Park Square, Milton Park,
Abingdon, Oxon, OX14 4RN UK

© 2023 by Apple Academic Press, Inc.

Apple Academic Press exclusively co-publishes with CRC Press, an imprint of Taylor & Francis Group, LLC

Library and Archives Canada Cataloguing in Publication

Title: Climate change and microbial diversity: advances and challenges / edited by Suhaib A. Bandh, PhD, Javid A. Parray, PhD, Nowsheen Shameem, PhD.
Names: Bandh, Suhaib A., editor. | Parray, Javid Ahmad. editor. | Shameem, Nowsheen, editor.
Description: First edition. | Includes bibliographical references and index.
Identifiers: Canadiana (print) 20220174954 | Canadiana (ebook) 20220174997 | ISBN 9781774637821 (hardcover) | ISBN 9781774637838 (softcover) | ISBN 9781003302810 (ebook)
Subjects: LCSH: Microorganisms—Climatic factors. | LCSH: Microbial ecology. | LCSH: Climatic changes. | LCSH: Sustainable agriculture.
Classification: LCC QR100 .C55 2022 | DDC 579/.17—dc23

Library of Congress Cataloging-in-Publication Data

Names: Parray, Javid Ahmad, editor. | Bandh, Suhaib A., editor. | Shameem, Nowsheen, editor.
Title: Climate change and microbial diversity : advances and challenges / edited by Suhaib A. Bandh, PhD, Javid A. Parray, PhD, Nowsheen Shameem, PhD.
Description: First edition. | Palm Bay, FL : AAP/Apple Academic Press, [2022] | Includes bibliographical references and index.
Identifiers: LCCN 2022008017 (print) | LCCN 2022008018 (ebook) | ISBN 9781774637821 (hbk) | ISBN 9781774637838 (pbk) | ISBN 9781003302810 (ebk)
Subjects: LCSH: Microorganisms--Climatic factors. | Microbial ecology. | Climatic changes.
Classification: LCC QR100 .C54 2022 (print) | LCC QR100 (ebook) | DDC 579/.17--dc23/eng/20220325
LC record available at https://lccn.loc.gov/2022008017
LC ebook record available at https://lccn.loc.gov/2022008018

ISBN: 978-1-77463-782-1 (hbk)
ISBN: 978-1-77463-783-8 (pbk)
ISBN: 978-1-00330-281-0 (ebk)

About the Editors

Suhaib A. Bandh, PhD

Suhaib A. Bandh, PhD, is an Assistant Professor in the Higher Education Department of Jammu and Kashmir at Government Degree College, D.H. Pora, Kulgam, India, where he teaches courses on environmental science and disaster management at the graduate level. He is the President of the Academy of EcoScience in addition to being a life member of the Academy of Plant Sciences, India, and the National Environmental Science Academy. He is a recipient of many awards and has participated in a number of national and international conferences organized by reputed scientific bodies in India and abroad. He has a number of scientific publications to his credit, published in some highly reputed and impacted journals. Dr. Bandh has recently edited and authored a number of books, including *Freshwater Microbiology: Perspectives of Bacterial Dynamics in Lake Ecosystems; Environmental Management: Environmental Issues, Awareness and Abatement,* and *Environmental Perspectives and Issues.* Dr. Bandh is an academic editor of a number of internationally reputed Hindawi journals.

Javid A. Parray, PhD

Javid A. Parray, PhD, is an Assistant Professor in the Higher Education Department at Government Degree College Eidgah, Srinagar, India, where he teaches the subject of environmental science. He has published many research articles in reputable refereed international and national journals. He had authored books on different themes, including *Approaches to Heavy Metal Tolerance in Plants; Sustainable Agriculture: Biotechniques in Plant Biology;* and *Sustainable Agriculture: Advances in Plant Metabolome and Microbiome.* He was awarded the "Emerging Scientist of Year Award" by the Indian Academy of Environmental Science for the year 2015 in addition to being awarded with an international travel grant for participating in international conferences. Dr. Parray completed a fast-track project entitled "Molecular characterization and metabolic potential of rhizospheric bacteria from *Arnebia benthamii* across North

Western Himalaya" at CORD, University of Kashmir, India. He finished his postdoctorate research associateship on a DBT-funded project entitled "Tissue culture-based network programme on saffron." He earned a PhD in Environmental Science with a specialization in plant microbe interaction on the topic "Evaluating role of Rhizospheric bacteria in saffron culture" from the University of Kashmir, India.

Nowsheen Shameem, PhD

Nowsheen Shameem, PhD, is an Assistant Professor in the Department of Environmental Science at Cluster University Srinagar, India. She has worked as a project associate on the DBT-sanctioned project on "Spawn production for the entrepreneurs of Kashmir Valley" at CORD, University of Kashmir, India. She has also worked as a group leader for drafting a number of research proposals and ideas. She has published many research articles in reputed, refereed international and national journals of Springer, Elsevier, and Hindawi. She has also presented her research work at international and national conferences. Dr. Shameem finished her doctorate through the University of Kashmir on "Phytochemical analysis and nutraceutical value of some wild mushrooms growing in Kashmir Valley" in 2017.

Contents

Contributors

Sheikh Firdous Ahmad
Scientist, ICAR-National Research Centre on Pig, Rani, Guwahati, Assam, India

Mahajabeen Akhter
Department of Bioresources, University of Kashmir, Srinagar, Jammu & Kashmir, India

Tawseef Rehman Baba
Division of Fruit Science, SKUAST-K, Shalimar, Jammu and Kashmir, India

Bharat Bhushan
Professor and Principal Scientist, Division of Animal Genetics,
ICAR-Indian Veterinary Research Institute, Izzatnagar, Bareilly 243122, UP, India

Manishankar Chakraborty
Department of Biotechnology, Assam University Silchar, India

Waqas Mohy UD Din
Institute of Soil and Environmental Science, University of Agriculture, Faisalabad 38040, Pakistan

Triveni Dutt
Joint Director (Academic), ICAR-Indian Veterinary Research Institute, Izzatnagar,
Bareilly 243122, UP, India

Zia UR Rahman Farooqi
Institute of Soil and Environmental Science, University of Agriculture, Faisalabad 38040, Pakistan

Huma Habib
Islamia College of Science and Commerce, Srinagar, India

Anita V. Handore
P. G. Department of Microbiology, HPT Arts and RYK Science College, Nashik, India

D. V. Handore
Research and Development Division, Sigma Winery Pvt. Ltd. Sinnar, Nashik, India

Musheerul Hassan
Clybay Private Limited, Bangalore, India

Shiekh Marifatul Haq
University of Kashmir, Srinagar, India

Muhammad Mahroz Hussain
Institute of Soil and Environmental Science, University of Agriculture, Faisalabad 38040, Pakistan

Rajib Karmakar
Department of Agricultural Chemicals, Bidhan Chandra Krishi Viswavidyalaya,
Directorate of Research, Research Complex Building, Nadia, West Bengal, India

Nafeesa Farooq Khan
Biological Invasion Lab, University of Kashmir, Jammu and Kashmir, India

S. R. Khandelwal
H. A. L. College of Science & Commerce, Nashik, India

Irshad A. Lone
Department of Chemistry, Govt. Degree College Handwara, Higher Education Department,
Jammu & Kashmir, India

P. B. Mazumder
Department of Biotechnology, Assam University Silchar, India

Debayan Nandi
Department of Biotechnology, Assam University Silchar, India

Gulzar Ahmed Rather
Sathyabama Institute of Science and Technology, Chennai, India

Taniya Sengupta Rathore
Faculty of Life Science, Mandsaur University, Mandsaur, Madhya Pradesh, India

Zafar A Reshi
Biological Invasion Lab, University of Kashmir, Jammu and Kashmir, India

Lakkakula Satish
Department of Biotechnology Engineering, Ben-Gurion University of the Negev, Beer Sheva, Israel

Amrina Shafi
Education Department, Government of Jammu & Kashmir, Srinagar, India

Manzoor Ahmad Shah
Biological Invasion Lab, University of Kashmir, Jammu and Kashmir, India

Khurshid Ahmad Tariq
Department of Zoology, Islamia College of Science & Commerce (UGC Autonomous), Srinagar,
Jammu & Kashmir, India

Udaya Kumar Vandana
Department of Biotechnology, Assam University Silchar, India

Uzma Zehra
Biological Invasion Lab, University of Kashmir, Jammu and Kashmir, India

Abbreviations

AGBD	aboveground biomass density
AGP	aboveground phytomass
BEF	biomass expansion factor
BGBD	belowground biomass density
CO	carbon monoxide
CO_2	carbon dioxide
CUE	carbon utilization efficiency
CT	cutting
DBH	diameter at breast height
DG	degradation
ESI	electrospray ionization
FR	forest fire
GSVD	growing stock volume density
GZ	grazing
HMT	heavy metal tolerant
IPCC	Intergovernmental Panel on Climate Change
IVI	importance value index
LD	low-disturbed
LOC	line of control
LP	lopping
MALDI	matrix-assisted laser desorption/ionization
NE	northeast
NW	northwest
OC	organic carbon
OM	organic matter
PCA	principal component analysis
PGP	plant-growth-promotion
PHC	petroleum hydrocarbon
RC	road connectivity
SE	southeast
SSB	single-stranded binding proteins

SW   southwest
TBD   total biomass density
TCD   total carbon density
TPC   triphenyl tetrazolium chloride
VFAs   volatile fatty acids

Preface

Modernization has forced mankind and other life forms into a number of problems. Poverty, climate change, inequality, and destruction of the atmosphere are some of the global issues that are shared. The United Nation's SDGs (sustainable development goals) are a blueprint for an improved and supportable proposal for all. The SDGs are designed to address the worldwide problems such as poverty, injustice, hunger, horrible change of climate, destruction of the environment, peace, and justice confronting us. Several reports suggest that nothing has been done and the SDGs would be unacceptable if a new inclusion plan is not put in place. The ongoing global climate change triggered by greenhouse gas growth constitutes one of twenty-first century's main scientific and political challenges. According to available information, the carbon dioxide buildup in the atmosphere is currently occurring in the atmosphere at an unprecedented rate.

The temperature-based microbial dynamics and pCO_2 are still not completely understood. Soil microbiologists play a vital role in the issue of global warming not only with regard to carbon dioxide, but also in relation to nitrous oxide and particularly methane. They should stress the services that the safe soil microbiota provides to society. It is well known that the production and removal of these gases are involved in microbial ecology or environmental microbiology. Rumen microorganisms are responsible for almost 30% of methane emissions, and a new approach to the shortening of rumen methanogens is underway by offering alternative electron receivers such as herbal components. However, the role of methanotrophs in soil for the reduction of methane cannot be ignored. On an average, they scavenge between 800 and 1000 kg of methane per hectare per year.

In the global fluxes of key greenhouse gases like methane, carbon dioxide, and nitrous oxide, the microbial processes play a key role, and these processes likely respond to any change in these gases including the dynamic interactions that occur between microorganisms and other abiotic and biotic factors. The climate change mitigation ability through the management of terrestrial microbial processes by cutting greenhouse gas emissions is an attractive future prospect. Microorganisms are generally acknowledged to have played a significant role in assessing greenhouse gas atmospheric concentrations. This kind of environmental issue is solved by the main reaction mechanism for climate change by the change of its microbial population structure and composition. Simply by using cycling methods for nutrients and stimulating their genetic functional content to degrade and remove contaminants or gases that result in global warming. Synergistically, the biogeochemical cycles and microbial communities act as decent climate change resolution mechanism. The use of greenhouse gases as energy sources and building up their cells is very important to microorganisms.

Indian climate change research and feedback on measuring biogeo-chemical processes and defining the source–decrease relationship used the information to construct predictive climate models. Extensive studies have been performed on the process of microbial function and related microbial diversity. However, our understanding of the microbial climate change response has remained limited in expert opinion, and our knowledge of climate change and input from terrestrial microorganisms must be improved. Knowledge on microbial communities' structure and biological patterns, including functional interactions between microor-ganisms and plant communities, must be created as a matter of urgency.

CHAPTER 1

Microbial Responses Under Climate Change Scenarios: Adaptation and Mitigations

ZIA UR RAHMAN FAROOQI, WAQAS MOHY UD DIN, and
MUHAMMAD MAHROZ HUSSAIN*

Institute of Soil and Environmental Science, University of Agriculture, Faisalabad 38040, Pakistan

Corresponding author. E-mail: hmahroz@gmail.com

ABSTRACT

Climate change is a disastrous change in the current weather conditions at a larger scale which alters the temperature and suitable conditions for all living organisms including microbes, the tiny creatures which are found nearly everywhere on the Earth. Soil microbes are also part of this regime, which exist at different temperatures and conditions. They require an optimum temperature range to survive, which when altered, disturbs the community and results in an imbalance in the soil ecology and its processes. However, soil microbes possess some response and adaptation mechanisms to survive the harsh environmental conditions. All these mechanisms with future research needs are explained in detail in this chapter.

1.1 INTRODUCTION

Anthropogenic activities and their impact on the environment have led to the extinction of a long range of animals and plants on Earth. But extensive research and documentation have been carried out vis-à-vis the floral and faunal species, society, and habitat losses. In contrast, microbes have not been debated that extensively in the perspectives of climate change (especially the climate change impact of microbes). Although invisible to the naked eye, the density and diversity of bacteria and other microbes play a key role in retaining a healthy worldwide ecosystem. In short, microbial ecosystems are the lifeblood of the planet Earth. Though the impacts on human microorganisms are less evident and, of course, not very typical, the main problem is that alterations in biodiversity and microbial activity involve the compatibility of all other creatures and thus climate change.

Microorganisms play an important role in the nutrient networks, living conditions, agriculture and worldwide food networks. Microbes exist in almost any environment on Earth as, for example, in deep terrain and severe weather conditions they are the only existent life forms. They were observed on Earth about 3.8 billion years ago and are expected to survive beyond the future. Though microorganisms play an important role in climate change management, no much attention has been given to microbial climate change research and strategy development. The extremely diverse molecular responses of their ecological variables complicate the description of their role in this ecosystem. Microbial communities are flexible formations that adapt quickly to changing environmental conditions. In the experimental forests, the sensitivity of soil microbial respiration was reduced due to experimental soil movement in mobile areas (Bradford et al., 2008), leading to a decline in ecosystem level in a few years (Conant et al., 2011).

Recent research has shown that cooler regions are more sensitive to soil mineralization rates than warmer regions (Dessureault-Rompré et al., 2010). In marine systems, planktonic bacterial populations show maximum nutrient use effectiveness when they come closer to the in situ temperature (Hall et al., 2009), while planktonic bacteria correlate negatively and the effective phosphorus content is reduced (Hall et al., 2011). Microbial communities present in marine environments have exhibited indication that functions with lower optimal temperature values in higher latitudes adapt to local ambient temperatures (Simon et al., 1999). It has also been noted that microbial communities adjust to instabilities in the redox state (DeAngelis et al., 2010) and adapt to changing rainfall patterns (Szukics et al., 2010; Evans and Wallenstein, 2012). These illustrations show that microbial variation is a universal phenomenon that happens in different environments and can greatly affect the function of ecosystems in the upcoming climate system (Allison et al., 2010).

The term "adaptation" is habitually used at the level of organisms or populations and is a general term that describes the adaptability of organisms to the environment (Rose et al., 1996). Though it is frequently used to define certain processes, such as changing gene frequencies on the population level, the period itself does not indicate a particular method (Hoegh-Guldberg and Bruno, 2010). Bradford et al. (2008) suggest that it can also be used at the community level to explain alterations in the overall functioning of microbial communities in response to changing climate. The variations seen at the community level are due to changes in the comparative characteristics of the microbial population as a whole. We test the relevance of microbial communities to the process by which the degree of observation of the characteristics of society is determined by the current environmental conditions. Because microbial communities are compatible with the local environment, microbial variation is frequently not achieved at all. For instance, several experiments found no indication of thermal variation in thermal experimentations (Hartley et al., 2008; Rinnan et al., 2009).

In oceans, microbial communities from cold climates usually exhibit optimal growth above in situ temperatures (Johnson et al., 2006) implying that the microbial adaptation is largely limited. This may be due to rate constraints (i.e., time delay or delay between changes in environmental factors and the physiology of the normal microphone community) or a major physical trade-off that is not fully integrated. This chapter briefly explained the classification of microbes and their adaption under different climate change scenarios.

1.2   CLASSIFICATION OF MICROORGANISMS

Microorganisms, on the basis of their functions, are of three types: reducers, neutralizers, and decomposers. All these types of microbes are extremely profitable although the third type is more beneficial. They act as synergists and enhance the biological, chemical and physical properties of soil, water, and sediments. The recovery type can make the neutral type useful. Both of these types of microorganisms breakdown organic matter to obtain increased crops and reduced pollution (Zhou et al., 2009). Beneficial microorganisms include the ecology of microorganisms (including microorganisms, bacteria, fungi, and viruses). They can retain ecological balance; prevent the spread of pests, and decay of harmful chemicals into the environment, such as *Bacillus thuringiensis* in nature.

Russian scientist and evolutionary leader Eli Meteknikov first proposed some bacteria. He indicated that in the early 20th century it was feasible to replace infectious plants with beneficial microorganisms. Collat first coined the term "probiotic" in 1953. Probiotic is a technical term in microbiology, which means that microbes and their metabolism contribute to the balance of the full microbes in animal intestines. Probiotics are highly antibiotic, especially preventable. Probiotics usually involve bacteria, cyanobacteria, microalgae, and fungi, but microalgae are not included in English literature (Zhou et al., 2009). Today, the term "microbiological medicine" for human health is very popular in the food and livestock industries of China. There are three types of microbiological agents: probiotics, prebiotics, and synbiotics. Though there are many microbial agents, they mainly consist of bacilli, lactobacilli, bifidobacteria, bacteroids, and yeast (Fu et al., 2005). Current research focuses on probiotics and prebiotics. Probiotics are direct microbial foods that have been used for some time and are found in most foods, mainly milk (Mussatto and Mancilha, 2007; Kesarcodi-Watson et al., 2008; McCue, 2010).

Microorganisms are important to the environment because they play a role in the Earth's original cycle, carbon and nitrogen. Microbes are involved in the "purity" of the soil from the production of oxygen, the control of biomass and dead organic matter. Microorganisms include microalgae, viruses, fungi, and viruses. Microalgae absorb carbon dioxide mainly through photosynthesis and oxygen supplied to marine animals. The main function of bacteria and fungi is to get rid of dirty substances in the air, thereby keeping the water quality clean. In fact, because of the lack of oxygen, the unbalance can cause anaerobic digestion or washing, producing harmful gases such as hydrogen sulfide and methane. There are two food chains in the aquatic ecosystem: the detritus chain and the first food chain. In most marine environments, the smaller organisms are the larger organisms in the picture. Aerobic conditions (e.g., wastewater) are dominated by algae, diatoms, and cyanobacteria. Some photosynthetic bacteria settle under anaerobic conditions (contaminated or eutrophic water). There is a huge difference between the water and the environment, especially in the case of deep water and the fishing economy and large-scale planting. The association of microbial ecosystems, global and agricultural systems and their impact on climate change are listed below.

1.3 ROLE OF TERRESTRIAL MICROBES AND CLIMATE CHANGE IMPACTS

Terrestrial biomass is ten times larger than sea level biomass, and the plants on the terrestrial system makeup a huge portion of the world's biomass. The whole net product is 30,000 plants. The Earth contains five tons more carbon, than all the fossil fuels in the environment. Soil microorganisms limit the amount of carbon collected into the soil and release it into the air and indirectly store carbon in nutrients and soil by providing nitrogen and nitrogen to monitor reproduction (Ju et al., 2017). Plants provide high levels of carbon to mycorrhizal fungi. In many ecosystems, mycorrhizal fungi cause plants to absorb large amounts of nitrogen and phosphorus. Plants remove carbon dioxide from the environment through photosynthesis and produce organic substances to stimulate the Earth's atmosphere. In contrast, autotrophic respiratory plants and various concentration viruses produce CO_2 in the environment (Yazdanpanah et al., 2016). Temperature affects the balance between processes, which in turn affects the recording and storage (now a quarter) of anthropogenic carbon in the Earth's atmosphere. The release of carbon into the warm environment is expected to accelerate. The first 10 cm of soil is estimated to be 143 cm, and a total of 100 cm of soil structure (including old carbon 144) indicates high carbon loss in warming areas. Large amounts of nutrients are needed to account for the variations in carbon variability among different soil layers (excluding organic matter, temperature, precipitation, pH, and content) (Reinhart et al., 2016). In fact, global warming from a warming response shows that global warming from warming causes positive changes and increases the amount of climate change, especially in warm and warm climates soil. Most of the carbon is collected in the soil. Climate change affects the composition and diversity of infectious environments (e.g., climate and temperature) or indirectly (e.g., farmland, planting, and root system). Soil variability affects crop diversity and is important for the functioning of soil conditions (including carbon cycling).

Climate change has a direct or indirect effect on the microbial community, which affects warming, rainfall, soil quality, and crop abundance. Due to the low carbon content of desert soil microorganisms, enhanced carbon uptake from plants stimulates the use of nitrogen complexes, microbial biomass, variety (such as fungal diversity), enzymatic activity, and often organic matter. These changes improve the soil fertility and carbon loss, but preferentially vacuum and semi-arid areas act as carbon sinks. A clearer sense of the reaction of atmospheric biomass to CO_2 levels and the rainy season requires a better understanding of the response and function of infected areas. The magnitude of the rate of microbial activity is a portion of how the organic matter of biomass is decomposed. Higher performance means additional carbon is discharged into the atmosphere. Laboratory studies show that increasing temperature increases virus, but virus growth does not change, and temperature is expected to promote carbon accumulation in soil (Hicks et al., 2019).

An 18-year-old field study found that as soil temperature increased, viral load decreased, hard-matrix dry heads grew over time, and carbon dioxide was lost. Within 6 months of exposure to elevated growth temperatures, Antarctic Peninsula and Arctic benthic Conventus increased cyanobacterial diversity and toxin production. Transmission of toxic species or transmission of toxins from existing pollutants affects the ocean, and cyanobacteria are often the major producers of benthic organisms. Climate change can raise the frequency, strength, and extent of cyanobacteria in many neutrophil ponds, barns, and ducks. Floral cyanobacteria cause a variety of neurotoxins, liver and skin toxins in birds and mammals, comprising waterfowl, cows, and dogs, and can be harmful to recreational drinking water (Liu et al., 2019).

1.4   MICROORGANISM IN MARINE ENVIRONMENT AND EFFECT OF CLIMATE CHANGE

Marine biomass affects 70% of the Earth's crust from coastlines to open sea reefs and coral reefs. Photosynthetic microorganisms use solar energy on the surface of 200 millimeters of water, and marine life in the southern hemisphere uses organic chemicals and socks. It also affects the formation of the lakes region. Elevating the population will not merely alter the natural processes, but also lessen the amount of water, reduce the quantity and circulation, and affect the distribution of violence and food transport. Rainfall, salinity, and precipitation affect mixing and rotation. Nutrients from the air, rivers, and fauna can also affect the structure and function of interesting places, and climate change affects all of these natural substances. By dissolving carbon, nitrogen, and minerals, marine organisms form the basis of the oceans, creating global carbon and biological material (Fotedar et al., 2016; Collins et al., 2019).

The oil, impregnation, and deposition of the solid balloon into organic substances in water vapor are essential for long-term processes to separate CO_2 from the atmosphere. The balance between carbon dioxide and regenerative processes is achieved through the reconstruction and burial of the ocean floor, thereby determining its impact on climate change. Disadvantages of marine burial and impact of greenhouse gases that focus on atmospheric temperature, acidization, mixing, thermal cycling, nutritional, radiation, and extreme weather events affecting marine ecology, including fertility and important change, influence Sea nets, carbon export, marine burial. Climate change affects racial integration, transformation, mobility, and retaliation, or coercion. Heat, acidization, eutrophication, and ocean extremes (such as fishing and tourism) all have to do with coral pollution and the migratory nature of macro and cyanobacterial mats present in the bottom. The ability of corals to adjust to climate change is greatly enhanced by the response of related pathogens, comprising microalgae symbiosis, and bacteria (Robinson et al., 2018).

Thousands of coral-inhabiting organisms maintain health and restore waste, provide vitamins, vital nutrients, and support the invulnerable system to fight infection. Consequently, environmental degradation and coral contamination can quickly reverse the coral infection. This change will undoubtedly affect the climate and resilience of the coral microbial system, affecting the efficiency and extent of coral regeneration to coral and climate change, and its relationship to other aspects of the coral reef ecosystem. Generally, viral infections are more widespread than in large organisms. Consequently, many types of infection have a history of life differences, and their distribution, lifestyle (e.g., labor organizations), and environmental factors may have an impact on geographical location and function. Waves, which control the oil and return to it, are vital to the marine environment. If you cannot go to a happy place, evolution is the only way to survive. Microorganisms are closely related to a large population and to rapid periods of reproduction such as bacteria, archaea, and microglia (Torda et al., 2017; Patel et al., 2020).

Only a handful of studies have examined the effect of environmental change on variability related to acid acidization and other climate change. Similarly, molecules influence the behavior of the body and the effect of this change on the effect of inorganic chemicals. Ocean acidization raises significantly higher pH conditions than previous reports on marine mammals, affecting the stable pH of cells. Inorganic forms of internal pH are highly susceptible. Components such as organism size, flocculation status, metabolic activity, and particle size can change the control of the process. Many environmental and physical factors are responsible for virus responses and overall competition in their natural environment. For example, obesity increases protein synthesis in eukaryotic phytoplankton and reduces cellular ribosome density (Webster et al., 2016).

Sea temperature is seen as a contributing factor to the small amount of plankton on large plankton. This can lead to changes in the biochemical flow as cell flow. Increasing the seawater temperature, acidity and decreased nutritional value are expected to increase the production of outer phytoplankton dissolved organic king, which increases the production of viruses but not fruits through changes in the circulatory system. Heating iron-catalyzed cyanobacteria also reduces iron degradation. This could adversely affect future combustion of nitrogen provided by the Seafood Web. Particular attention must be paid to the general and limited biodiversity responses to climate change and stress related to climate change. Therefore, effective transitional forces in the region, such as carbon remineralization and changes in carbon sequestration, as well as building cycling remain a major concern (Zaneveld et al., 2016; Balamurugan et al., 2020).

1.5 ROLE OF MICROBES IN AGRICULTURE: A CLIMATE CHANGE SCENARIO

Conferring to the World Bank (World Bank data on the farm), 40% of the world's climate is used for agriculture. This amount is expected to rise, making significant alterations in the flow of carbon, nitrogen, phosphorus, and additional nutrients into the soil. Then again, these changes are related to the obvious loss of biodiversity (including infection). There is growing fascination in using animal pathogens to improve agricultural security and reduce the impact of climate change on food security. However, this requires a better understanding of the effect of climate change on biodiversity. On the surface of fossil fuels producing methane, methane and methane are synthesized in nature and made-up of anaerobic environments (air, paddy fields, tracts of animals (especially animals), seawater, and biogas). The main precipitation of CH_4 is the atmospheric oxidation and oxidation of 20–90 sulfides in soil, soil and water. Atmospheric CH_4

biomass has increased significantly in recent years (2014–2017), while fossil fuel methane sources and / or the air industry have increased or decreased in the atmosphere, but the cause remains unclear. CH_4 oxidation is a major threat to methane regulation and global warming (Kashyap et al., 2018; Thomashow et al., 2019).

Hydrothermal fossil fuels and fertilizer applications have significantly increased the use of estrogen sites, adversely affecting terrestrial systems and adversely affecting the ecosystem. Agriculture is the largest source of N_2O, a powerful greenhouse gas emitted by microbes that are invisible to nitrogen. N_2O reductase can convert into rhizobia (inside the nose), and certain viral infections can also convert N_2O to N_2 (without greenhouse gas). Climate change affects microbial nitrogen exchange (degradation, mineralization, denitrification, fixation) and N_2O production rates. We must understand the impact of climate change and further human events on the nitrogen oxide micro-exchange. The combined effect of climate change and fertilizer eutrophication can be both significant and perhaps unexpected for contagion networks. For example, the attacks often cause Algal blooms, but different results have been observed in the Lake Zurich (Baggio et al., 2020).

Reducing phosphorus levels in fertilizer reduces eukaryotic phyto-plankton flow, but adds nitrogen to the phosphorus, aiding nitrogen-producing cyanobacterial plankton. In the lack of efficient screening, annual hybridization plays an essential role in managing cyanobacterial population. In turn, increasing the temperature increases the temperature of the fuel and decreases the concentration, which keeps the toxicity of cyanobacteria constant. Crop planting can range from excessive cultivation (small labor, fertilizer, capital) to extreme farming (heavy labor). High temperatures and drought are adversely affecting crop growth. In widespread agriculture (such as grass), trap-based soils are more soil and drought-resistant than bacterial traps commonly found in concentrated systems (such as wheat) (Cao et al., 2016).

Global external studies show that soil fungi and bacteria live in certain locations and react in a different way to precipitation and soil pH. Climate change can increase drought and reduce the diversity and prevalence of viruses and fungi in arid regions of the world. Reducing soil biodiversity reduces the overall effectiveness of cropping areas and reduces their ability to contribute to crop growth.

1.6   ADAPTATION OF MICROBES UNDER DIFFERENT ENVIRONMENTAL CONDITIONS

Traditionally, the term "adapter" refers to the size of a living thing or population. This is a generic term for the body's organs to improve their strength in a certain area. It is frequently used to define a particular event, such as a change in gene density, but the term itself does not mean any other reference (Hochachka and Somero, 2002). It is also recommended here that it can be used at a regional level to explain changes in the overall functioning of small areas and environmental changes. Changes in service performance are seen at the local level as a result of changes in the contribution of the relative risk of infection to all activities in the community. Microbial environment changes are defined as behavioral processes that have been monitored by levels of specific human pathogens that are appropriate for the current climate.

1.6.1   COMMUNITY-LEVEL ADAPTATION

Although microorganisms can adapt to environmental changes via genetic progression (Bennett and Lenski, 2007; Angilletta Jr and Angilletta, 2009), transfer of horizontal gene (Trevors et al., 1987; Nemergut et al., 2013), and physiological malleability (Schimel et al., 2007). We believe that changes in community composition and the comparative impact of different taxa to overall activities are key processes to promote adaptation to climate change at the ecosystem level. Changes in the relative frequency of the population due to differences in growth or mortality (referred to as species types) are mainly due to regional environmental circumstances (e.g., resources, temperature) (Van der Gucht et al., 2007). Microbial population structure can also be disturbed by immigration burden or immigration (e.g., flushing) populations, collectively known as collective impact (Leibold et al., 2004). At some point, the composition and composition of the microbial community (component survival and relative abundance) depend on the classification of species and the relative impact of large-scale effects. Similarly, alterations in the function of existing microbial communities depend on the relative rate at which community members are physiologically favorable and the rate at which new genes can be acquired by mutation (Pörtner et al., 2006).

The physical adjustment can be done via a variety of processes, comprising regulation of intracellular solutes (Csonka and Hanson, 1991; De Vrieze et al., 2012) and variations in lipid composition (Hall et al., 2010). These two technologies describe the functioning of existing environments and the potential impact of infected areas in reaction to climate change to new activities that have been lost and lost as a result of climate change. In some cases, the functionality of the virus depends on the work of various members of the community, as they only work in a particular area (Lennon and Jones, 2011). In many cases, the composition of the community is closely related to altering environmental gradients on comparatively little seasonal time period. In lakes (Shade et al., 2007), streams (Crump and Hobbie, 2005), and marine ecosystems (Fuhrman et al., 2006), seasonal changes in aquatic microbial communities are closely related to temperature. It has been suggested that these changes in the social media environment due to the selection of features fit better for each community. These changes are rooted in an environment that is detrimental to global environmental performance.

For example, if different areas of infection destroy the same amount of waste, the formation of the environment can change the rates of respiration by 20% (Strickland et al., 2009). In a new study, the metabolic content of aspirin leaf barn degradation depends on how the microbial community pollutes the virus (Wallenstein et al., 2010). The formation of toxic geoenvironmental changes leads to reduced microbial metabolic processes that is needed to be unravelled under various environments.

1.6.2 THERMAL ADAPTATION OF MICROBIAL COMMUNITIES

As the climate changes, infected regions change by changing the temperature. As mentioned above, their rate of evolution is determined by the different characteristics of the technology, the availability of existing facilities and resources. As the region of expertise moves beyond the decorative zone, important physical changes (such as breathing) can occur, which can directly affect the behavior of an ecosystem (such as CO_2). Ultimately, these changes can lead to discipline and competition in the community. Consequently, in a healthy population, the population is still out of the total cell population, causing a mutation (Hall et al., 2008). Burnout is a clear example of a regional climate that forms an infected environment and is influenced by climate change. In short, heat regulates the rate of the metabolic process (Schmidt et al., 2011). Over time, heat can affect the human immune system and physiology (Adams et al., 2010; Saggar et al., 2013).

Schindlbacher et al. (2011) observed that extremely hot temperatures followed a similar pattern in MAT, compared with three large-area soil organisms. That is, each region is more active within this initial temperature range. Consequently, the significant relationship among temperature and metabolic processes is not constantly directly related to the environmental level. For instance, some experiments have observed that there is no relationship among the MT transition time and soil organic matter due to geoclimatic inclines (Fissore et al., 2009; Binkley and Fisher, 2013). Consequently, the temperature dependency of microbial metabolism on the marine environment is usually considered to be relatively straightforward (Arístegui et al., 2009; Kirchman et al., 2009). This difference may be part of the cross-system analysis of unmodified microbial adapter capacity (i.e., different regions of the region with different temperatures to respond to curves corresponding to your temperature system). The mechanism by which the infected environment responds to the temperature will affect the time and temperature of the room. On a broader scale, global warming and terrestrial climate must be sensitive to climate change.

Ambient temperature not only affects seasonal mean and temperature, but also daily decreases and spatial variation. Because of the high temperature of the water, the temperature of the water shortage changes gradually. The annual temperature of the major river basins is approximately the same, but the annual temperature of the Alps river may be lower that showed temporary decline in temperature of the area as aquatic biota may be higher in the tropics than in the deep underground environment. This is important because areas with low temperature (such as heat experts) are more susceptible to environmental and infection than infectious areas than the general population (such as heat experts). We hypothesized that if climate change only modifies temperature by raising the minimum, width, and average temperature, sweetness fluctuations will be reduced by larger adjustments (such as travel parameters). In contrast, regions with a microclimate have a greater degree of heat in warmer climates and have to adapt to a faster rate of global warming through imaging.

Therefore, warming has a greater effect on the stress response in hot and humid climates than in temperate regions. At intermediate levels, changes in the abundance of existing entities can affect the biochemical rate activation. Heat specialists can deal with severe weather in hot climates, near the equator, between vortices, or in tight, stable areas.

1.6.3   MICROBIAL ADAPTATION TO ALTERED PRECIPITATION REGIMES

The rules governing the process of adapting to changes in temperature can be extrapolated to changes in certain climate conditions that are affected by climate change. These parameters are the major factor in the biochemical cycle promoted by the virus. For example, soil-contaminated soil moisture is important for poor differentiation in biodiversity (Wilson and Griffin, 1975; Snyder et al., 2009). Soil moisture has an influence in controlling the rate of microbial action. Increasing soil moisture increases the rate of aerobic processes (oxygen is limited). Consequently, the history of soil moisture degradation in local contaminants can also be used as a means of controlling existing taxation (Evans and Wallenstein). Local factors that affect health under climate change include microorganisms forced in response to water scarcity, and swift alterations in redox status and osmotic and matrix problems.

Different systems can be used for soil moisture pulses (Schimel et al., 2007). First, bacteria can "block" pulses. That is, bacteria can maintain function independently of the immune system. At the local level, an average level of tenderness affects the outcome of improving coordination, including the cost of integration associated with this process. The climatic conditions found in tropical soils and their performance (respiration, methanogenesis, N_2O production, and iron deficiency) are closely related to this water supply (DeAngelis et al., 2010). All solutions to these waters require the provision of resources including large macronutrients carbon, nitrogen, and phosphorous (P). Consequently, the mechanism of microbial biomass can be observed in the precipitation gradient.

We study two studies here and their results are in line with our normal body system. First, when the annual rainfall over Mongolian grasslands passes through atmospheric gradients, soil microbial metabolic fraction (qCO_2; respiration/microbial biomass) decreases (Liu et al., 2010). In additional words, a decrease in carbon utilization efficiency (CUE) and a decrease in soil moisture indicate higher precautionary measures than low soil moisture. In desert areas, carbon has microbial biomass: Nitrogen increases with increasing annual precipitation, driving revenue into rich tanks with minimal soil moisture (Gallardo and Schlesinger, 1992). These changes in C:N and CUE can be attributed to changes in the local environment. Recent culture-based studies have shown that C:fungal biomass, rather than bacteria, can cause bacteria from fungal to CUE ratio. These examples indicate that c-block is an independent soil with few virulence factors to increase the amount of water. Similarly, changes in temperature can alter virus-regulated phosphorous cycling in the wet environment, depending on the ability of de novo to form new phospholipids in response to changes in warming conditions (Hall et al., 2009). there are. Research in Maine proves that to be the case. Phosphatides have been replaced by thio-lipids by the deep phosphorus inhibition of some cyanobacteria (Lambers et al., 2011). Therefore, changes in the warm climate may increase the utilization of phosphorus by current microbial communities and alter the phosphorous cycling at the ecological level. These instances indicate that microbial community interactions and microbial activity may significantly alter the biochemical process.

1.7 INTEGRATION OF MICROBIAL ADAPTABILITY AND BIOGEOCHEMISTRY

The availability of resources must determine the rate and extent of environmental change, and thus the degree of chemical change due to climate change. In the resource-poor environment, the cost of controlling the body determines the correlation (i.e., phenotypic plasticity) of the existing regions. Changes in the social environment have occurred gradually due to the burden of factors that bring about population. In an economy with high levels of diversity, different affiliations can increase toxicity, but different societies are more susceptible to environmental change.

Like other climates, biological activity that affects climate change depends on the degree of climate change that is adapted to local adaptation. To better determine the effect of climate change on ecosystems, research questions should focus on the relationship among climate change and current climate change patterns in this region. The main difference between these relationships, the greater the likelihood is that climate change will adversely affect the Earth's environmental conditions.

In more temperate climatic conditions per year compared to rapid climate change, a low level of risk in the virus can lead to severe chemical changes. In a stable environment, the level of flexibility is limited to large-scale faults, forcing current community members to work beyond the relevant niche, and to replace existing biochemical cycles. Climate dispersal throughout the year must be adapted to rapid climate change through a series of adaptations to its population (e.g., changes in regional composition).

1.8 CONCLUSIONS AND FUTURE PERSPECTIVES

First, a simple method for measuring local-level structures is used to measure the relationship that can occur in favorable conditions or networks in different field conditions. The ratio should be equal to your current concentration. For example, instead of measuring nitration efficiency under optimal conditions, a more effective method is used in laboratory tests, as it is better to measure the stabilization rate globally and at warmer temperatures. Then, researchers need to measure changes in functionality locally with climate change. This is possible by calculating the time taken to recover part of the original activity (e.g., respiration) after disturbances in the region (e.g., heat intensity). Third, new biogeochemical models ought to be built to provide high-quality functional features at the regional level. Combining the active elements of existing systems with a better understanding of how and how biological processes will affect the ecosystem process in future climates.

There is a dire need to improve understanding of global biodiversity and marine composition. To critically understand the role of microbes in global climate change and its complex nature more advanced research is needed to be done by using modern analytical and scientific approaches. We need the quantitative information to monitor the effects of these organisms. Although models can be compared to models, many of these models have not identified air and land markets. The reason for this error is not due to the mathematical formula of such models, but the absence of evolution and evolutionary data does not preclude reliable predictions of biological effects on climate change. Surface sowing is an important way to implement a system and requires measuring, measuring and comparing the current and future state of the Earth system.

KEYWORDS

- soil ecology
- high temperature
- carbon sequestration
- adaptation responses

REFERENCES

Adams, H. E.; Crump, B. C.; Kling, G. W. Temperature Controls on Aquatic Bacterial Production and Community Dynamics in Arctic Lakes and Streams. *Environ. Microbiol.* **2010,** *12,* 1319–1333.

Allison, S. D.; Wallenstein, M. D.; Bradford, M. A. Soil-Carbon Response to Warming Dependent on Microbial Physiology. *Nat. Geosci.* **2010,** 3, 336–340.

Angilletta, Jr, M. J.; Angilletta, M. J. *Thermal Adaptation: A Theoretical and Empirical Synthesis*; Oxford University Press, 2009.

Arístegui, J.; Gasol, J. M.; Duarte, C. M.; Herndld, G. J. Microbial Oceanography of the Dark Ocean's Pelagic Realm. *Limnol. Oceanography.* **2009,** *54,* 1501–1529.

Baggio, M.; Chavas, J. P.; Di Falco, S.; Hertig, A.; Pomati, F. The Effect of Anthropogenic and Environmental Factors in Coupled Human-Natural Systems: Evidence from Lake Zürich. *Nat. Resour. Model.* **2020,** *33,* e12245.

Balamurugan, S.; Sivakumar, K.; Sivasankar, P.; Radhakrishnan, M.; Balagurunathan, R.; Anilkumar, N.; Subramanian, M. Metagenomic 16S RDNA Amplicon Data of Microbial Diversity and Its Predicted Metabolic Functions in the Southern Ocean (Antarctic), 2020.

Bennett, A. F.; Lenski, R. E. An Experimental Test of Evolutionary Trade-Offs during Temperature Adaptation. *Proc. Natl. Acad. Sci.* **2007,** *104,* 8649–8654.

Binkley, D.; Fisher, R. F. *Ecology and Management of Forest Soils*; Wiley Online Library, 2013.

Bradford, M. A.; Davies, C. A.; Frey, S. D.; Maddox, T. R.; Melillo, J. M.; Mohan, J. E.; Reynolds, J. F.; Treseder, K. K.; Wallenstein, M. D. Thermal Adaptation of Soil Microbial Respiration to Elevated Temperature. *Ecol. Lett.* **2008,** *11,* 1316–1327.

Cao, X.; Wang, Y.; He, J.; Luo, X.; Zheng, Z. Phosphorus Mobility among Sediments, Water and Cyanobacteria Enhanced by Cyanobacteria Blooms in Eutrophic Lake Dianchi. *Environ. Pollut.* **2016,** *219,* 580–587.

Collins, S.; Boyd, P. W.; Doblin, M. A. Evolution, Microbes, and Changing Ocean Conditions. *Ann. Rev. Marine Sci.* **2019,** *12,* 181–208.

Conant, R. T.; Ryan, M. G.; Ågren, G. I.; Birge, H. E.; Davidson, E. A.; Eliasson, P. E.; Evans, S. E.; Frey, S. D.; Giardina, C. P.; Hopkins, F. M. Temperature and Soil Organic Matter Decomposition Rates–Synthesis of Current Knowledge and a Way Forward. *Global Change Biol.* **2011,** *17,* 3392–3404.

Crump, B. C.; Hobbie, J. E. Synchrony and Seasonality in Bacterioplankton Communities of Two Temperate Rivers. *Limnol. Oceanography*. **2005**, *50*, 1718–1729.

Csonka, L. N.; Hanson, A. D. Pro karyotic Osmoregulation: Genetics and Physiology. *Annu. Rev. Microbiol*. **1991**, *45*, 569–606.

De Vrieze, J.; Hennebel, T,; Boon, N.; Verstraete, W. Methanosarcina: The Rediscovered Methanogen for Heavy Duty Biomethanation. *Bioresour. Technol*. **2012**, *112*, 1–9.

DeAngelis, K.M.; Silver, W. L.; Thompson, A. W.; Firestone, M. K. Microbial Communities Acclimate to Recurring Changes in Soil Redox Potential Status. *Environ. Microbiol*. **2010**, *12*, 3137–3149.

Dessureault-Rompré, J.; Zebarth, B. J.; Georgallas, A.; Burton, D. L.; Grant, C. A.; Drury, C. F. Temperature Dependence of Soil Nitrogen Mineralization Rate: Comparison of Mathematical Models, Reference Temperatures and Origin of the Soils. *Geoderma*. **2010**, *157*, 97–108.

Evans, S. E.; Wallenstein, M. D. Soil Microbial Community Response to Drying and Rewetting Stress: Does Historical Precipitation Regime Matter? *Biogeochemistry*. **2012**, *109*, 101–116.

Fissore, C.; Giardina, C. P.; Swanston, C. W.; King, G. M.; Kolka, R. K. Variable Temperature Sensitivity of Soil Organic Carbon in North American Forests. *Global Change Biol*. **2009**, *15*, 2295–2310.

Fotedar, R.; Zeyara, A.; Al Malaki, A.; Fell, J. W.; Stoeck, T.; Filker, S.; Boekhout, T.; Kolecka, A.; Bukhari, S. J.; Abdel-Moati, M. A. Marine Microbes and Climate Change-a Qatari Prospective. Paper Read at Qatar University Life Science Symposium 2016: Biodiversity, Sustainability and Climate Change, with Perspectives from Qatar, 2016.

Fu, D.; Yu, X.; Wang, F. The Effects of Compound Microorganisms on the Growth and Water Quality in Carps. *Feed Res*. **2005**, *8*, 43–45.

Fuhrman, J.A.; Hewson, I.; Schwalbach, M. S.; Steele, J. A.; Brown, M. V.; Naeem, S. Annually Reoccurring Bacterial Communities Are Predictable from Ocean Conditions. *Proc. Natl. Acad. Sci*. **2006**, *103*, 13104–13109.

Gallardo, A.; Schlesinger, W. H. Carbon and Nitrogen Limitations of Soil Microbial Biomass in Desert Ecosystems. *Biogeochemistry*. **1992**, *18*, 1–17.

Hall, E. K.; Dzialowski, A. R.; Stoxen, S. M.; Cotner, J. B. The Effect of Temperature on the Coupling between Phosphorus and Growth in Lacustrine Bacterioplankton Communities. *Limnol. Oceanography*. **2009**, *54*, 880–889.

Hall, E. K.; Neuhauser, C.; Cotner, J. B. Toward a Mechanistic Understanding of How Natural Bacterial Communities Respond to Changes in Temperature in Aquatic Ecosystems. *ISME J*. **2008**, *2*, 471–481.

Hall, E. K.; Singer, G. A.; Kainz, M. J.; Lennon, J. T. Evidence for a Temperature Acclimation Mechanism in Bacteria: An Empirical Test of a Membrane-Mediated Trade-Off. *Funct. Ecol*. **2010**, *24*, 898–908.

Hall, E.; Maixner, F.; Franklin, O.; Daims, H.; Richter, A.; Battin, T. Linking Microbial and Ecosystem Ecology Using Ecological Stoichiometry: A Synthesis of Conceptual and Empirical Approaches. *Ecosystems*. **2011**, *14*, 261–273.

Hartley, I. P.; Hopkins, D. W.; Garnett, M. H.; Sommerkorn, M.; Wookey, P. A. Soil Microbial Respiration in Arctic Soil Does Not Acclimate to Temperature. *Ecol. Lett*. **2008**, *11*, 1092–1100.

Hicks, L. C.; Meir, P.; Nottingham, A. T.; Reay, D. S.; Stott, A. W.; Salinas, N.; Whitaker, J. Carbon and Nitrogen Inputs Differentially Affect Priming of Soil Organic Matter in Tropical Lowland and Montane Soils. *Soil Biol. Biochem.* **2019,** *129,* 212–222.

Hochachka, P. W.; Somero, G. N. *Biochemical Adaptation: Mechanism and Process in Physiological Evolution*; Oxford University Press, 2002.

Hoegh-Guldberg, O.; and Bruno, J. F. The Impact of Climate Change on the World's Marine Ecosystems. *Science.* **2010,** *328,* 1523–1528.

Ju, C.; Xu, J.; Wu, X.; Dong, F.; Liu, X.; Tian, C.; Zheng, Y. Effects of Hexaconazole Application on Soil Microbes Community and Nitrogen Transformations in Paddy Soils. *Sci. Total Environ.* **2017,** *609,* 655–663.

Kashyap, P.L.; Srivastava, A. K.; Tiwari, S. P.; Kumar, S. 2018. *Microbes for Climate Resilient Agriculture*; John Wiley & Sons, 2018.

Kesarcodi-Watson, A.; Kaspar, H.; Lategan, M. J.; Gibson, L. Probiotics in Aquaculture: The Need, Principles and Mechanisms of Action and Screening Processes. *Aquaculture.* **2008,** *274,* 1–14.

Kirchman, D. L.; Morán, X. A. G.; Ducklow, H. Microbial Growth in the Polar Oceans— Role of Temperature and Potential Impact of Climate Change. *Nat. Rev. Microbiol.* **2009,** *7,* 451–459.

Lambers, H.; Brundrett, M. C.; Raven, J. A.; Hopper, S. D. Plant Mineral Nutrition in Ancient Landscapes: High Plant Species Diversity on Infertile Soils Is Linked to Functional Diversity for Nutritional Strategies. *Plant Soil.* **2011,** *348,* 7.

Leibold, M. A.; Holyoak, M.; Mouquet, N.; Amarasekare, P.; Chase, J. M.; Hoopes, M. F.; Holt, R. D.; Shurin, J. B.; Law, R.; Tilman, D. The Metacommunity Concept: A Framework for Multi-Scale Community Ecology. *Ecol. Lett.* **2004,** *7,* 601–613.

Lennon, J. T.; Jones, S. E. Microbial Seed Banks: The Ecological and Evolutionary Implications of Dormancy. *Nat. Rev. Microbiology.* **2011,** *9,* 119–130.

Liu, M.; Ma, J.; Kang, L.; Wei, Y.; He, Q.; Hu, X.; Li, H. Strong Turbulence Benefits Toxic and Colonial Cyanobacteria in Water: A Potential Way of Climate Change Impact on the Expansion of Harmful Algal Blooms. *Sci. Total Environ..* **2019,** *670,* 613–622.

Liu, Z.; Fu, B.; Zheng, X.; Liu, G. Plant Biomass, Soil Water Content and Soil n: P Ratio Regulating Soil Microbial Functional Diversity in a Temperate Steppe: A Regional Scale Study. *Soil Biol. Biochem.* **2010,** *42,* 445–450.

McCue, M. D. Starvation Physiology: Reviewing the Different Strategies Animals Use to Survive a Common Challenge. *Comp. Biochem. Physiol. A Mol. Integr. Physiol.* **2010,** *156,* 1–18.

Mussatto, S. I.; Mancilha, I. M. Non-Digestible Oligosaccharides: A Review. *Carbohy. Poly.* **2007,** *68,* 587–597.

Nemergut, D. R.; Schmidt, S. K.; Fukami, T.; O'Neill, S. P.; Bilinski, T. M.; Stanish, L.F.; Knelman, J. E.; Darcy, J. L.; Lynch, R. C.; Wickey, P. Patterns and Processes of Microbial Community Assembly. *Microbiol. Mol. Biol. Rev.* **2013,** *77,* 342–356.

Patel, N. P.; Shimpi, G. G.; Haldar, S. Evaluation of Heterotrophic Bacteria Associated with Healthy and Bleached Corals of Gulf of Kutch, Gujarat, India for Siderophore Production and Their Response to Climate Change Factors. *Ecol. Indic.* **2020,** *113,* 106219.

Pörtner, H.O.; Bennett, A. F.; Bozinovic, F.; Clarke, A.; Lardies, M. A.; Lucassen, M.; Pelster, B.; Schiemer, F.; Stillman, J. H. Trade-Offs in Thermal Adaptation: The Need for a Molecular to Ecological Integration. *Physiol. Biochem. Zool.* **2006,** *79,* 295–313.

Reinhart, K. O.; Dangi, S. R.; Vermeire, L. T. The Effect of Fire Intensity, Nutrients, Soil Microbes, and Spatial Distance on Grassland Productivity. *Plant Soil.* **2016**, *409*, 203–216.

Rinnan, R.; Rousk, J.; Yergeau, E.; Kowalchuk, G. A.; Bååth, E. Temperature Adaptation of Soil Bacterial Communities Along an Antarctic Climate Gradient: Predicting Responses to Climate Warming. *Global Change Biol.* **2009**, *15*, 2615–2625.

Robinson, C.; Wallace, D.; Hyun, J.-H.; Polimene, L.; Benner, R.; Zhang, Y.; Cai, R.; Zhang, R.; Jiao, N. An Implementation Strategy to Quantify the Marine Microbial Carbon Pump and Its Sensitivity to Global Change. *Natl. Sci. Rev.* **2018**, *5*, 474–480.

Rose, M.; Nusbaum, T.; Chippindale, A. *Laboratory Evolution: The Experimental Wonderland and the Cheshire cat Syndrome.* Adaptation edited by: Rose, M. R.; Lauder, G. V.; Academic Press, 1996.

Saggar, S.; Jha, N.; Deslippe, J.; Bolan, N.; Luo, J.; Giltrap, D.; Kim, D.-G.; Zaman, M.; Tillman, R. Denitrification and n2o: N2 Production in Temperate Grasslands: Processes, Measurements, Modelling and Mitigating Negative Impacts. *Sci. Total Environ.* **2013**, *465*, 173–195.

Schimel, J.; Balser, T. C.; Wallenstein, M. Microbial Stress-Response Physiology and Its Implications for Ecosystem Function. *Ecology* **2007**, *88*, 1386–1394.

Schindlbacher, A.; Rodler, A.; Kuffner, M.; Kitzler, B.; Sessitsch, A.; Zechmeister-Boltenstern, S. Experimental Warming Effects on the Microbial Community of a Temperate Mountain Forest Soil. *Soil Biol. Biochem.* **2011**, *43*, 1417–1425.

Schmidt, M. W.; Torn, M. S.; Abiven, S.; Dittmar, T.; Guggenberger, G.; Janssens, I. A.; Kleber, M.; Kögel-Knabner, I.; Lehmann, J.; Manning, D. A. Persistence of Soil Organic Matter as an Ecosystem Property. *Nature.* **2011**, *478*, 49–56.

Shade, A.; Kent, A. D.; Jones, S. E.; Newton, R. J.; Triplett, E. W.; McMahon, K. D. Interannual Dynamics and Phenology of Bacterial Communities in a Eutrophic Lake. *Limnol. Oceanography* **2007**, *52*, 487–494.

Simon, M.; Glöckner, F. O.; Amann, R. Different Community Structure and Temperature Optima of Heterotrophic Picoplankton in Various Regions of the Southern Ocean. *Aquat. Microb. Ecol.* **1999**, *18*, 275–284.

Snyder, C. S.; Bruulsema, T. W.; Jensen, T. L.; Fixen, P. E. Review of Greenhouse Gas Emissions from Crop Production Systems and Fertilizer Management Effects. *Agric. Ecosyst. Environ.* **2009**, *133*, 247–266.

Strickland, M. S.; Lauber, C.; Fierer, N.; Bradford, M. A. Testing the Functional Significance of Microbial Community Composition. *Ecology.* **2009**, *90*, 441–451.

Szukics, U.; Abell, G. C.; Hödl, V.; Mitter, B.; Sessitsch, A.; Hackl, E.; Zechmeister-Boltenstern, S. Nitrifiers and Denitrifiers Respond Rapidly to Changed Moisture and Increasing Temperature in a Pristine Forest Soil. *FEMS Microbiol. Ecol.* **2010**, *72*, 395–406.

Thomashow, L. S.; Kwak, Y.S.; Weller, D. M. Root-Associated Microbes in Sustainable Agriculture: Models, Metabolites and Mechanisms. *Pest Manage. Sci.* **2019**, *75*, 2360–2367.

Torda, G.; Donelson, J. M.; Aranda, M.; Barshis, D. J.; Bay, L.; Berumen, M. L.; Bourne, D. G.; Cantin, N.; Foret, S.; Matz, M. Rapid Adaptive Responses to Climate Change in Corals. *Nat. Clim. Change* **2017**, *7*, 627–636.

Trevors, J.; Barkay, T.; Bourquin, A. Gene Transfer Among Bacteria in Soil and Aquatic Environments: A Review. *Can. J. Microbiol.* **1987**, *33*, 191–198.

Van der Gucht, K.; Cottenie, K.; Muylaert, K.; Vloemans, N.; Cousin, S.; Declerck, S.; Jeppesen, E.; Conde-Porcuna, J.-M.; Schwenk, K.; Zwart, G. The Power of Species Sorting: Local Factors Drive Bacterial Community Composition Over a Wide Range of Spatial Scales. *Proc. Natl. Acad. Sci.* **2007**, *104*, 20404–20409.

Wallenstein, M. D.; Hess, A. M.; Lewis, M. R.; Steltzer, H.; Ayres, E. Decomposition of Aspen Leaf Litter Results in Unique Metabolomes When Decomposed under Different Tree Species. *Soil Biol. Biochem.* **2010**, *42*, 484–490.

Webster, N.; Negri, A.; Botté, E.; Laffy, P.; Flores, F.; Noonan, S.; Schmidt, C.; Uthicke, S. Host-Associated Coral Reef Microbes Respond to the Cumulative Pressures of Ocean Warming and Ocean Acidification. *Sci. Rep.* **2016**, *6*, 19324.

Wilson, J. M.; Griffin, D. Water Potential and the Respiration of Microorganisms in the Soil. *Soil Biol. Biochem.* **1975**, *7*, 199–204.

Yazdanpanah, N.; Mahmoodabadi, M.; Cerdà, A. The Impact of Organic Amendments on Soil Hydrology, Structure and Microbial Respiration in Semiarid Lands. *Geoderma.* **2016**, *266*, 58–65.

Zaneveld, J. R.; Burkepile, D. E.; Shantz, A. A.; Pritchard, C. E.; McMinds, R.; Payet, J. P.; Welsh, R.; Correa, A. M.; Lemoine, N. P.; Rosales, S. Overfishing and Nutrient Pollution Interact with Temperature to Disrupt Coral Reefs Down to Microbial Scales. *Nat. Commun.* **2016**, *7*, 11833.

Zhou, Q.;, Li, K.; Jun, X.; Bo, L. Role and Functions of Beneficial Microorganisms in Sustainable Aquaculture. *Bioresour. Technol.* **2009**, *100*, 3780–3786.

CHAPTER 2

Microbial Manipulation for Reduced Greenhouse Gas Emissions

SHEIKH FIRDOUS AHMAD[1*], TRIVENI DUTT[2], and
BHARAT BHUSHAN[3]

[1]Scientist, ICAR-National Research Centre on Pig, Rani, Guwahati,
Assam, India

[2]Joint Director (Academic), ICAR-Indian Veterinary Research Institute,
Izzatnagar, Bareilly 243122, UP, India

[3]Professor and Principal Scientist, Division of Animal Genetics,
ICAR-Indian Veterinary Research Institute, Izzatnagar,
Bareilly 243122, UP, India

*Corresponding author. E-mail: firdousa61@gmail.com

ABSTRACT

Climate change, for which the green house gases (GHGs) are primarily responsible, refers to the continually evolving global temperature and precipitation patterns. Carbon dioxide, methane, and nitrous oxide with differing potential to induce global warming are the three common GHGs from different natural and artificial sources. A notable percentage of atmospheric greenhouse gases is contributed by agriculture and livestock production, among anthropogenic sources. The livestock sector accounts for 58% of the GHG emissions from the agriculture and allied sectors. While ruminants makeup the bulk of GHG emissions, non-ruminants like pigs, chickens, etc., contribute to a minor amount. Approximately 18% of total global GHG emissions and 37% of anthropogenic methane gas are derived globally from ruminant processing systems. Meat and milk production account for 35 and 30% of GHG emissions from the livestock sector, respectively. During ruminant stomach fermentation, methane is released through a multistep process with considerable global warming potential. One of the exciting developments that can significantly reduce GHG emissions is microbial modulation. Several approaches ranging from dietary manipulation to sophisticated genetic biotechnological techniques, have been attempted to manipulate the microbial environment, to reduce GHG emissions, mainly methane. Sustainable practices to mitigate methane in ruminants are felt to be the most necessary without affecting production. To minimize GHG emission, ionophores, organic acids, and oils are part of food manipulation. Tannins and saponins also help to reduce the emissions of ruminant-stomach methane. However, each intervention has one or more pitfalls, and any single process cannot give a simple, lasting remedy. This chapter aims to address GHG emissions, mainly methane, from agriculture and allied industries, and possible mitigation measures to facilitate sustainable production with minimal environmental impact.

2.1   INTRODUCTION

Various problems of differing magnitudes and meanings confront the current world. Besides population explosions and food shortages, climate change is one of the prime threats facing human society. By the year 2050, the human population will reach 9.6 billion, with a 70% rise in agricultural and livestock production (FAO, 2009). Due to unavoidable differences, both problems cannot be dealt with simultaneously. However, prioritization of goals both inside and around the tasks is essential. A mechanism of sudden changes in precipitation patterns and other climatic environments is referred to as climate change. Terming it as a recent origin problem would be unjust because it has happened since the early days, but its pace has been unfairly fast in the last few decades. One example of worldwide climate change is extreme warming. Global temperature change, including carbon dioxide (CO_2), methane (CH_4), nitrogen oxides (mainly N_2O), and other gases, is primarily the repercussion of GHGs. These gases vary in their ability to trigger the climate change phenomena as the capacity of CO_2, CH_4, and N_2O the three primary global warming gases is scaled to 1, 23 (or 25), and 298, respectively (Solomon et al., 2007; Ribeiro et al., 2015).

The release of greenhouse gases into the atmosphere results from numerous natural and anthropogenic sources (Yue and Gao, 2018). Wild forest fires, permafrost, decomposition, water vapors, etc., are the natural sources of greenhouse gas pollution, while anthropogenic sources include automobile emissions, mining, and fossil fuel combustion, etc. (Fig. 2.1). The Intergovernmental Panel on Climate Change (IPCC), convened by the United Nations, stressed that human activities had contributed significantly to global greenhouse emissions in recent decades, contributing to the increased release of infrared radiation from the atmosphere, causing global warming. The deposition of GHG in the atmosphere has eventually contributed to sudden shifts in climatic conditions with elevated temperatures (rise of 0.6°C–0.7°C) (Lascano and Cárdenas, 2010) and increased occurrence of floods and droughts.

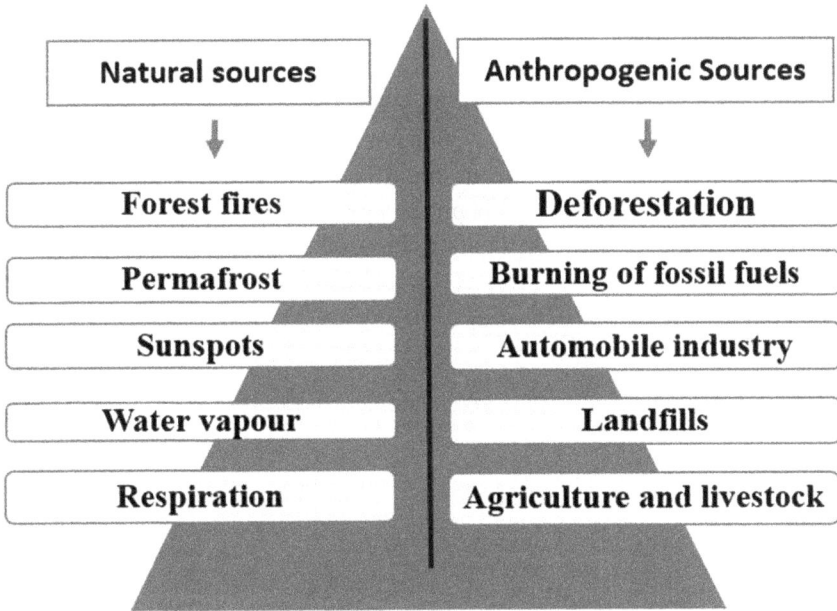

FIGURE 2.1   Natural and anthropogenic sources of greenhouse gases.

Agriculture and livestock industries add substantial quantities of GHGs to the environment, particularly CH_4 and N_2O, among anthropogenic sources. Out of the total GHG emissions from agriculture and associated sectors, livestock accounts for about 58% of total emissions, while other agricultural activities account for the remainder. GHGs are primarily derived from rice, wheat, and ruminant animals like buffalo, sheep, and goats, contributing nearly 80% of GHG emissions. In contrast, non-ruminants like pigs and chickens, etc., contribute the remaining (Haque, 2018). A high percentage of the energy consumed by ruminants is converted into various gases, including methane, a portion of which is released to the atmosphere. Approximately 18% of total global GHG emissions and 37% of anthropogenic methane gas are produced globally from ruminant processing systems (Sejian et al., 2016). Meat and milk production contribute 35 and 30% to GHG emissions from the general livestock sector (Haque, 2018), respectively. The carbon footprints for milk production are 2.8, 3.4, and 6.5 kg CO_2-eq/kg FPCM (fat and protein-corrected milk), respectively, in terms of GHG emission indicators for cattle, buffalo, and small ruminants. Similarly, carbon footprints are 46.2, 53.4 kg CO_2-eq/kg meat for beef/carabeef and little ruminant beef, respectively.

CO_2 emitted from livestock is not used in the net calculation of GHGs emitted from livestock, although they consume plants that use CO_2 for photosynthesis (Steinfeld et al., 2006). Thus, only CH_4 and N_2O are the significant GHGs released from the livestock sector, with substantial climate change potential. N_2O is produced mainly due to the fermentation of manure and chemical fertilizers to be used as livestock feed (Grossi et al., 2019). Figure 2.2 summarizes the proportionate input to GHG emissions from various practices related to livestock rearing.

SOURCES OF GHG EMISSIONS FROM LIVESTOCK

■ Enteric fermentation ■ Feed porduction ■ Manure storage ■ Processing and transportation

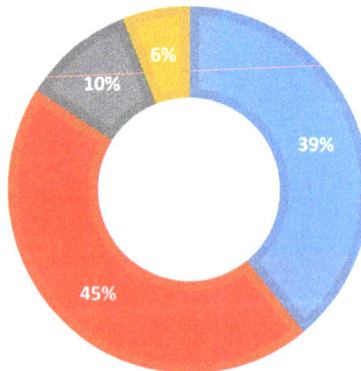

FIGURE 2.2   Summary of different sources of GHGs from livestock.

Different livestock types differ in their input to ambient GHG pollution. The four-chambered stomach types of ruminant species like the horse, buffalo, sheep, and goat are the key contributors (Table 2.1). As a result of the stomach fermentation process and livestock manure (Lascano and Cárdenas, 2010), methane is created (Table 2.2). Further, by feed preparation and field control, they are responsible for indirect GHG releases into the atmosphere.

TABLE 2.1 Species-Wise Contribution to GHG Emissions into the Atmosphere

Livestock type	GHG contribution (%)
Sheep and goat (small ruminants)	6.7
Poultry	8.0
Buffalo	8.7
Pigs	9.0
Cattle	67.6

TABLE 2.2 Activity-Wise Contribution to GHG Emissions into the Atmosphere.

Activity type	GHG contribution (%)
Manure fermentation	6.0
Enteric fermentation	39.0
Feed production activities	45.0
Other	10.0

2.2 RUMINANT DIGESTIVE MECHANISM

We need to understand the digestive process of ruminants contributing to methane production and eventual emissions into the atmosphere before any further. It reaches their stomach via the oesophagus when animals consume food. With microbiota assistance, the ruminant stomach's rumen component is mainly responsible for the fermentation phase. There is a symbiotic relationship between rumen microbiota and ruminants in which later provide habitat to the former and receive final fermentation products and, in return, microbial proteins (Castillo-González, et al., 2014). Ruminants are adapted for digestive grasses and roughage (fiber rich feed) abundant in cellulose and hemicellulose. These animal groups have a significant capacity to digest the polysaccharides (compounds high in polymeric carbohydrates) (Karasov and Douglas, 2013), a reason for the abundance of microbial populations in their rumen. Bacteria, protozoa, and fungi responsible for breaking down complicated feed materials into simpler molecules are mostly found in the rumen microbiota and eventually form new compounds. Fermentation results in new combinations primarily comprising volatile fatty acids (VFAs) and digestive gaseous products, methane and hydrogen (Moss et al., 2000). VFAs are short-chain fatty acids that ruminants use as energy sources. They travel through the wall of the rumen and, through circulation, enter the liver. The transformation of complex polysaccharides gives animals the ability to digest cellulose and hemicellulose.

2.2.1   METHANOGENESIS IN RUMINANTS

Livestock, in particular, the ruminant production systems face the twin problems of improving food production on one side while attempting to reduce GHG emissions (Grossi et al., 2019) at the same time. Food supply-demand rises each passing day with the population boom; therefore, there is an urgent need to make the livestock industry profitable by using different methods to minimize GHG emissions. Several variables, including feed intake, feed type, and density, fermentation-related rumen and microbiota, and others, affect the ultimate development of methane in ruminant stomachs (Shibata and Terada, 2010). Hydrogen gas does not concentrate in the rumen, since various microbiota groups use it. Methane gas, however, accumulates and acts as a sink of hydrogen. Depending on different dietary aspects, ruminants are estimated to incur 2–12% of the feed's gross energy (Beauchemin et al., 2008). Due to methane development in the rumen, it is no longer beneficial to the host (Meale et al., 2012). Most of the methane (87%) in ruminants results from the rumen, and little is produced in the large intestines (13%). Methanogenesis contributes to energy loss, rendering it useless for further use for body metabolism, thus lowering feed use effectiveness (Gastelen, 2017). In short, the ruminants' digestive system function so that the gaseous digestive materials, mainly methane, are produced, contributing to inefficiency in energy use and ultimately leading to GHG emissions, climate change, and global warming.

Ruminants degrade complex and polymeric carbohydrates into monomeric units which are in turn converted into volatile fatty acids. The metabolism of volatile fatty acids contributes to producing an abundant number of chemicals, including hydrogen and carbon dioxide, which (Hook, 2010) acts as a virtual sink for methane. In a rumen system where carbon dioxide is decreased with protons' aid (Ellis et al., 2008), methanogenesis is responsible for methane formation. Methanogens, belonging to the archaea community, perform the conversion of substrates into methane. Methanogenesis is a complex and collaborative mechanism focused on hydrogen ions and the type of microbiota (Patra, 2017). Methanogenesis and methane emissions contribute between 40 and 45% of the global GHG emissions from ruminants and about 90% of enteric fermentation GHG emissions (Meale et al., 2012). Methanogenesis is favorable to ruminants as it favors the removal of hydrogen from rumen which otherwise inhibits the development of volatile fatty acids when in excess quantities. In other words, to retain a desirable rumen climate and successful, productive output of ruminants, the phenomenon of methanogenesis is required. The reported inhibition mechanism prevents reduced co-factors (NADPH, NADH, and FADH) from re-oxidizing (Ungerfeld, 2020).

In contrast, other possible mechanisms in the rumen are responsible for methane production, the most prevalent and vital being the formation by reducing CO_2 with hydrogen ions. The gaseous products of digestion are extracted by the eructation process from the animal stomach (Leng, 2018). The production and release of methane are an evolutionary adaptation mechanism aimed at eliminating excess amounts of hydrogen from ruminants' guts. If for some cause, the eructation process is halted or slowed, gases may accumulate, and digestion of the fiber in the stomach may stop (McAllister and Newbold, 2008). The amount of methane production in the rumen depends on different variables, such as the degree of feed ingestion, the nature of feed carbohydrates, and microflora. The average output of methane in cattle is around 250–500 L/day, contributing to energy inefficiency and environmental emissions. Therefore, meticulous conservation techniques may aid in the effective use of feed resources and make the ecosystem healthy.

2.2.2 OTHER METHANE OUTPUT SOURCES

In addition to the production of methane in large quantities in the ruminant stomach, other sources including fermentation during the preparation of manure, fertilizer use (organic and inorganic), agricultural mechanization, transportation, manufacturing, etc. (Kumari et al., 2020). Among them, manure's fermentation is the primary cause of GHG pollution from animals, while others are the secondary causes. During manure handling, the direct emissions of GHGs come in the form of CH_4 and N_2O. However, emissions rely on several variables, including manure quality and environmental conditions (predominantly the temperature during storage). The organic matter and nitrogen content of manure, which primarily influence its GHG emissions, are among the various composition aspects (Ren et al., 2017). Under anaerobic conditions contributing to GHG production, the organic matter in manure is partly decomposed. Abundant gases are created if the liquid slurry is made from manure and processed under anaerobic conditions. However, N_2O gas production is preferred under mixed aerobic and anaerobic conditions. Thus, the formation of CH_4 and N_2O is mainly preferred by solid dung and liquid slurry forms. With extended storage times, the amount of gases released increases under hot and humid conditions (Brouček and Čermák, 2015). N_2O is developed at manure stages by nitrification and denitrification. The nitrification process leading to the production of N_2O from ammonia and ammonium compounds is preferred in aerobic conditions. Denitrification, leading to the production of N_2O from nitrates, is followed under anaerobic conditions. The fertilizer/manure application on pastures is also responsible for significant atmospheric emissions of GHGs (Grossi et al., 2019). Acceptable practices in manure management may help to reduce the release of GHG from livestock rearing into the atmosphere. Feed production processes contribute significantly to GHG emissions from livestock, especially from non-ruminants and account for 60–80% of GHG emissions from non-ruminants, while the output from ruminants is 35–45%.

2.2.3 *MINIMIZING THE METHANOGENESIS PHENOMENON*

Methanogens are mainly responsible for the production of methane in ruminants among the various groups of rumen microbiota. They are, thus, the primary targets of experimentation aimed at reducing the production of methane. Methanogens are found in high abundance in ruminants (Meale et al., 2012), and the content of methanogens also varies based on the diet the ruminant has consumed (Hook, 2010). Thus, dietary modulation can help to minimize the rumen methane output and the resulting emissions into the outer world.

The most visible way to limit methanogenesis in the rumen is to divert hydrogen supply to other electron acceptors. The system, however, is thermodynamically unfavorable and the whole process returns to the standard route. Different forms of forages vary in their structure, including soluble and insoluble carbohydrates, together with lignification status. These forage properties affect the rumen fermentation process and subsequently on various gaseous materials in ruminants. Emission of GHGs may be decreased if grasses are cut at early growing stages and fed as such or after conversion into silage, attributable to the lower lignin content and higher soluble carbohydrate content favoring lower microbial growth and lower emissions of GHGs. Forage digestibility is also an essential factor affecting GHG emissions from ruminants as enhancing feed digestibility (Olijhoek et al., 2018) by various methods, including cutting, chaffing, soaking, grinding, and steam treatment, can help to minimize the GHG emission (Knapp et al., 2014). The development of acetogenesis, which entails converting acetate to acetyl-CoA and CO_2, is a potential means of lessening methanogenesis (Attwood et al., 2011). Figure 2.3 indicates the various routes available where the output of GHG can be undervalued.

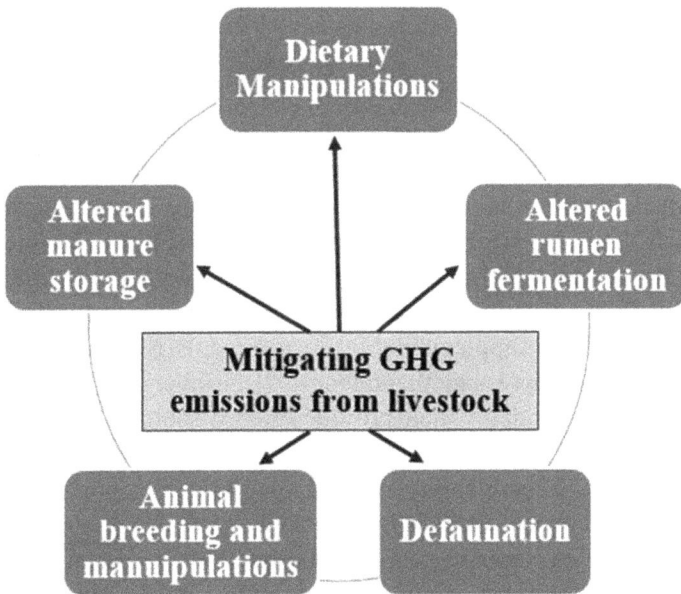

FIGURE 2.3 Summary of different available pathways wherein the GHG productions can be reduced.

2.3   DIETARY MANIPULATIONS

2.3.1   FEEDING FOCUS

Feeding focus is an essential factor vis-à-vis the GHG emission. Feeding extra concentrate to animals can help to increase the digestibility and reduce the emission of GHGs. It, however, would require large margins to raise concentrate volumes. Furthermore, animals may threaten medical conditions such as metabolic acidosis, metabolic alkalosis, etc. However, there is a rather complicated connection between GHG emissions and the balance of carbohydrate–protein in livestock feed. Feeding high-carbohydrate feed contributes to increased methane production, whereas feeding extra proteins results in elevated ammonia and nitrous oxide into the atmosphere (Haque, 2018). For such a strategy, an issue remains that such manipulations compromise multiple animal species' productivity. Besides, feeding additional concentrates as a mitigation technique against GHG emissions may prove to be inexpensive. There is a great need to examine various forage processing/manipulations to reduce rumen's gas output (Lascano and Cárdenas, 2010).

2.3.2   FEEDING FATS, OILS, AND LIPIDS

Words that refer to various substances but are mostly used interchangeably are fats, oils, and lipids. Fats' high-energy feed elements are solid at room temperature, whereas oils at room temperature are liquid. On the other side, lipids are high-energy plant-origin feed substances. The integration of these high-energy compounds into livestock feed helps to minimizing enteric methane emission by reducing organic matter and carbohydrate fermentation. Lipid supplementation in ruminants is considered one of the best ways to suppress methanogenesis (Martin et al., 2010). Fats, the energy-dense molecules which provide energy sources do not undergo metabolism in the rumen and, therefore, have no capacity for methane production. The effectiveness of lipid supplementation in minimizing GHG emissions in ruminants is influenced by multiple factors, including the degree of incorporation, source of fat and fatty acid content, and the physical type of supplementation (Meale et al., 2012). Fatty acids with carbon numbers ranging from 8 to 14 are most effective in reducing methane production in the rumen (Haque, 2018). The capacity of lipids in ruminants to minimize methanogenesis is proportional to the amount of long-chain fatty acids present therein (Beauchemin and McGinn, 2006). For ruminants, diets with fat/lipid content ranging from 5 to 8% are prescribed as more fat disturbs their digestive system. To eliminate GHG emissions from ruminants (Grainger and Beauchemin, 2011), a similar concentration of fat in livestock diets is proposed to further minimize up to 15% per unit of FPCM (Knapp et al., 2014). However, rapid food changes have been reported to be harmful to their digestive system by introducing fats to livestock feed. An additional downside is that the production of VFAs is simultaneously limited, along with GHG emissions from the use of essential oils. The detrimental impact on feed intake and digestion is an additional downside of feeding unnecessary lipids in the ruminant diet. These adverse effects can be negated by ensuring that the rumen's lipids are released slowly or by pre-processing feed before feeding (Meale et al., 2012).

2.3.3   THE ADDITIVES IN FEED

To decrease methane emissions in ruminants, supplementation of feed with compounds such as ionophores, organic acids, antibiotics, chemicals, and other such substances of plant or animal origin is essential, as they have been seen to reduce methane emissions (Castillejos et al., 2007). Ionophores, especially monensin, have been extensively studied in cattle for their mitigating effect on GHG emissions. Lasalocid, salinomycin, nigericin, and gramicidin are other ionophores used to minimize methane release in ruminants. The possible mechanism of action involves increased feed conversion efficiency, and altered fermentation favoring decreased the methane production in the rumen (Lascano and Cárdenas, 2010). However, ionophores' effect on GHG development in rumen does not last long as microbes respond to changed conditions in a significantly short period (Beauchemin et al., 2008). In addition to other factors, ionophores' mitigating impact on greenhouse gas concentrations depends on the type of ionophore used and the ionophore's dosage size.

Condensed tannins are complex compounds that bind to proteins and polysaccharides in the feed, decreasing the ability of feed to be digested in the rumen. Condensed tannins modify the rumen's metabolism, reducing the development of essential metabolites that, in turn, alter the methanogenesis rate, thus helping to mitigate ruminant GHG emissions.

However, excessive feeding of tannins can affect the body's metabolism and reduce the efficiency of animals. A drop-in fiber digestibility reduces hydrogen production in the rumen and is a possible mechanism of action of condensed tannins. However, the reduction of condensed tannin GHG emissions varies with various legumes and depends on tannins' concentration and composition.

Probiotics (Lactobacillus, Saccharomyces, and Enterococcus) and prebiotics are currently used in ruminants to improve their health status. To help reduce the development of GHGs in the rumen, probiotics, and prebiotics are reported; however, studies on this aspect are limited (Lascano and Cárdenas, 2010).

Organic acids, including malate, fumarate, pyruvate, etc., help to reduce the hydrogen production in the rumen, thus minimizing methane generation. These acids improve the concentration of propionate, which decreases the production of hydrogen, resulting in less methane formation in the rumen. However, direct feed supplementation with organic acids appears to be mostly uneconomical and is thus not favored.

2.3.4   MANIPULATION OF ANIMALS

Reducing the number of cattle being reared would lead to decreased emissions of GHGs. However, it can decrease production and hamper the achievement of goals. GHG emissions are lowered as livestock productivity increases. Therefore, productivity needs to be increased urgently so that demand stays steady and emissions of GHGs decrease. Besides, animal productivity is multifactorial, mainly based on an animal's genetic structure and its environment. Therefore, animal productivity can be increased either by using elite animals that are genetically superior or by creating environmental conditions that guarantee a specific animal's optimal potential. However, there is an association between an animal's genetic structure and the external factors to which it is subjected. Therefore, to increase animals' productivity to the highest possible standards, the criteria for the best genetic makeup and ideal environmental conditions should be maintained. In terms of their GHG emissions, human variation also occurs between species. Better breeding plans are required to reproduce elite animals for full productivity. Breeding would have fewer food needs for elite animals higher in efficiency, and GHG emissions will also be lowered (Lascano and Cárdenas, 2010). An efficient animal is more likely to retain more protein in its body than voiding it out via urine and feces. In recent years, broilers' production quality has dramatically led to a decline in poultry GHG emissions (Williams et al., 2015). However, the need for their nutrition and management is also growing for highly profitable animals, which may not be viable under the production systems currently being practiced, especially in developing countries. In addition to genetic selection and animal improvement by effective breeding plans, enhanced animal health status improved reproductive efficiency. It decreased mortality figures provide a low to medium potential for minimizing greenhouse gas discharges into the environment from livestock rearing (Grossi et al., 2019). Improved animal fertility will substantially reduce methane (10–24%) and nitrous acid (9–17%) pollution from the livestock industry (Grossi et al., 2019).

2.4   VACCINES AND DEFAUNATION

Defaunation registered as a process to help decrease GHG emissions in ruminants, refers to the mechanism by which some protozoan organisms in the rumen are removed. The altered population of methanogens (archaea bacteria) and decreased hydrogen production are likely the pathways to minimize GHG pollution. To minimize GHG emissions from animals, any feed additive that reduces the rumen's protozoal activity would help. Saponins are one of the significant compounds with anti-protozoan activity (defaunation). The degree of defaunation depends on saponin applied to the feed, the dose, and administration quality. Several plant extracts have been reported to reduce the activity of protozoa (Hess et al., 2003; Abreu et al., 2004; Agarwal et al., 2009). Similarly, to help reduce methane development in ruminants, immunogenic substances, including vaccines, also identified. In recent times, many immunogenic compounds that limit the production/emission of methane from livestock have been tried and patented.

2.5   MANURE STORAGE CONDITION MANIPULATION

One of the key features influencing GHG production from fermentation during the storage of manure is temperature. Lowering storage temperatures minimize the emission of GHGs. The pollution reduction is, however, still contingent on other factors. Methane and nitrous oxide emissions can be lowered to amounts of 55 and 41%, respectively, if the manure is removed from livestock farms at frequent intervals in the form of slurry (Mohan-kumar et al., 2018). Using slated and grooved floors, by this strategy, help to minimize pollution. Similarly, in modern poultry housing schemes, belt scrapers can help to eliminate manure from poultry farms and reduce GHG pollution routinely. During the preparation of manure (Grossi et al., 2019), the separation of manure into solid–liquid phases is stated to have a high potential to reduce methane emissions. The mitigating capacity of separation treatment is estimated to be 30% higher than that of untreated manure. GHG emissions can be altered by the modification of exposure to aerobic/anaerobic environments. During preparation and injection for land use, the slurry can reduce GHG emissions from manure (Holly et al., 2017). The use of manure for the processing of biogas serves two goals: the availability of energy supply and eliminating GHG emissions. Compared to conventional manure use processes, the anaerobic digestion process for biogas processing helps to minimize pollution by up to 30% (Battini et al., 2014).

2.6 CONCLUSIONS

Despite its GHG pollution estimates, agriculture and related industries especially the animal husbandry remains an essential part of modern society, primarily because of its capacity to assist the ever-growing population of the planet in maintaining nutritional stability. However, the meticulous execution of effective mitigation measures helps to reduce the emission of greenhouse gases from livestock. The management methods aimed at minimizing GHG release from livestock include dietary modulation, proper feeding of forages, enrichment of feed (condensed tannins, saponins, ionophores, organic acids), and secondary metabolites in feed, immunization, and introduction of biotechnological procedures. Different mitigating techniques vary in their ability, durability of results, economic feasibility, and adoption by farmers. However, before either of these methods is considered practically, its entire life cycle's impact needs to be studied through its metabolism and interactive analysis.

KEYWORDS

- **environment**
- **greenhouse gases**
- **production**
- **manipulation**
- **livestock**

REFERENCES

Abreu, A.; Carulla, J. E.; Lascano, C. E.; Díaz, T. E.; Kreuzer, M.; Hess, H. D. Effects of Sapindus Saponaria Fruits on Ruminal Fermentation and Duodenal Nitrogen Flow of Sheep Fed a Tropical Grass Diet with and without Legume. *J. Anim. Sci.* 2004. https://doi.org/10.2527/2004.8251392x.

Agarwal, N.; Shekhar, C.; Kumar, R.; Chaudhary, L. C.; Kamra, D. N. Effect of Peppermint (Mentha Piperita) Oil on in Vitro Methanogenesis and Fermentation of Feed with Buffalo Rumen Liquor. *Anim. Feed Sci. Technol.* **2009.** https://doi.org/10.1016/j.anifeedsci.2008.04.004.

Attwood, G. T.; Altermann, E.; Kelly, W. J.; Leahy, S. C.; Zhang, L.; Morrison, M. Exploring Rumen Methanogen Genomes to Identify Targets for Methane Mitigation Strategies. *Anim. Feed Sci. Technol.* **2011.** https://doi.org/10.1016/j.anifeedsci.2011.04.004.

Battini, F.; Agostini, A.; Boulamanti, A. K.; Giuntoli, J.; Amaducci, S. Mitigating the Environmental Impacts of Milk Production via Anaerobic Digestion of Manure: Case Study of a Dairy Farm in the Po Valley. *Sci. Total Environ.* **2014.** https://doi.org/10.1016/j.scitotenv.2014.02.038.

Beauchemin, K. A.; Kreuzer, M.; O'Mara, F.; McAllister, T. A. Nutritional Management for Enteric Methane Abatement: A Review. *Austr. J. Exp. Agri.* **2008.** https://doi.org/10.1071/EA07199.

Beauchemin, K. A.; McGinn, S. M. Methane Emissions from Beef Cattle: Effects of Fumaric Acid, Essential Oil, and Canola Oil. *J. Anim. Sci.* **2006.** https://doi.org/10.2527/2006.8461489x.

Brouček, J.; Čermák, B. Emission of Harmful Gases from Poultry Farms and Possibilities of Their Reduction. *Ekol. Bratislava* **2015.** https://doi.org/10.1515/eko-2015-0010.

Castillejos, L.; Calsamiglia, S.; Ferret, A.; Losa, R. Effects of Dose and Adaptation Time of a Specific Blend of Essential Oil Compounds on Rumen Fermentation. *Anim. Feed Sci. Technol.* **2007.** https://doi.org/10.1016/j.anifeedsci.2006.03.023.

Castillo-González, A. R.; Burrola-Barraza, M. E.; Domínguez-Viveros, J.; Chávez-Martínez, A. Rumen Microorganisms and Fermentation. *Archivos de Medicina Veterinaria.* **2014.** https://doi.org/10.4067/S0301-732X2014000300003.

Ellis, J. L.; Dijkstra, J.; Kebreab, E.; Bannink, A.; Odongo, N. E.; McBride, B. W.; France, J. Aspects of Rumen Microbiology Central to Mechanistic Modelling of Methane Production in Cattle. *J. Agric. Sci.* **2008**. https://doi.org/10.1017/S0021859608007752.

FAO (Food and Agriculture Organization of the United Nations). No Title. In: *Global Agriculture Towards 2050.* High Level Expert Forum Issues Paper; Rome, 2009.

Gastelen, S. van. *Predicting Methane Emission of Dairy Cows Using Milk Composition*; Wageningen University. 2017. https://doi.org/10.18174/425382

Grainger, C.; Beauchemin, K. A. Can Enteric Methane Emissions from Ruminants Be Lowered without Lowering Their Production? *Anim. Feed Sci. Technol.* **2011**. https://doi.org/10.1016/j.anifeedsci.2011.04.021.

Grossi, G., Goglio, P.; Vitali, A.; Williams, A. Livestock and Climate Change: Impact of Livestock on Climate and Mitigation Strategies. *Anim. Front.* **2019**, *9* (1), 69–76.

Haque, M. N. Dietary Manipulation: A Sustainable Way to Mitigate Methane Emissions from Ruminants. *J. Anim. Sci. Technol.*. **2018**. https://doi.org/10.1186/s40781-018-0175-7.

Hess, H. D.; Kreuzer, M.; Díaz, T. E.; Lascano, C. E.; Carulla, J. E.; Soliva, C. R.; Machmüller, A. Saponin Rich Tropical Fruits Affect Fermentation and Methanogenesis in Faunated and Defaunated Rumen Fluid. *Anim. Feed Sci. Technol.* **2003**. https://doi.org/10.1016/S0377-8401(03)00212-8.

Holly, M. A.; Larson, R. A.; Powell, J. M.; Ruark, M. D.; Aguirre-Villegas, H. Greenhouse Gas and Ammonia Emissions from Digested and Separated Dairy Manure during Storage and after Land Application. *Agric. Ecosyst. Environ.* **2017**. https://doi.org/10.1016/j.agee.2017.02.007.

Hook, S. E.; Wright, A. D. G.; McBride, B. W. Methanogens: Methane Producers of the Rumen and Mitigation Strategies. *Archaea.* **2010**. https://doi.org/10.1155/2010/945785.

Karasov, W. H.; Douglas, A. E. Comparative Digestive Physiology. *Compr. Physiol.* **2013**. https://doi.org/10.1002/cphy.c110054.

Knapp, J. R.; Laur, G. L.; Vadas, P. A.; Weiss, W. P.; Tricarico, J. M. Invited Review: Enteric Methane in Dairy Cattle Production: Quantifying the Opportunities and Impact of Reducing Emissions. *J. Dairy Sci.* **2014**. https://doi.org/10.3168/jds.2013-7234.

Kumari, S.; Fagodiya, R. K.; Hiloidhari, M.; Dahiya, R. P.; Kumar, A. Methane Production and Estimation from Livestock Husbandry: A Mechanistic Understanding and Emerging Mitigation Options. *Sci. Total Environ.* **2020**. https://doi.org/10.1016/j.scitotenv.2019.136135.

Lascano, C. E.; Cárdenas, E. Alternatives for Methane Emission Mitigation in Livestock Systems. *Rev. Bras. Zootec.* **2010**. https://doi.org/10.1590/s1516-35982010001300020.

Leng, R. A. Unravelling Methanogenesis in Ruminants, Horses and Kangaroos: The Links between Gut Anatomy, Microbial Biofilms and Host Immunity. *Anim. Product. Sci.* **2018**. https://doi.org/10.1071/AN15710.

Martin, C.; Morgavi, D. P.; Doreau, M. Methane Mitigation in Ruminants: From Microbe to the Farm Scale. *Animal* **2010**. https://doi.org/10.1017/S1751731109990620.

McAllister, T. A.; Newbold, C. J. Redirecting Rumen Fermentation to Reduce Methanogenesis. *Austr. J. Exp. Agric.* **2008**. https://doi.org/10.1071/EA07218.

Meale, S. J.; McAllister, T. A.; Beauchemin, K. A.; Harstad, O. M.; Chaves, A. V. Strategies to Reduce Greenhouse Gases from Ruminant Livestock. Acta Agriculturae Scandinavica A: *Anim. Sci.* **2012**. https://doi.org/10.1080/09064702.2013.770916.

Mohankumar Sajeev, E. P.; Winiwarter, W.; Amon, B. Greenhouse Gas and Ammonia Emissions from Different Stages of Liquid Manure Management Chains: Abatement

Options and Emission Interactions. *J. Environ. Qual.* **2018**. https://doi.org/10.2134/jeq 2017.05.0199.

Moss, A. R.; Jouany, J. P.; Newbold, J. Methane Production by Ruminants: Its Contribution to Global Warming. *Anim. Res.* **2000**. https://doi.org/10.1051/animres:2000119.

Olijhoek, D. W.; Løvendahl, P.; Lassen, J.; Hellwing, A. L. F.; Höglund, J. K.; Weisbjerg, M. R.; Noel, S. J.; McLean, F.; Højberg, O.; Lund, P. Methane Production, Rumen Fermentation, and Diet Digestibility of Holstein and Jersey Dairy Cows Being Divergent in Residual Feed Intake and Fed at 2 Forage-to-Concentrate Ratios. *J. Dairy Sci.* **2018**. https://doi.org/10.3168/jds.2017-14278.

Patra, A.; Park, T.; Kim, M.; Yu, Z. Rumen Methanogens and Mitigation of Methane Emission by Anti-Methanogenic Compounds and Substances. *J. Anim. Sci. Biotechnol.* **2017**. https://doi.org/10.1186/s40104-017-0145-9.

Ren, F.; Zhang, X.; Liu, J.; Sun, N.; Wu, L.; Li, Z.; Xu, M. A Synthetic Analysis of Greenhouse Gas Emissions from Manure Amended Agricultural Soils in China. *Sci. Rep.* **2017**. https://doi.org/10.1038/s41598-017-07793-6.

Ribeiro Pereira LG, Machado FS, Campos MM, Guimaraes Júnior R, Tomich TR, Reis LG, C. C. Enteric Methane Mitigation Strategies in Ruminants: A Review. *Rev. Colomb. Ciencias Pecu.* **2015**, *28* (2), 124–143.

Sejian, V.; Bhatta, R.; Malik, P. K.; Madiajagan, B.; Ali, Y.; Al-Hosni, S.; Sullivan, M. G. J. Livestock as Sources of Greenhouse Gases and Its Significance to Climate Change. *InTechOpen* **2016**. https://doi.org/DOI: 10.5772/62135.

Shibata, M.; Terada, F. Factors Affecting Methane Production and Mitigation in Ruminants. *Anim. Sci. J.* **2010**. https://doi.org/10.1111/j.1740-0929.2009.00687.x.

Solomon, S., D.; Qin, M.; Manning, Z.; Chen, M.; Marquis, K. B.; Averyt, M. T.; Miller HL; Solomon, S.; Qin, D.; Manning, M.; Chen, Z.; Marquis, M.; Averyt, K. B.; Tignor, M.; Miller, H. L. Summary for Policymakers. In *Climate Change 2007: The Physical Science Basis. Contribution of Working Group I to the Fourth Assessment Report of the Intergovernmental Panel on Climate Change.* Qin, D.; Manning, M.; Chen, Z.; Marquis, M.; Averyt, K.; Tignor, M.; Mill, H. L., Eds.; Cambridge University Press: New York, 2007. https://doi.org/10.1038/446727a.

Steinfeld, H.; Gerber, P.; Wassenaar, T.; Castel, V.; Rosales, M. H. C. Livestock's Long Shadow: Environmental Issues and Options; FAO: Rome, 2006.

Ungerfeld, E. M. Metabolic Hydrogen Flows in Rumen Fermentation: Principles and Possibilities of Interventions. *Front. Microbiol.* **2020**. https://doi.org/10.3389/fmicb.2020.00589.

Williams, A.; Chatterton, J.; Hateley, G.; Curwen, A.; Elliott, J. A Systems-Life Cycle Assessment Approach to Modelling the Impact of Improvements in Cattle Health on Greenhouse Gas Emissions. *Adv. Anim. Biosci.* **2015**. https://doi.org/10.1017/s20404700 14000478.

Yue, X. L.; Gao, Q. X. Contributions of Natural Systems and Human Activity to Greenhouse Gas Emissions. *Adv. Clim. Chang. Res.* **2018**. https://doi.org/10.1016/j.accre.2018.12.003.

CHAPTER 3

Biogeochemical Processes and Climatic Change

IRSHAD A. LONE[1*] and MAHAJABEEN AKHTER[2]

[1]*Department of Chemistry, Govt. Degree College Handwara, Higher Education Department, Jammu & Kashmir, India*

[2]*Department of Bioresources, University of Kashmir, Srinagar, Jammu & Kashmir, India*

Corresponding author. E-mail: drirshadlone@gmail.com

ABSTRACT

Climate change is presumably one of the most growing challenges that the world is globally facing today. It is a complex and multidimensional phenomenon, which refers to the changes over time in the average surface temperature of the earth and associated changes in its atmosphere, hydrosphere, cryosphere, and biosphere. Growing studies suggest most of the climatic challenge over the last several decades is directly attributed to anthropogenic activities and the subsequent release of gases, notably methane (CH_4), carbon dioxide (CO_2), and nitrous oxide (N_2O), responsible for global warming. The terrestrial and aquatic biogeochemical processes have been found as key components of the earth's climate system as they control atmospheric concentrations of notable greenhouse gases which exert a warming effect. Around half of the anthropogenic carbon emissions are being contained by terrestrial and aquatic ecosystems, thereby, significantly reducing the footprint of CO_2 gas in the atmosphere and limit the global climate challenge. Since microorganisms constitute an essential part for the sustenance of higher living forms in both terrestrial and aquatic ecosystems, their role in global climatic change cannot be ignored. In this chapter, we have documented the crucial role and responses of microorganisms in climatic change and other human activities. Through the existing literature, attempts have been made to assess the underlying biogeochemical mechanisms by which microbes discharge and take up greenhouse gases. The ensuing effects as well as feedbacks in both terrestrial and aquatic ecosystems have also been documented. This chapter shall also underline the scope of better understanding of the role of microorganisms in climatic change for a sustainable future.

3.1 INTRODUCTION

Today, the world is facing certain unprecedented changes in climate and environment in human history. This is being demonstrated at the global level by the melting of the polar ice caps and at the local level by the more extreme weather conditions, such as heavy rains and floods. The average global temperature is soaring up continuously, with around 1.5°F raised up during the last century alone (Dutta and Dutt, 2016). Experts suggest it will continue to rise by around 0.5–8.6°F in another 100 years (Climate Change: Basic Information. US EPA, 2015). This poses a serious global threat as slight fluctuations in average temperature could lead to catastrophic shifts in climatic change. Microbes, which remind the beginning of life on earth some 3800 million years ago, constitute a major portion of the earth's biosphere and hold great significance in this backdrop. They determine the effluxes of notable gases responsible for the greenhouse effect, such as CH_4, CO_2, and N_2O, which are considered to as essential components and major feedback responses to global changes (Falkowski et al., 2008). Microbes have the ability to sustain in almost all environmental conditions on the earth, such as deep subsurface and tough environments, and as a result, are crucial in regulating climatic change (Cavicchioli et al., 2019). They exhibit a critical role by both serving as producers and users of greenhouse gases in the atmosphere by recycling essential elements like carbon and nitrogen. Heterotrophic microbes on one side decompose organic matter to emit greenhouse gases; photosynthetic microbes on the other end utilize atmospheric CO_2. A balance between the two processes on either side is the key factor for determining the net carbon flux. Studies on the role of climate change on microbes and their biogeochemical processes are considered as crucial aspects to predict the future impact of climate change on all forms of life (Weiman, 2015). Having generation times as short as a few hours, microbes provide an ideal platform for better understanding of the effects of climate change on biological systems and the global biogeochemical cycles that microbes mediate. It would be prudent, therefore, to learn and understand much more about the complex and intricate microbial functions and biogeochemistry associated with it so that the future dynamics of global climate change could be precisely forecasted (Singh et al., 2010). In this chapter, we briefly illustrate the role of microbial contributions and responses to climatic changes across the global systems (terrestrial and aquatic ecosystems) and their tendency in broader sense to accelerate or decelerate the impact of anthropogenic climatic change. Attempts have been made to develop a better understanding of various biogeochemical mechanisms of microbes from the prism of climatic change. Microbial feedbacks and their consequent effects in climatic change mediation for a sustainable future have also been documented.

3.2 CLIMATIC CHANGE

Climatic change is defined as the change in the state of the earth's climate resulting from the combination of biotic processes, variations in solar radiation, volcanic eruptions, and human activity. It can be identified by shifts in the mean and/or the variability of its properties that persist for an extended period, typically decades or longer. Climate change may occur due to either natural internal processes or external forces, such as modulations of the solar cycles, volcanic eruptions, and persistent anthropogenic changes in the composition of the atmosphere or due to its inclination toward land use (Agard et al., 2014). The earth's climate is directly dependent on the latitude of the earth, the tilt of its axis, the movements of wind belts over it, the topography of the land and oceans, and their difference in temperatures. The earth nurses a favorable climate, which lets plants, animals, and microbes to live in. This is possible due to the thick layer of gases surrounded by the earth, which acts like a blanket and keeps the planet warm. Without this blanket, the earth would be 20°C–30°C colder and least appropriate for life to exist. Due to the rise in average temperature across the length and breadth of the world, the climatic change becomes inevitable. This leads to the heating up of the earth, and subsequently, the phenomenon of global warming wherein a rise in temperature of the global atmosphere occurs over a phase of time. The envelope of gases engulfing the earth gets much denser, and in this manner, traps more heat in the atmosphere tending the planet to warm up further.

The greenhouse effect is observed whereby the earth's atmosphere in the presence of gases such as CO_2, CH_4, and water vapor traps solar radiation by allowing the radiating sunlight to enter, but absorbs the heat given out by the earth's surface. This provides a blanketing effect in the lower strata of the earth's atmosphere, which is being further augmented due to human anthropogenic actions like industrializations, burning of fossil fuels, and agricultural usage (Olufemi et al., 2014). In recent years, emissions of greenhouse gases have witnessed a dramatic rise owing to human activity and other natural issues like volcanic eruptions. The notable greenhouse gases such as CH_4, CO_2, N_2O, and halocarbons accumulate in the earth's atmosphere, causing concentrations to increase with the passage of time. A significant upsurge in these gases with predominant changes in climate has been observed in both terrestrial and aquatic (fresh water and marine water) systems.

3.3 SCOPE AND RELEVANCE OF TERRESTRIAL AND AQUATIC ECOSYSTEMS IN MICROBIAL WORLD

3.3.1 TERRESTRIAL ECOSYSTEM AND MICROBES

The soil provides an enormous pool of dynamic carbon, and as a result, serves as a crucial factor in the development of global climate. An estimate of around 2000 billion tonnes of organic carbon is present in the soil, which is more than the combined pool of carbon in the atmosphere and vegetation (Singh et al., 2010). The exchange of CO_2 between soil and atmosphere serves as a major component of the global carbon cycle, and it has been estimated that about 10% of the CO_2 of the atmosphere percolates through soil annually. The CO_2 flux from the terrestrial surface, which generates emissions of 50–75 Pg (1 Pg = 1000 billion tonnes) of carbon per annum, acts as a chief contributor to the cycling of global carbon (Behrenfeld, 2014). There are about 10^{29} microbes in terrestrial atmosphere similar to the total number of microbes in maritime atmosphere (Flemming and Wuertz, 2019). Around 120 billion tonnes of carbon is being consumed up on an annual basis by autotrophic soil microbes, while heterotrophic soil microbes cumulatively give out about 119 billion tonnes of carbon (Singh et al., 2010). Plants through the process of photosynthesis take away CO_2 from the environment and generate organic stuff that provides energy to terrestrial ecological units. On the contrary, respiration (autotrophic) by plants and respiration (heterotrophic) by microbes discharge CO_2 back into the atmosphere (Ballantyne et al., 2017). Temperature plays a key role in the equilibrium between these conflicting forces, and as a result, the capability of the terrestrial ecological units to confine and stock up anthropogenic carbon emissions. Warming is found to step up carbon release into the atmosphere (Ballantyne et al., 2017). Forests cover, which constitutes about 30% of the earth surface, holds about 45% of global carbon and sequester about 25% of anthropogenic CO_2 (Pan et al., 2011). Non-forested, arid, and semiarid regions covering about 47% of the terrestrial surface respond in a different way to anthropogenic climate change than forested regions and are vital for global carbon change (Hovenden et al., 2019). Lakes that constitute about 4% of the non-glaciated land area are responsible for emission of a considerable amount of carbon in the form of CH_4 gas (Verpoorter et al., 2014). Decomposed plant litter, commonly called peat, covers about 3% of the land surface. Such peatlands, which contain about 30% of global soil carbon, serves as an important global carbon sink (Gallego and Prentice, 2013). Permafrost also constitutes the largest terrestrial carbon sink of organic matter (Hultman et al., 2015). Studies have shown that climate warming of around 1.5°C–2°C is enough to significantly reduce the permafrost by about 28%–53% (Hoegh et al., 2018), and as a result, puts huge carbon pools accessible for microbial respiration and greenhouse gas discharges. Forecasts from global assessments of responses to increased temperatures indicate that terrestrial carbon loss due to warming is yielding a positive feedback that will accelerate the rate of climate change especially in cold and temperate-carbon-rich soils (Crowther et al., 2016).

3.3.2   AQUATIC ECOSYSTEM AND MICROBES

The aquatic ecosystems cover up to about 70% of earth's surface, ranging from coastal estuaries, mangroves, and coral reefs to the open oceans, and as a result, have a critical role in the issue of climate change. The upper oceanic layer forms an interface in which CO_2 and other greenhouse gases have mounted in the course of these years, whereas the deep-sea layer has served as the most alarming carbon storehouse in the world. The microbes in the ocean ecosystem process around 6×10^{12} kg (6 gigatonnes) of carbon annually. Studies have shown that the oceans and the soil ecosystems together represent a world carbon sink of about 3×10^{12}

kg (3 gigatonnes) of carbon, among which CO_2 emissions of about 40% get absorbed from fossil fuel burning (Singh et al., 2010). Marine life in deeper zones uses organic and inorganic chemicals for energy, whereas phototrophic living systems particularly microbes utilize the sun's energy in the upper 200 m of the oceanic water column, which influence the composition and behavior of marine life communities, accordingly. The variation in water temperature (ranging from around −2°C in ice-covered oceans to >100°C in hydrothermal vents) also affects the composition and behavior of marine life systems (Sunagawa et al., 2015). Besides altering the biological processes, rising temperatures reduces water thickness and in this manner stratification and circulation, which subsequently affects organismal distribution and nutrient mobility. Through the fixations of carbon, nitrogen, and remineralization of organic stuff, marine microbes constitute the foundation of seafood webs and, consequently, global carbon and nutrient cycles (Azam and Malfatti, 2007). The sedimentation of carbon that is fixed in the form of organic matter to marine deposits is a plausible, long-term mechanistic approach for quenching CO_2 from the atmosphere. It strikes an equilibrium between restoration of CO_2 and nutrients by means of demineralization against burial in the ocean floor, and as a result regulates the impact of climate change. The acidity of oceans also serves as a key agent in forecasting the climate change. Oceans have recorded an acidification of around 0.1 pH units since preindustrial times, and further reductions of 0.3–0.4 pH units are estimated by the end of this century (Hurd et al., 2018).

3.3.3   MICROBES IN CLIMATIC CHANGE

The role of microbes is of paramount importance and has been a deter-mining factor in the atmospheric concentration of key greenhouse gases such as CH_4, CO_2, and N_2O in the streak of global history (Singh et al., 2010). Microbes generate and eliminate greenhouse gases since the time they evolved in the ocean some 3500 million years ago and later their shift to land some 2000 million years ago (Zimmer, 2010). They mediate chief biogeochemical processes responsible for greenhouse gas fluctuations among soils and the atmosphere (Falkowski et al., 2008). Microbes have responded to numerous alterations in global climate, which has too, in return, impacted them (Zimmer, 2010). In reality, the effect of climate change has led to a plethora of changes in microbial world.

From the change in diversity and composition of microbes to reduction in their biomass, the climatic change exerts direct and indirect changes in microbial metabolic activity even resulting in their death eventually. Microbes demonstrate negative or positive feedbacks in terms of their physiology and greenhouse gases emission against the climatic change. As the temperature rises, microbial community structures register change in their processes of respiration, fermentation, and methanogenesis. For biotic and abiotic components, the impact of climate change resulted in manifestations of unprecedented floods, intense storms, wildfires, heat waves, extreme heat conditions, drought, substandard air quality, and other natural disasters. The responses of microbes across different kingdoms (such as viruses, bacteria, and fungi) in climate change have also been documented. From accelerating global warming by means of organic material decay to the upsurge in the fluctuation of CO_2 concentrations in the air, microbes lead to a positive feedback to rising global temperatures (Baron et al., 2018). Microbes register a change in their enzyme activity and physiological property on account of varying temperatures. Finally, the efficiency of microbes in utilizing carbon decides the carbon response to climate change.

3.4 BIOGEOCHEMICAL PROCESSES OF MICROBES IN CLIMATIC CHANGE (A MECHANISTIC APPROACH)

In 1926, the term "biogeochemistry" was first introduced by V. Vernadsky, which involves the merging or linking of three scientific disciplines, namely biology, geology, and chemistry. The discipline intends to understand intricate processes, preferably microbial-mediated processes, which transform and recycle both organic and inorganic substances in soils, sediments, and waters. Catalyzed by bacteria and archaea, such processes with diverse and highly evolved cellular mechanisms maintain the entire biosphere. Microbial processes have played a central role as drivers and responders of climate change. They indulge in the global changes of the chief biogenetically generated greenhouse gases, such as CO_2, CH_4, and N_2O, and are expected to react swiftly to climate change. The net effect of microbes on biogeochemical cycles has been diverse over the years, and microbes have both contributed to and mitigated changes in the earth's climate. The potential to mitigate climate change by plummeting greenhouse gas emissions through regulating terrestrial and aquatic microbial processes is an exciting prospect for the future (Endeshaw et al., 2018). By altering their community structures and composition, microbes provide an effective feedback reaction mechanism for climatic change. In this way, environmental crisis is resolved by employing the processes of nutrient cycling and invigorating their well-designed genetic material for mortifying and eradicating substances or gases that promotes global warming (Endeshaw et al., 2018). Biogeochemical cycles and microbial communities when connected together can serve as an effective mechanistic tool to solve climate change. In the following section, we describe the impact of microbes on these cycles. The impact of climatic change directly and indirectly on microbial communities (terrestrial and aquatic) and their corresponding biogeochemical processes have also been illustrated.

3.4.1   MICROBES: BIOGEOCHEMICAL PROCESS AND GREENHOUSE GASES

3.4.1.1   CARBON DIOXIDE (CO_2) CYCLE

The existing levels of atmospheric CO_2 largely depend upon the balance between photosynthesis and respiration. In marine ecosystems, photosynthesis is principally carried out by phytoplankton, whereas autotrophic and heterotrophic respiration recovers much of the carbon taken up during photosynthesis to the dissolved inorganic carbon pool (Arrigo, 2005). On the contrary, the intake of CO_2 from the atmosphere in ecosystems of terrestrial world is showing dominance of plants. Microbes add significantly to net exchange of carbon by means of disintegration processes and heterotrophic respiration. They also contribute indirectly to it by acting as plant pathogens or symbionts and by amending nutrient accessibility in the soil. Microbes have the ability to either consume inorganic carbon as CO_2 through photosynthesis or chemosynthesis and storing it as organic matter in the form of cell mass, or consume organic matter in the course of respiration. Such phenomena ultimately produce CO_2 or CH_4 as metabolic byproducts that are released as gases into the ocean or atmosphere. CO_2 subsists as a gaseous substance in the air. Before getting incorporated into living systems, it must be converted into usable organic form. The conversion process through which CO_2 is utilized from the aerial pool and fixed in the form of organic matter is termed as carbon fixation. The well-known illustration of carbon fixation is photosynthesis, a phenomenon by which energy obtained from sunlight is exploited to synthesize organic molecules. Important microbes called photosynthetic algae show such a type of carbon fixation. Chemoautotrophic microbes, such as archae and bacteria, are proficient in converting CO_2 into sugar type and thereby making it accessible for cell formation. During respiration, a part of organic carbon is released back into the atmosphere in the form of CO_2. The remaining organic carbon is cycled among organisms through the food chain. After the death of an organism, the decomposition process starts with the help of bacteria during which carbon of the dead organism is discharged into the soil or the atmosphere. The CO_2 produced in this way gets recycled by dissolved in the water and reaches soil where algae, plants, and bacteria convert it into organic carbon again. Carbon may then transfer among organisms from producers to consumers. In the decaying process, their tissues are finally broken down by means of bacterial degradation and CO_2 is freed back to the atmosphere or ocean (Reay and Grace, 2007; Carney et al., 2007). In aquatic habitats, the carbon cycling is also established by a variety of fungi and bacteria. This carbon cycling entails the conversions of carbon in the presence or absence of oxygen. Algal involvement is an aerobic process of carbon conversion; while fermentation, where microbes cycle the carbon compounds to yield energy, is an anaerobic process of carbon conversion. The deep mud of lakes, ponds, and other water bodies provides the regions for anaerobic conversions of carbon cycle. Microbes, such as green and purple sulfur bacteria, are also able to participate in the cycling of carbon. They degrade carbon compounds by utilizing the energy gained from degradation of hydrogen sulfide (H_2S). Other microbes like *Thiobacillus ferrooxidans* degrade carbon by utilizing the energy retrieved from the removal of an electron from compounds containing iron. The carbon degradation anaerobically is, thus, a joint effort of numerous bacteria, such as *Clostridium butyricum, Bacteroides succinogenes,* and *Syntrophomonas* species. Such bacterial group effort is named as *inter-species hydrogen transfer*, which is ultimately accountable for bulk of CO_2 release into the atmosphere.

3.4.1.2 *METHANE (CH$_4$) CYCLE*

This cycle entails the transformation of sugars (organic residues) into CH_4 by the processes of methanogenesis, which is accomplished under anoxic conditions by a specific group of archaea called *methanogens*. However, a large amount of CH_4 liberated into soils is used as an alternative source of energy by *methanotrophs*. This occurs generally through an aerobic process, where the accessibility of oxygen acts as a rate-limiting factor. Some atmospheric CH_4 is also oxidized by methanotrophs. The CO_2 released in this manner of methane oxidation then makes its way into the CO_2 cycle. Microbes, thus, regulate global emissions of CH_4 presumably more directly than they regulate the emissions of CO_2. Studies have shown that annual natural emissions of around 2.5×10^2 million tonnes of CH_4 are mediated by microbial process of methanogenesis, a biogeochemical phenomenon performed by a cluster of archaea in anaerobic environments of wetlands, rumens, termite guts, and oceans. Conversely, such natural habitations are surpassed by emanations from anthropogenic actions such as fossil fuel extraction, rice cultivation, livestock farming, and landfill, in which around 3.2×10^2 million tonnes of CH_4 are emitted per annum. In recent years, atmospheric CH_4 levels have shown a sharp rise (2014–2017), but the reasons are ambiguous so far, despite the fact that increased emissions have been attributed to methanogens and/or fossil fuel industries and/or reduced atmospheric CH_4 oxidation, as a result posturing a severe threat in containing climate warming (Nisbet et al., 2019). Bacterial methanotrophs act as an important buffer to the vast quantities of CH_4 released in a few of these environments. Type I methanotrophs or low-affinity methanotrophs, which remain instrumental merely at > 40 parts per million (ppm) concentration of CH_4, can often use a huge proportion of the CH_4 liberated into soils prior to its escape to the atmosphere. Such methanotrophs mainly belong to the class Gamma proteobacteria. Bacterial methanotrophs may also serve as a sink for CH_4 already existing in the air. Type II methanotrophs or high-affinity methanotrophs, which remain instrumental merely at < 12 ppm concentration of CH_4, can eliminate about 30 million tonnes of CH_4 from the air per annum. Such methanotrophs mainly belong to the class α-proteobacteria.

3.4.1.3   NITROGEN (N₂) CYCLE

As with the carbon cycle, the earth's nitrogen pool is continually recycled. All organisms need nitrogen to grow, but only a few can use the gaseous form of nitrogen (N_2) that is in the atmosphere. Nitrogen being the major component in the atmosphere in elemental form constitutes almost 78% of the gases in the atmosphere of the earth. Other forms of nitrogenous gases such as NH_3, NO, and N_2O also exist in the atmosphere. In the atmosphere, N_2 exists as an extremely stable molecule that remains unusable by animals and plants without fixation. Nitrogen fixation is the method by which atmospheric nitrogen is changed into usable chemical forms for living beings. Specific microbes "fix" nitrogen from atmosphere into forms (NH_4^+ or NO_3^- or N_2O) that other organisms can utilize. For instance, Rhizobium bacteria fix nitrogen by the formation of nodules on the roots of legumes, such as soybeans and alfalfa. Such bacteria adhere to the root hair of a plant and in reaction the plant makes an empty thread that leads to the root. By means of this infection thread, bacteria start growing with the preliminary development of a nodule onto the root. Bacteria grow through this infection thread and eventually initiate the formation of a nodule on the root. The nodule weight of the root constitutes about 30% due to bacteria. Through the phenomenon of symbiotic nitrogen fixation, the plant provides nutrients and energy to bacteria and fungus, which, in return, provides nitrogen from the atmosphere in a usable form to the plant (Bardgett and Wardle, 2010). Some bacteria are specific for certain plants; that is why the species that infects alfalfa will not infest soybeans. *Rhizobium trifolium* and other specific bacteria have endogenous nitrogenase enzymes, which have the ability to fix atmospheric nitrogen into ammonium ion structure. The plants, in turn, as a part of the symbiotic relationship convert the fixed ammonium ion to amino acids and nitrogen oxides that are subsequently used for the synthesis of proteins and other nonnutritive substances, such as alkaloids. The major source of anthropogenic substrate is agricultural application of nitrogen fertilizers and manure. In soil, plants and microbes use a considerable amount of NH_4^+, and the remaining portion is transformed into NO_3^- by NH_3 oxidizing bacteria and archaea through the process of nitrification. Most NO_3^- is then converted into N_2 via various nitrogen oxides (including N_2O) by denitrification processes (carried out by denitrifying bacteria), and these eventually escape into the atmosphere. Just like emissions of CH_4 and CO_2, global emissions of N_2O are largely originated from microbes. Autotrophic ammonia (NH_3)-oxidizing bacteria, which belong to the class of β-proteobacteria, are found to be largely responsible for the production of N_2O by means of nitrification process. On the other hand, some Achaea has also been observed to play an essential part in nitrification process. For natural or artificial deposition of each tonne of reactive nitrogen species (mainly fertilizer) on the surface of earth, 10–50 kg of nitrogen is emitted in the form of N_2O. The environmental availability of nitrogen has registered a significant rise owing to the utilization of fertilizers and the burning of fossil fuels, thereby perturbing global biogeochemical processes and threatening ecosystem sustainability (Steffen et al., 2015). With the help of microbial oxidation and reduction of nitrogen, agriculture has been found as the chief producer of the intoxicating greenhouse gas of N_2O (Cavicchioli et al., 2019). On the contrary, microbes such as Rhizobacteria (in root nodules) contain enzymes such as N_2O reductase, which have the capability of transforming N_2O to non-greenhouse gas of N_2. The nitrogen transformations processes via microbes, such as mineralization, decomposition, nitrification, denitrification, and fixations, are perturbed by climate change, which ultimately results in the release of N_2O (Greaver et al., 2016). So, it is crucial to find out about the impact of climate change and other anthropogenic actions on microbe-mediated conversions of nitrogen species. The detailed steps of N_2 cycle are further elaborated in Box 3.1.

BOX-3.1: General steps of N_2 cycle

1. **Nitrogen fixation:** This is the initial step in the phenomenon of transformation of nitrogen into utilizable forms for plant life. Such transformations are performed with the help of microbes, which transform nitrogen into usable ammonium ionic forms. Two types of microbial bacteria have been found responsible for nitrogen fixations. The first type involves the nonsymbiotic and free-living bacteria. For instance, anabaena, cyanobacteria or blue-green algae, nostoc, Azotobacter, Clostridium, and beijerinckia. The second type involves the mutualistic symbiotic bacteria. For instance, Rhizobium bacteria of leguminous plants. Free-living and symbiotic microbes performs nitrogen fixation in a more effective manner. Such bacteria with the help of nitrogenase enzyme mix nitrogen with hydrogen to generate ammonia and are further transformed with the assistance of bacteria into other useful forms.

2. **Nitrification:** This is the phenomenon that involves the conversion of ammonium ions into nitrates by the intervention of microbes. Such conversion of ammonia to nitrate is accomplished by soil-living bacteria and other nitrifying bacteria in various stages. In the first step of nitrification, the ammonium ion (NH^{4+}) is oxidized into nitrites (NO^{2-}) by a specific bacterial species such as the Nitrosomonas. In the second stage, bacterial species like Nitrobacteria are accountable for the oxidation of the nitrite species into nitrate (NO^{3-}) species. It is noteworthy to mention that ammonia gas is toxic to plants, and as a result, it is converted to

nitrates or nitrites. However, ammonium ion is a useful source of energy to microbes involved in various biogeochemical processes. Nitrite is also toxic for plants and animals if not utilized properly, which is why it is immediately converted into nitrate ions by various microbial species.

3. **Assimilation**: This step is the process by which plants and other living systems incorporate nitrate and ammonia formed through nitrogen fixation and nitrification. Plants take up nitrates from these forms of nitrogen from soil through their root hairs and incorporate them in production of cellular components, such as nucleic acids, amino acids, and chlorophyll. A portion of nitrogen in plants having symbiotic relationship with Rhizobium is incorporated in the form of ammonium ions from the nodules directly. Other living forms also assimilate nitrogen by means of food chain structures.[39]

4. **Ammonification**: During the process of assimilation, the organic nitrogen is produced in bulk quantities in living systems in the form of proteins, amino acids, and nucleic acids. At a time when living systems die and decay, decomposers like fungi and bacteria turn this organic nitrogen into ammonia. Such a process of conversion of organic nitrogen into ammonia is called ammonification. Several types of enzymes are involved in this process. For instance, *Glu Synthase* (Plastid and Cytosolic), *Glu Dehydrogenase*, and *Glu 2-oxoglutarate aminotransferase* (Ferredoxin and NADH dependent). During the process of ammonification, ammonia exists in soil in the form of the positively charged ammonium ion (NH^{4+}). This charge has the tendency to adhere the nitrogen to soil clay minerals due to which the nitrogen is prevented from getting lost by runoff or leaching; however, there is also a disadvantage that nitrogen cannot be readily mobilized to reach plant roots for its uptake.

5. **Denitrification**: This marks the end of nitrogen cycle in which additional nitrogen molecules in the soil are recycled back to the atmosphere. Denitrification process involves the reduction of nitrates into gaseous nitrogen. The process only occurs at places where there is little or no oxygen, such as in deep waterlogged soils. Pseudomonas and Clostridium, which belong to a specific faction of bacteria, are responsible for this kind of process. Denitrifying bacteria use nitrate in the soil as an electron acceptor instead of oxygen to perform respiration, and as a result, produce nitrogen gas, which is inert and unavailable to plants. Eventually, nitrate is transformed into nitrogen gas, which recycles back to atmosphere.

3.4.2   IMPACT OF CLIMATIC CHANGE ON BIOGEOCHEMICAL PROCESS OF MICROBES

Climate change can impact the diversity and structure of microbial communities in a direct manner by temperature and seasonality or in an indirect manner by plant litter, plant composition, and root exudates. Diversity of microbes in soil impacts diversity of plant in terrestrial environment, and thereby, plays a vital role in ecosystem functions, including carbon cycling (Jing et al., 2015; Delgado-Baquerizo et al., 2016). Climate can disturb interactions between species, and as a result, species may adapt, migrate, and be replaced by others or go extinct (Hutchins and Fu, 2017). Ocean warming, acidification, eutrophication, and overuse by fishing and tourism together show a profound impact on microbial world (Ford et al., 2018). Following are shown some of the implicit and explicit impacts of climate change upon microbial world of terrestrial communities and their associated biogeochemical phenomenon.

3.4.2.1   CARBON DIOXIDE CYCLE

It is widely believed that an increase in the levels of CO_2 leads to an alteration in the discharge of labile amino acids, organic acids, and carbohydrates from roots of plants that in turn can encourage microbial activity (Bardgett et al., 2009). Stimulation in microbial growth and activity can subsequently disturb the CO_2 fluxes in accordance with the availability of nutrients, such as nitrogen. Elevated levels of carbon released by the roots may guide soil nitrogen immobility due to enhanced levels of microbial biomass. This results in less quantity of nitrogen available to plants and as a result limits plant growth. An increase in soil carbon to nitrogen ratio followed by fungal dominance and diversity is also observed (Hurd et al., 2018). Fungi which possess a virtue of having higher carbon assimilation efficiencies than bacteria, and whose cell walls are made up of chitin and melatin, become much more resistant to decomposition than those in bacterial cell membranes and walls made up of phospholipids and peptidoglycan. This is the main reason why we have exceptionally low respiration rates at places dominated by fungi, which ultimately augments the potential for carbon sequestration. Permafrost is a special environment in the global biome. A major portion of carbon of the terrestrial globe is entrapped in the earth's permafrost. It has been estimated that permafrost contains as much carbon as is found in the environment and in plants together. It acts as a giant carbon freezer. The predicted rise in average global temperature may lead to thawing of permafrost. As a result, the microbial activity gets enhanced due to which carbon molecules in the soil are metabolized. The microbial degradation of molecules ultimately leads to the production of greenhouse gases, such as CO_2 and CH_4 into the air, which can impel extra warming. All bacteria, whether in the permafrost or anywhere else in the world, require carbon to exhibit growth and generate cellular biomass. Microbes under anaerobic conditions in aqueous environment have the capability to reduce iron for its energy. This provides the mechanism to make us realize how microbes continue to exist and grow slowly in such extreme temperature conditions with slight or no oxygen. At places of high salt concentrations and sub-zero temperature, the water remains in liquid form due to depression in the freezing point. This facilitates the active iron-reducing bacteria to live in extreme brine climate conditions. The microbial effects of climatic change can be mitigated by agricultural solution of Biochar, a thermo-chemically transformed biomass product obtained during depleted atmospheric conditions. It stabilizes and accumulates organic material in soils rich in iron content (Weng et al., 2017). Biochar has the tendency to improve organic material detention by declining microbial-mediated mineralization and downplaying the impact of root exudate over producing organic matter from minerals, thus encouraging growth of grasses and sinking the release of carbon (Weng et al., 2017).

3.4.2.2   METHANE (CH_4) CYCLE

CH_4 constitutes the second most vital human-generated greenhouse gas in terms of net climate challenge. Utilization of CH_4 by microbes in terms of methanotrophy operates as the major terrestrial sink. So, for better understanding of emission of CH_4, it would be important to identify the response of CH_4 flux to climate change. Recent studies suggest that climate warming, predominantly at high latitudes, may lead to a considerable rise in net CH_4 emissions from permafrosts and wetlands, which as a result will serve as a prominent positive feedback to global climate warming (Schuur et al., 2015). Various studies have suggested a significant 30% decline in CH_4 uptake by soil microbes due to elevated levels of CO_2, which results in increase in CH_4 efflux. The mechanistic approach, which leads to declined uptake of CH_4, remains elusive. Plant-mediated increases in soil moisture have also been observed to be responsible for reductions in CH_4 consumption. Enhanced levels of CO_2 may have an indirect impact upon CH_4 emissions in terms of microbial physiology and activity. The increase in soil moisture due to plants under the influence of enhanced levels of CO_2 may guide to more anoxic situation, which may lead to an increase in methanogenesis process and fall in methanotrophy. On the other hand, rise in temperature and decline in moisture are presumed to raise net uptake of CH_4 through terrestrial ecological units, as they perpetually boost diffusion paces of gases and microbial contact to atmospheric CH_4 and oxygen. From the studies, it could be inferred that CH_4 uptake increases due to enhanced temperature, whereas it decreases due to an increase in levels of CO_2 (Dutta and Dutt, 2016). Alterations in flux of CH_4 are also linked to fluctuations in the community structure and loads of methanotrophs. Some studies have suggested a decline in the abundance of type II methanotrophs in response to increasing precipitation and temperature. Regardless of the extensive reaction of the microbe-mediated flux rates of CH_4 to predictable variations in atmospheric and climate conditions, more studies on the temperature-sensitive microbes of various factions of methanotrophs and methanogens, and their relations with moisture and levels of CO_2 across diverse ecological units, have to be carried out to more precisely forecast the future dynamics of fluxes of terrestrial CH_4.

3.4.2.3 *NITROGEN (NITROUS OXIDE) CYCLE*

The impact of changes in climate and atmospheric composition on fluxes of N_2O mediated through microbes has been observed less prominent compared to the impact of fluxes of CH_4 and CO_2. The general emissions of N_2O have been found largely reliant upon the availability of reactive species of nitrogen, such as N_2O and NH_3. Over the past century, there has been a three- to five-fold rise in the discharge of reactive species of nitrogen on account of mushrooming of industries and other anthropogenic discharges of N_2O from the burning of fossil fuels, the spiraling of agricultural affairs, and allied emissions of NH_3. Such enhanced accessibility of reactive species of nitrogen in ecosystems of terrestrial world is expected to result in an increase in nitrification and denitrification processes and consequently elevated N_2O production (Greaver et al., 2016). Agriculture has been observed as the largest emitter of potent greenhouse gas N_2O that is being released by microbial oxidation and reduction of nitrogen (Cavicchioli et al., 2019). The progress in understanding the biogeochemical processes and ecophysiology of microbes in agriculture, which causes reduction of N_2O to harmless elemental N_2, offers alternative avenues for mitigating emissions (Bakken and Frostegard, 2017). Employing bacteria types with elevated activity of N_2O reductase has considerably reduced emissions of N_2O from soya varieties, with the result such genetically and naturally modified types having elevated activity of N_2O reductase offer alternating options for extenuating N_2O emissions (Bakken and Frostegard, 2017). Influxes of fixed nitrogen from human activities as a result will accelerate production and emission of oxidized nitrogen gases that add dramatically to the greenhouse gas budget. Increasing CO_2 concentrations may lead to higher rates of nitrogen fixation by some microbes. While some studies suggest a decline in denitrification process with enhanced CO_2 levels, others suggested no effect at all. Further, at the time of availability of surplus mineral nitrogen, enhanced emissions of N_2O were observed under elevated levels of CO_2. Alternatively, it infers that when availability of reactive species of nitrogen is little, an ecosystem will observe reduced emission of N_2O under increased levels of CO_2, and vice versa. Contradicting reports have been observed regarding the sensitivity of temperature on nitrifying and denitrifying communities of microbial world. One school of thought suggested a predictable rise in temperature would not significantly influence nitrifying or denitrifying enzymes; while, another school of thought suggested that the proportion of the net N_2O flux linked with nitrification process declines at elevated temperatures, which was connected to an alteration in structure of the community of NH_3-oxidizing bacteria (Barnard et al., 2005). There is a need for further investigation over this area which would spotlight on reviewing production of N_2O with the help of nitrifying and denitrifying microbes in reaction to a varying accessibility of reactive species of nitrogen, and subsequently its relations with variations in climate and environmental composition.

3.5 MICROBIAL FEEDBACK

The response of microbes as feedback due to change in climate is calibrated in the form of greenhouse gas fluctuation which could either increase (feedback as positive) or lessen (feedback as negative) the pace of climate change (Singh et al., 2010). It has been observed that climate change leads to both direct and indirect effects on behavior of microbes and accordingly stir up subsequent feedbacks, respectively (Bardgett and Wardle, 2010). The former effects may result due to extreme climatic conditions, such as rise in temperature, drought, floods, or freezing, whereas the latter effects may occur due to climatic sponsored variations in plant productivity, diversity and allied modifications. Such direct and indirect impacts were employed as important tools to figure out the responsibility of microbes and their mode of action in feedbacks of carbon cycle during climate change. Adopting this approach puts two different microbial feedback mechanisms of carbon cycle forward.

3.5.1   DIRECT FEEDBACK

3.5.1.1   TEMPERATURE ELEVATION

Rise in temperature will lead to a rise in microbial respiration. This will promote mobilization of carbon from terrestrial soil to atmosphere. An estimated loss of 5% of total soil carbon has been already documented with around 20–200 ppm upsurge in concentration of CO_2 in the atmosphere by the end of 21st century. Alternatively, climatic warming will yield positive feedback on climatic change by enhancing the carbon flow from soil to the environment (Friedlingstein et al., 2006). The decomposition dynamics rate of relatively stable organic material would take place by climatic warming. This would, in longer term, facilitate a more efficient community of microbes and as a result will show positive feedback to climate (Frey et al., 2013). Permafrost also serves as an important feedback for climatic change. Studies have shown that by the end of 21st century permafrost would thaw as much as 25%. As a result, the accessibility of the preserved organic material for decomposition by microbes would get enhanced, thereby exercising a positive feedback on cycling of carbon and subsequent climatic change. Studies have shown that 4×10^4 kg of permafrost carbon in organic form has turned out to be susceptible to increased rates of disintegration (Dutta and Dutt, 2016). Other important indicators of climatic change are the transformations in the structure, diversity, and strength of the communities of microbes. A rise in temperature for instance will climb up the relative abundance of β-proteobacteria that promotes respiration manifesting in positive feedback, a feature not ubiquitous to other types of bacteria (Frey et al., 2013). Due to the cumulative effect of such uncertainties, unsuccessful models have resulted which are incapable of forecasting the mechanistic feedbacks of the soil carbon in terms of the disintegration prototypes.

3.5.1.2 EXTREME CONDITIONS

The soil microbial activity is altered due to stress conditions, such as drought and freezing; however, their post effects vary across different ecosystems (Bardgett et al., 2009). For instance, microbial activity is suppressed on account of severe drought conditions, which generate a negative feedback upon microbial decay and loss of soil carbon, owing to decline in the action of the phenol oxidase enzyme responsible for organic material disintegration. On the contrary, microbial activity is accelerated in areas of drought conditions where water table gets lowered that results in anaerobic soil microbes getting exposed to oxygen. This gives positive feedback for cycling of carbon due to the enhanced activity of phenol oxidases. It is noteworthy to mention here that drought-hit areas at the same time suffer a decline in methane emissions due to the sensitivity of methanogens to desiccation and toxicity of oxygen (Bardgett et al., 2009). In the broader context, however, the tendency of global warming is to increase methane emissions. This is corroborated by the fact that enhanced net output under the influence of climate change increases emissions of methane whereas declined rainfall and soil drought conditions possess conflicting impacts (Christensen et al., 2003).

3.5.2 INDIRECT FEEDBACK

Indirect feedbacks of microbe communities from climatic change are intervened via plant growth and constitution of the foliage of the ecological unit (Bardgett et al., 2009). Two different mechanistic routes are adopted for such kinds of feedbacks that are discussed below.

Route I: The first mechanistic approach of climatic-change-driven feedback remains linked to enhancement in atmospheric concentrations of CO_2, an outcome of enhanced photosynthesis of plants. The enhanced CO_2 levels result in transport of photosynthetic carbon from the body of plants to its roots and related mycorrhizal fungi (Johnson et al., 2005). Finally, the carbon reaches heterotrophic soil microbes by means of exudation of organic compounds via roots resulting in the augmented flow of carbon into the soil.

This provides soil microbes an increase in the level of carbon content available that ultimately leads to an increase in respiration of microbes and the subsequent discharge of carbon content to the air. A part of allied carbon gets mobilized back into the soil by getting dissolved in water (Zak et al., 1993).

Route II: The second mechanistic microbial feedback route of climate change needs a little moment to swift into an action as it relies on the modifications in diversity of foliage and the functional structure of an ecosystem. Owing to the warming of climate, alterations in the plant distribution species and their functional groups have been documented. This has the potential to modify the bodily environment of the soil through its variation in the depth of root and its structure. Alternatively, such changes carry a considerable change in the qualitative and quantitative production of litter, which serves as a key component of soil organic carbon (Bardgett et al., 2009). The disparity in leaf litter distribution causes the changes in the plant functional groups, thereby accounting for the heterotrophic respiration of soil microbes and their rate of decomposition. Evergreen shrubs, which are slow-growing plants, generate substandard litter (rich in intractable compounds such as phenolic acid and lignin), which due to impeded microbial activity does not experience fast disintegration. The abundant occurrence of this kind of vegetation provides a negative feedback on carbon exchange and climate change. On the contrary, legumes and graminoids, which exhibit fast growth, produce superior litter that undergoes swift microbial decomposition. The abundant occurrence of this kind of vegetation provides a constructive feedback on activity of microbes and mineralization of carbon. From the above discussion of various mechanistic feedbacks of microbes, it could be inferred that the feedback of carbon cycle does not work independently but are dependable upon several biological and climate variables (Singh et al., 2010).

3.6 CONCLUSION

Microbes play a crucial role in shaping the atmospheric concentration of greenhouse gases. Through biogeochemical processing, they act as drivers of climatic change. Microbes break down organic matter and add significantly to emissions of greenhouse gas of CH_4, CO_2, and N_2O through heterotrophic mode of methanogenesis, respiration, and denitrification, respectively. Several factors affect the balance of microbial greenhouse gas emission versus capture, together with the aquatic and terrestrial biomes, the neighboring atmosphere, food web responses, and interactions, and in particular, human-sponsored climate change and other anthropogenic activities. Man-made activities that unswervingly impact microbes encompass emissions of greenhouse gases (CO_2, CH_4, and N_2O), contamination through eutrophication, population growth, and agriculture land usage. All these positive feedbacks speed up global climate change. But, on the other end, microbes sequester different greenhouse gases emissions and decelerate or thwart climate change by conversions to more utilizable organic forms. In this way, the role of microbes in climate change is multifaceted, challenging, and incomplete to understand because appropriate interpretations for fundamental reasons and effects of anthropogenic- or human-related climate impact on living systems still subsist. Simultaneously, however, there is no denying the fact that human activity is contributing to climate change which is a worrying concern for the entire globe. An urgent, unrelenting, and rigorous endeavor is required to explicitly incorporate microbes into research, technological advancement, policy and decision-making initiatives. The prospective contributions of these microbes can be seized by taking into account the intricate impact of climate change upon them and their consequent long- and short-term feedbacks. If exploited judiciously, microbes could serve as essential natural repository tools for regulating global climate challenges. They not only participate in the rate of climatic change but can also serve as effective tools for its mitigation and adaptation. Conversely, if appropriate consideration is not given, microbes may possibly act as the most rising accelerators to the crisis. It is a lofty occasion, therefore, to intensely learn this side and realize its acting biogeochemical mechanistic approaches more accurately, so that microbes could be suitably utilized for effective developing solutions.

KEYWORDS

- climate change
- nutrient cycles
- nitrogen
- phosphorus
- microbial feedback
- direct feedback

REFERENCES

Agard, J.; Schipper, ELF.; Birkmann, J.; Campos, M.; Dubeux, C; Nojiri, Y.; Olsson, L.; Osman-Elasha, B.; Pelling, M.; Prather, M. J.; Rivera-Ferre, M. G.; Ruppel, O. C.; Sallenger, A.; Smith, K. R.; St Clair A. L.; Mach, K. J.; Mastrandrea, M. D.; Bilir, T. E. Annex II: Hlossary, (ed).; IPCC **2014**, 1757–1776.

Arrigo, K. Marine Microorganisms and Global Nutrient Cycles. *Nature* **2005**, *437*, 349–355.

Azam, F.; Malfatti, F. Microbial Structuring of Marine Ecosystems. *Nat. Rev. Microbiol.* **2007**, *5*, 782–791.

Bakken, L. R.; Frostegard, A. Sources and Sinks for N_2O, Can Microbiologist Help to Mitigate N_2O Emissions. *Environ. Microbiol.* **2017**, *19*, 4801–4805.

Ballantyne, A., et al. Accelerating Net Terrestrial Carbon Uptake during the Warming Hiatus Due to Reduced Respiration. *Nat. Clim. Change* **2017**, *7*, 148–152.

Bardgett, R. D.; Wardle, D. A. *Aboveground–Belowground Linkages*; Oxford University Press: Oxford, **2010**.

Bardgett, R. D.; De Deyn, G. B.; Ostle, N. J. Plant–Soil Interactions and the Carbon Cycle. *J. Ecol.* **2009**, *97*, 838–839.

Barnard, R.; Leadley, P.; Hungate, B. Global Change, Nitrification, and Denitrification: A Review. *Global Biogeochem. Cycl.* **2005**, *19*, GB1007.

Baron, Y. M.; Phillips, R.; Milo, R. The Biomass Distribution on Earth. *Proc. Natl Acad. Sci. USA.* **2018**, *115*, 6506–6511.

Behrenfeld, M. J. Climate-Mediated Dance of the Plankton. *Nat. Clim. Change* **2014**, *4*, 880–887.

Carney, K. M.; Hungate, B. A.; Drake, B. G. Megonigal, J. P. Altered Soil Microbial Community at Elevated CO_2 Leads to Loss of Soil Carbon. *Proc. Natl Acad. Sci. USA* **2007**, *104*, 4990–4995.

Cavicchioli, R., et al. Scientists Warning to Humanity: Microorganisms and Climatic Change. *Nat. Rev. Microbiol.* **2019**, *17*, 569–586.

Christensen, T. R., et al. Factors Controlling Large Scale Variations in Methane Emissions from Wetlands. *Geophys. Res. Lett.* **2003**, *30*, 10–13.

Climate Change: Basic Information. US EPA **2015**, http://www3.epa.gov/climatechange/basics/.

Crowther, T. W., et al. Quantifying Global Soil Carbon Losses in Response to Warming. *Nature* **2016**, *540*, 104–108.

Delgado-Baquerizo, M., et al. Microbial Diversity Drives Multifunctionality in Terrestrial Ecosystems. *Nat. Commun.* **2016**, *7*, 10541.

Dutta, H.; Dutt, A. The Microbial Aspect of Climatic Change. *Energ. Ecol. Environ.* **2016**, *1* (4), 2019–2232.

Endeshaw, A.; Birhanu, G.; Zerihun T.; Genene T. Microbial Function on Climate Change – A Review. *Environ. Pollut. Climate Change* **2018**, *2* (1). DOI: 10.4172/2573-458X.1000147

Falkowski, P. G.; Fenchel, T; Delong, E. F. The Microbial Engines That Drive Earth's Biogeochemical Cycles. *Science* **2008**, *320*, 1034–1039.

Falkowski, P. G.; Fenchel, T.; Delong, E. F. The Microbial Engines That Drive Earth's Biogeochemical Cycles. *Science* **2008**, *320*, 1034–1039.

Flemming, H. C.; Wuertz, S. Bacteria and Archaea on Earth and Their Abundance in Biofilms. *Nat. Rev. Microbiol.* **2019**, *17*, 247–260.

Ford, A. K., et al. Reefs under Siege: The Rise, Putative Drivers, and Consequences of Benthic Cyanobacterial Mats. *Front. Mar. Sci.* **2018**, *5*, 18.

Frey, S. D.; Lee, J; Melillo, J. M.; Johan S. J. The Temperature Response of Soil Microbial Efficiency and Its Feedback to Climate. *Nat. Clim. Change* **2013**, *3*, 395–398.

Friedlingstein, P; Bopp, L; Rayner, P., et al. Climate-Carbon Cycle Feedback Analysis: Results from the C4MIP Model Intercomparison. *J. Clim.* **2006**, *19*, 3337–3353

Gallego, S. A. V.; Prentice, I. C. Blanket Peat Biome Endangered by Climate Change. *Nat. Clim. Change* **2013**, *3*, 152–155.

Greaver, T. L., et al. Key Ecological Responses to Nitrogen Are Altered by Climate Change. *Nat. Clim. Change* **2016**, *6*, 836–843.

Hoegh, G. O., et al. Special Report: Global Warming of 1.5°C. *IPCC* **2018**, Ch. 3.

Hovenden, M. J., et al. Globally Consistent Influences of Seasonal Precipitation Limit Grassland Biomass Response to Elevated CO_2. *Nat. Plants* **2019**, *5*, 167–173.

Hultman, J., et al. Multi-Omics of Permafrost, Active Layer and Thermokarst Bog Soil Microbiomes. *Nature* **2015**, *521*, 208–212.

Hurd, C. L.; Lenton, A.; Tilbrook, B.; Boyd, P. W. Current Understanding and Challenges for Oceans in a Higher- CO2 World. *Nat. Clim. Change,* **2018**, *8*, 686–694.

Hutchins, D. A.; Fu, F. X. Microorganisms and Ocean Global Change. *Nat. Microbiol.* **2017**, *2*, 17508.

Jing, X., et al. The Links between Ecosystem Multi-Functionality and Above and Below Ground Biodiversity Are Mediated By Climate. *Nat. Commun.* **2015**, *6*, 8159.

Johnson, D; Kresk, M.; Stott, A. W.; Cole, L.; Bardgett, R. D.; Read, D. J., et al. Soil Invertebrates Disrupt Carbon Flow through Fungal Networks. *Science* **2005**, *309*,1047.

Nisbet, E. G., et al. Very Strong Atmospheric Methane Growth in the Four Years 2014–2017: Implications for the Paris Agreement. *Global Biogeochem. Cycl.* **2019**, *33*, 318–342.

Olufemi, A.; Reuben, O.; Olufemi, O. Global Climate Change. *J. Geosci. Environ. Protect.* **2014**, *2*, 114–122.

Pan, Y., et al. A Large and Persistent Carbon Sink in the World's Forests. *Science* **2011**, *333*, 988–993.

Reay, D. S.; Grace, J. *Greenhouse Gas Sinks*; Reay, D. S., et al., Eds.; CABI Publishing: Oxfordshire, 2007; pp 1–10.

Schuur, E. A. G., et al. Climate Change and the Permafrost Carbon Feedback. *Nature* **2015**, *520*, 171–179.

Singh, B. K.; Bardgett, R. D.; Smith, P.; Reay, D. S. Microorganisms and Climate Change: Terrestrial Feedbacks and Mitigation Options. *Nat. Rev. Microbiol.* **2010**, *8*, 779–790.

Steffen, W., et al. Sustainability. Planetary Boundaries: Guiding Human Development on a Changing Planet. *Science* **2015**, *347*, 1259855.

Sunagawa, S., et al. Structure and Function of the Global Ocean Microbiome. *Science* **2015**, *348*, 126–359.

Verpoorter, C.; Kutser, T.; Seekell, D. A.; Tranvik, L. J. A Global Inventory of Lakes Based on High- Resolution Satellite Imagery. *Geophys. Res. Lett.* **2014**, *41*, 6396–6402.

Weiman, S. Microbes Help to Drive Global Carbon Cycling and Climate Change. *Microbe Mag.* **2015**, *10*, 233–238.

Weng, Z. H., et al. Biochar Built Soil Carbon Over a Decade By Stabilizing Rhizodeposits. *Nat. Clim. Change* **2017,** *7*, 371–376.

Zak, D. R.; Pregitzer, K. S.; Curtis, P. S.; Teeri, J. A.; Fogel, R.; Randlett, D. L. Elevated Atmospheric CO_2 and Feedback between Carbon and Nitrogen Cycles. *Plant Soil* **1993,** *151*, 105–11.

Zimmer, C. The Microbe Factor and Its Role in Our Climate Future. *Yale Environ.* **2010**. http://e360.yale.edu/feature/the_microbe_factor_and_its_role_in_our_climate_future/2279/.

CHAPTER 4

Seasonal Dynamics of Bacterial and Fungal Lineages in Extreme Environments

NAFEESA FAROOQ KHAN[1*], UZMA ZEHRA[1], ZAFAR A RESHI[1], MANZOOR AHMAD SHAH[1], and TAWSEEF REHMAN BABA[2]

[1]*Biological Invasion Lab, University of Kashmir, Jammu and Kashmir, India*

[2]*Division of Fruit Science, SKUAST-K, Shalimar, Jammu and Kashmir, India*

Corresponding author. E-mail: khnnafisa@gmail.com

ABSTRACT

Globally, bacterial and fungal communities are important factions present in all major ecosystems with a significant role to play in driving various ecosystem functionalities. Given the differences in life style of these micro-organisms and their occurrence in all ecosystems, these pervasive micro-organisms tend to adapt/withstand harsh environmental conditions such as extreme highs and lows in abiotic/physical characteristics. Besides the potential of exhibiting physiological plasticity (special ability to utilizing non-photosynthetic substrates like methane, sulfur, and iron), in order to evolve and survive in extreme environments, they also respond to trivial changes in environment by exhibiting different patterns in abundance and diversity. Although, many studies have centered focus on studying micro-organisms in hostile conditions but the role of seasonal dynamics in shaping these microbial communities in extreme environments remains an open question. Of late, the different trends in abundance of fungi and bacteria and their functional diversity in the face of seasonal dynamics is gaining focus. Here we have discussed how bacterial and fungal assemblages are continually evolving into different adaptations in extreme conditions. The chapter explains contributions of abiotic conditions (temperature, pressure, pH, oxygen saturation, salinity) and biotic conditions (low nutrition, high population density, or low prey availability) in temporal and spatial dynamics of bacterial and fungal lineages in extreme environments.

4.1 INTRODUCTION

Adapting to extreme, unchanging or evolving ecosystems, which are not generally hospitable for life forms, became a subject of increasing interest recently. Dohm and Maruyama (Dohm and Maruyama, 2015) presented the "Habitable Trinity" definition, harmonious state of coexistence with all essential components, that is, atmosphere, ocean, and land regulated by the light, fostering material and energy flow. Microbial species offer vital ecological resources, in this sense; they are the only biological components capable of prospering under severe adverse environmental conditions. They have a strong adaptive ability and physiological versatility that enables them to inhabit a variety of ecosystems. They have distinguished capacities to tolerate extreme conditions. In the case of terrestrial ecosystems, microbial communities are important for bulk decomposition and mineral content (Bardgett, 2005; Haubensak et al., 2002). Furthermore, seasons induce a dramatic shift in variables such as humidity, temperature, vegetation, and concentrations of nutrients that are vital to microorganisms' survival (Lazzaro et al., 2015). Their community richness, dominance, and vulnerability are significantly influenced by temporal and seasonal changes in abiotic factors such as precipitation, in environments often where microorganisms might be at or close their physiological level of tolerance (Hullar et al., 2006; Lipson and Schmidt, 2004; Schadt et al., 2003; Waldrop and Firestone, 2006). It has also been shown that in warm, fairly precipitation-limited pre-monsoon seasons, bacterial and fungal species have a distinct composition from those in the rainy seasons (Cregger et al., 2012). Knowing the variability of microbes in aquatic habitat and their possible reactions to environmental factors also remains a major challenge (Jackson et al., 2014). However, a study on Dongjiang River bacterioplankton showed varied composition and spread of proteobacteria, actinobacteria, and dry-to-wet seasonal bacteroidetes phyla, relating to seasonal variations with microbiome occurrence (Sun et al., 2017).

In the past few decades, the unique organisms that live in extreme environments have interested the scientific community. The majority of organisms which survive in stressful conditions are primarily prokaryotes. Among three life domains (i.e., bacteria, archaea, and eukarya), bacteria and fungi are the commonly known extremophiles, highly adapting to myriad extreme conditions. In microbial ecological studies, "extreme" refers to adverse environmental conditions that detriment the capacity of species to function. Extremophiles fall into two specific categories:

1. Extremophilic species that need harsh conditions for growth.
2. Extremotolerant species that can withstand extreme physicochemical parameters although they successfully grow under normal conditions.

In fact, extremophiles are recognized with respect to their astonishingly diverse way of life they live such as thermophiles prefer a growth temperature of min 60°C and max 80°C, hyperthermophiles prefer a temperature of 80°C or above, psychrophiles prefer a temperature of 15°C or less as a temperature beyond 20°C is nontolerating for them. Similarly, acidophiles perfectly grow at a pH range of 1–5, alkaliphiles grow beyond a pH of 9, barophiles/piezophiles grow in much higher hydrostatic pressure, halophiles-choicest environments to grow are higher salt medium, excellent growth of oligotrophiles occurs in nutrient-deprived regions; endolithophiles grow inside hard rock minerals and xerophiles flourish in xeric (water limited) regions (Hassan et al., 2016; Tiquia-Arashiro et al., 2016). With each microbe having its own characteristic adaptations to grow and survive, poly-extremophiles can operate in one or more physiochemical parameters, each type with extreme values (Yamagishi et al., 2018), redefining our view on nature of inception and evolution of life (Bertrand et al., 2015). For example, halo-alkali thermophile, *Natranaerokjbius thermophilus* has a mandatory requirement of three extreme conditions (Preiss et al., 2015; Banciu and Muntyan, 2015). Phylogenetic diversity of extremophiles is characteristically strong but its research is rather complicated (Rampelotto, 2013) while some others consist of both extremophiles and mesophiles. Pabulo (Rampelotto, 2013) reported a few orders or genera that exclusively contain extremophiles. The measures through which different types of organisms make adjustments to harsh environments help to reflect on the main attributes of cellular mechanisms, like those of biomolecular stabilization processes and genome guidance for the construction of biomolecules under all extreme environmental conditions. Such organisms pose abilities for colonizing intense environments combined with an incredibly large and flexible metabolic range through which they can utilize methane, sulfur, and iron-like substrates, although the understanding of adapted macromolecules and physiological functions at the molecular level in extreme environments is still unclear. The stability and activity of extremophiles are important for biotechnological purposes, for example, extremozymes are key enzymes obtained from these microorganisms which can tolerate harsh conditions in production plants which otherwise are protein destructive (Elleuche et al., 2014). Most such prospective enzymes have been evaluated from extremophiles that could sustain extreme temperatures, salinity, and pH, etc.

Notably, poly-extremophilic enzymes exhibit poly-extremophilicity which renders them more promising in biotechnological applications. Pertaining to evolutionary time scale and phylogenetics, (poly)extremophiles are at the root of evolutionary tree sketch and hyperthermophiles are closest to the early pioneers of life on the planet (Bell et al., 2015), which is why deeper insights into extremophile ecologies might very well help to predict model organisms in the galaxy. Extremophiles can flourish in harsh and stressful conditions, as is reported from the hyper-arid Atacama Desert and the Dry Valleys of Antarctica where water activity stays quite low (Goordial et al., 2016; Schulze-Makuch and Guinan, 2016; Schulze-Makuch et al., 2015). Equally, high temperature loving microorganisms mostly in natural hot springs, fumaroles or even some sea areas could imitate probable forms of life in many other outer space situations (Kelley et al. 2001; Corliss et al., 1981; Mulkidjanian et al., 2012). In the past few years, the emergence of relatively new methodologies using spectroscopic analysis and diverse extremophilic organisms as predictive models has established a further point of view that can be quite beneficial in astrobiology.

The rapid application of molecular tools over the last few years has resulted in increased advancement in the area, facilitating everyone to uncover unsettling insights into the existence of extremophiles with astonishing accuracy. Distinctly, high-throughput sequencing technology has transformed methods of exploring microbes in extremes and unveils microbial ecosystems with elevated amounts of richness and intricacy. In essence, some basic grasp of the biology of microorganisms can complement genetic mechanisms and must not be substituted by any other approach. A recent study using analysis of esters of fats through PLFA (phospholipid fatty acid analysis) procedure created correlation between the microbiome structures at varying soil depth in different seasons, to improve the understanding of the function of these microorganisms (Luo et al., 2019). A combined fine-tune among culture based and present day HTS techniques help in a better comprehension on part of microbial survivalism with its associated functionality under aggravated environments, for example, microbial communities identified from samples taken from four seasons gave better estimates of abundance using TRFLP, following it is quantitative PCR—bacteria 16S rRNA;18S rRNA fungi (Lazzaro et al., 2015). Focusing on such scientific-technological advances and breakthroughs, which mostly support futuristic evolutionary biology concepts, has perhaps made the occurrence of microbes in extreme environments the most enticing areas of research.

In this chapter, there is a summary of an overview of the microbial diversity of both fungi and bacteria in hostile planetary environments, their adaptive mechanisms, and new methods adopted to study the extremophiles. In terms of seasonality, the composition of some extremophiles like those of halophiles, psychrophiles, thermophiles, acidophiles, etc., is also discussed. At last, specified extremophile biotechnology and industrial uses are also discussed.

4.2 MICROBIAL LIFE IN EXTREME ENVIRONMENTS: TYPES OF EXTREME HABITATS

Irrespective of their small size, microbes play a critically important role in the biodiversity of all ecosystems in addition to the function of biogeochemical processes. While life happens to be 3.5 billion years old with ambient environmental requirements of temperature, pH, nutrients, and many other similar factors as a fundamental necessity to thrive. The planet earth is made of extreme habitats like freezing deep oceans, hot springs, glaciers, ocean depths, highland levels in all continents. Broadly speaking, though organisms seem to be susceptible to intense variations in the environment, and some microbes not only embrace certain circumstances, they also involve other requirements for their existence. For instance, these microbes have been reported in Hassan et al. (2016):

TABLE 4.1 Types of Extremophiles and Related Environmental Parameters.

No.	Stressor	Type of environment	Domain	Type of extremophile	Examples/References
1	High temperature	45°C and 122°C	**Bacteria**	**Thermophiles**	*Geobacillus stearothermophilus, Thermoanaerobacter Ethanolicus, Caldicellulosiruptor bescii, Saccharolyticus, Thermincola ferriacetica, Thermus scotoductus, Ureibacillus thermosphaericus* (Hassan et al., 2016)
			Fungi		*Humicola* sp. (Hassan et al., 2016)
2	Low temperature	−15°C to −20°C	**Bacteria**	**Psychrophiles**	*Exiguobacterium, Paenibacillus tylopili, Lysinibacillus, Pontibacillus, Desemzia incerta, Staphylococcus, Virgibacillus., Bacillus licheniformis, B. muralis, Sporosarcina globispora* Yadav et al. (2016) *Pseudomonas proteolytica, Morganella psychrotolerans, Arthrobacter gangotriensis, Arthrobacter kerguelensis, Pseudomonas meridian* (Tiquia-Arashiro et al., 2016).
			Fungi		*Penicillium* and *Cladosporium Aspergillus* spp., *Cylindrocarpon, Candida Glomerella, Cryptococcus, Golovinomyces, Phoma* (Gupta et al., 2014) *Thelebolus microspores, Mortierella* sp., *Antarctomyces psychrotrophicus, Thielavia antarctica* and *Apiosordaria antarctica, Penicillium* and *Alternaria* (Hassan et al., 2016)
3	High pressure	Require pressure 120 MPa	**Bacteria**	**Piezophiles**	*Shewanella, Methanococcus, Pyrococcus* and *Moritella* (Gupta et al., 2014)
			Fungi		*Aspergillus versicolor* (Hassan et al., 2016; Tiquia-Arashiro et al., 2016)
4	Salinity	At least 0.2 M concentrations of salt	**Bacteria**	**Halophiles**	*Halobacterium salinarum, Halarsenatibacter silvermanii* (Arora, 2019) *Halococcus hamelinensis, Haloarcula argentinensis, Haloferax alexandrines, Haloferax larsenii,*

TABLE 4.1 *(Continued)*

No.	Stressor	Type of environment	Domain	Type of extremophile	Examples/References
					Haloferax volcanii, Halolamina pelagic, Halostagnicola kamekurae, Haloterrigena thermotolerans, Natrinema sp., *Natronoarchaeum mannanilyticum* (Nath and Subbiimah)
			Fungi		*Wallemia ichthyophaga Hortaea werneckii*
5	Alkaline systems	pH>9	**Bacteria**	**Alkalophiles**	*Spirulina platinensis, Spirulina platinensis, Bacillus* sp. *Nostoc* sp., *Thermomonospora* sp.*, *Thermomonospora* sp. (Chang and Kang, 2004), *Streptomyces* sp., *Pseudomonas alcaliphila Bacillus licheniformis* (Tiquia-Arashiro et al., 2016)
			Fungi		*Saccharomyces cerevisiae*
6	Acidic systems	pH<3	**Bacteria**	**Acidophiles**	*Thiobacillus, Sulfolobus* and *Thermoplasma, Halomonas campisalis, Picrophilus oshimae,* and *Picrophilus torridus* (Arora, 2019) *Acidithiobacillus ferrooxidans Lactobacillus acidophilus Pilimelia columellifera, Thiobacillus thioparus, Actinobacteria,Streptacidiphilus* sp., *Arthrobacter nitroguajacolicus,Verticillium* sp.; *Bipolaris nodulosa, Pichia jadinii, Fusarium oxysporum* (Tiquia-Arashiro et al., 2016)
			Fungi		*Aspergillus niger, A. temeri, A. terreus, A. oryzae, Aspergillus tubingensis, Penicillium atramentosum, P. fellutanum, P. nalgiovense, P. citrinum, P. chrysogenum, P. purpurogenum, P. nalgiovense, P. aurantiogriseum, P. citrinum, P. Waksmanii* (Tiquia-Arashiro et al., 2016; Prasad, 2016)
7	Radiation	UV radiation/ ionizing radiations	**Bacteria** **Fungi**	**Radiophiles**	*Acinetobacter* sp., *Rhodobacter* sp. (Pérez et al., 2017) *Deinococcus radiodurans* and *Cryptococcus neoforman* (Gupta et al., 2014; Cox et al., 2010)

1. Interiors of waterbodies like lakes and oceans with a pressure range of nearly 110 MPa and a depth of tens of kilometers.
2. Within Earth's crust at around $6^{1/2}$ km deep.
3. Proton-rich and proton-deficient regions.
4. Cold waters at nearly −20°C and Fumaroles mostly at 122°C.

The broad number of extremophiles currently known is mentioned in Table 4.1. In this section, we discuss why the systematic perspective, rendered possible by the subjective studies on microbial ecology, is important to supervene dynamics in microbial populations.

4.2.1 HIGH-TEMPERATURE STRESS: THERMOPHILES AND HYPERTHERMOPHILES

Temperature is among the most important environmental factors that can affect microorganism's survival ability (Maheshwari et al., 2000). The term "Thermophiles" refers to extremophilic species with elevated temperatures tolerance potential between 55°C and 121°C including hyperthermophiles which are highly strong thermophilic, having a temperature requirement of more than 80°C (Maheshwari et al., 2000). A few thermophilic fungi modestly show progressive maturation in a temperature range of 45°C–61°C and it has been considered that truly thermophilic fungi do not mature in temperatures less than 20°C whereas purely develop at or above 50°C. Mostly, all fungi can grow at 45°C but based on temperature requirements they are distinguished among thermophilic and thermo-tolerant (Cooney and Emerson 1964) fungal species, where the latter grows well at low temperatures. Of all characterized fungal species, only about 30 confirmed thermophiles have been listed so far (Maheshwari et al., 2000). Thermophiles are reportedly the members of Sordariales, Eurotiales, Onygenales, and Mucorales orders (Morgenstern et al., 2012). *Thermomyces, Aspergillus, Thermoascus, Rhizomucor, Chaetomium,* and *Mycothermus* are among the most widely known thermophilic and thermotolerant taxonomic groups (Maheshwari et al., 2000; Natvig et al., 2015). Thermophily of fungus, as such, is not regarded to be quite radical albeit observed in individuals of bacteria and archaea. This is because of their unique and moderate preference of habitat and temperature in contrast to bacteria and archaea (Özdemir and Uzel, 2020).

Hot areas such as volcanoes that are heavily influenced by water vapors and deep-sea hydrothermal vents are characteristically densely populated by extremophiles (McDermott et al., 2018; Rampelotto, 2013). Thermophiles are chemoautotrophic utilizing carbon, iron, or sulfur compounds (Hassan et al., 2016). A lot of heat-loving microbes also show high acidic adaptation, for example, *Picrophilus* sp. and *Methanopyrus kandleri* vigorously flourish at a temperature of 60°C, and 121°C, respectively, and near a pH of 7 (Elleuche et al., 2014; Goordial et al., 2016; Kashefi and Lovley, 2003; Takai et al., 2008). Quite a few hyperthermophiles (growth temperature > 80°C) tend to expand at a high pressure (29.8 MPa) by maintaining a liquid state along with elevated temperatures (assumed upper limit of 407°C) (McDermott et al., 2018). Hyper-thermopiezophilic microorganisms, such as *M. kandleri* (Takai et al., 2008) and *Geogemma barossii* (Kashefi and Lovley, 2003), competently keep-up cell's physical strength against the highs of pressure and heat stress (Sharp et al., 2014).

4.2.2 ADAPTATIONS OF THERMOPHILES

The genetic basis to highlight adaptive responses to high temperatures is researched more intensively among all hostile environments. Among biomolecules (DNA, proteins, and lipids), the most deliberate aspect of adaptation for thermophiles is found in proteins. Proteins lose their activity (denature) at high temperatures and eventually lead to a considerable rise in the membrane fluidity and thus alter the metabolic activities. Thermophiles display a range of cellular alterations to avoid diminishing properties, in view of essential metabolic proteins and in them, the frequency of (double and single) esters are more (Schulze-Makuch and Guinan, 2016); thus enhancing the membrane mobility to a full extent. Unlike other organisms, heat stable proteins apropos of thermophiles are very small and alkaline (Bell et al., 2015). The thermostability is possibly a result of the following interactions: (1) hydrogen bonds; (2) reduced loop lengths of proteins; (3) increased secondary bonding propensity; (4) core hydrophobicity; (5) week interactions (van der Waals force); and (6) enhanced ionicity and protein packaging (Brininger et al., 2018). The enzyme's rigidity and noncompliance to remove complex packaging, predominated by rising temperature correspond by and large to escalated bi-sulfide bonding, joining two Cysteines (Reed et al., 2013; Siliakus et al., 2017).

In addition to electrostatic interactions and other physical properties for improving the stability of proteins, histones have displayed an intense decrease in destabilization of deoxyribonucleic acid via structurally similar H2A/H2 B, H3, and H4 histone core proteins present in eukarya (Stetter, 1999; Dohm and Maruyama, 2015). DNA of thermophiles often have thermal tolerance to positive super twists introduced by reverse gyrase (Jamroze et al., 2014). Mono- and di-valent salts (KCl and $MgCl_2$) also stabilize nucleic acids protecting DNA from hydrolyzing and depurination (Kelley et al. 2001). Definite stability to DNA is imparted by single-stranded binding (SSB) proteins and also by condensed volume super-structuring of genetic material (Luo et al., 2019).

The issue over the metabolic fundamentals of thermophilic fungi is far from settled. Other than this, reports of rejected arguments about growth, survival, and ultimate metabolism in thermophilic fungi also exist. Several analyses of enhanced protein retaining capacity are crucial for thermophilic fungi to live at extreme temperatures. Also, releasing thermostable polysaccharide digesting enzymes for the absorption of sugars is essential to thermophilic organisms as is depicted in *Thermomyces lanuginous* and *Penicillium duponti,* to use di-saccharide sucrose and glucose simultaneously at 50°C. Fungal thermophiles also make use of HSPs (heat shock proteins) to become thermotolerant. Using the pulse-labeling technique followed by SDS-PAGE separation, eight HSPs' synthesis was reported during heat shock, three HSPs were dominant in response. The role of these HSPs in *Thermomyces lanuginosus* around 60°C–62°C (Oberson et al., 1999) is in line with this definition. Thus, it was considered that a constitutive, abundantly expressed protein, HSP60, was essential in thermophilic adaptation.

4.2.3   PSYCHROPHILES

Eighty percent of the Earth's biosphere, featuring polar areas, deep-sea, frozen lakes, winter soils, coastal ocean sediments, and large mountain glaciers consist of cold temperatures < 5°C (Pikuta et al., 2008; Huston, 2008)]. These predominantly cold environments of earth harbor a number of cold-adapted psychrophiles, with a preferred temperature range of 20°C–25°C to meet their vital needs of biochemical processes (Schröder et al., 2020). Often the terms eurypsychrophile (wide) and stenopsychrophile (narrow) are used to classify the psychrophiles (Raymond-Bouchard et al.,

2018; Cavicchioli, 2006). Some well-documented fungal psychrophiles are: *Thelebolus microsporus, Mucor strictus, Phoma herbarum, Humicola marvinii, Pseudogymnoascus destructans, Mrakia psychrophila, Mrakia frigid, Mrakiella cryoconiti, Mrakiella aquatic, Mrakiella niccombsii, Sclerotinia borealis, Microdochium nivale, and Coprinus psychromorbidus* (Wang et al., 2015).

4.2.4 ADAPTATIONS

Microbes exposed to cold temperatures consequently find ways to respond to such temperature restrictions and to counter the detrimental impact of cold temperatures, physiological adaptations such as: (1) production of induced antifreeze proteins, (2) mediating chaperones, (3) increased flexibility of tRNA, (4) increased membrane fluidity (semifluid state), and (5) active cold secondary metabolites and pigments (Casanueva et al., 2010; De Maayer et al., 2014; Gupta et al., 2014; Yadav et al., 2016) are developed by such organisms. The mobility in psychrophilic membrane is controlled through absolute increasable fatty acids and also curtailed fatty acid chains or at times both. Few solute particles act as antifreeze proteins/compounds (cryoprotectants) to reduce the cytoplasm's freezing point (Celik et al., 2013; Kawahara, 2002). Considering the structural conformation, proteinaceous substances are altered through temperature. Proteins do down with two obstacles exclusive at low temperatures: (1) Cold denaturation (2) A slower rate of reaction. Less activity enforced at cold temperatures is often attributed to very adjustable cell components and solvent connection (Siddiqui and Cavicchioli, 2006; Merlino et al., 2010). Thermophiles potentially cling to aggregate of solvate, showing ditto adaptedness in enzymes found in salt stress (Karan et al., 2012) and therefore, overcoming the thermodynamic barriers (Santiago et al., 2016). Also, their enzymes are more flexible with less charged amino acid surfaces as compared to enzymes of the mesophiles (Schröder et al., 2020).

4.2.5 ACIDOPHILES

Acidophiles are identified in all habitats, but exclusively in acid mine drainage sites, sulfate fields, acid thermal hot springs and fumaroles, bioreactors, and coal spoils (Tiquia-Arashiro et al., 2016). These organisms, with few outliers, have pH homeostatic mechanisms, retaining pH close to neutral through pH maintenance on each side of the plasma membrane (Tiquia-Arashiro et al., 2016). They are present in bacterial and archaeal domains (Baker-Austin and Dopson, 2007), contributing to the reduction of iron and sulfur in extremely acidic environments (Druschel et al., 2004). Frequently reported acidophiles belong to bacteria but some acidic ecosystems reported fungal presence also, for example, *Aspergillus fumigates, Penicillium* sp.*, Pithomyces* sp. (Stierle et al., 2003). Fungi have developed specialized adaptation in hyphal structures to broaden their ecological amplitude.

4.2.6   ADAPTATIONS

Acidic environments altogether cause significant harm to the elementary structure of proteins leading to cell death (Selbmann et al., 2013) and to overcome pH-related issues, acidophiles have evolved some cellular mechanisms such as impermeability of protons (Konings et al., 2002). Complex ester-linked lipid cell membranes of bacteria and eukaryotes consist of a large core and different head group lipids whereas Archaea has tetra-ether lipids in the cell membranes (Baker-Austin and Dopson, 2007). Illustration of a permeability barrier in the cell membrane is revealed in *Sulfolobus solfataricus, Acidophilus thermoplasma, Picrophilus oshimae, Ferroplasma acidiphillum, Ferroplasma acidarmanus* (Shimada et al., 2002; van de Vossenberg et al., 1998a, 1998b; Macalady and Banfield, 2003). More or less acid loving microbiome co-opt omega(ω)-ali-cyclic esters of fats to stabilize lipid bilayer by reducing the membrane fluidity and improving the acid resistance (Chang and Kang, 2004). Membrane channels studied in an acidophile, *Acidithiobacillus ferrooxidans*, have recorded a curtailed pore size with upregulation of Omp40 (Amaro et al., 1991) and a reduced H^+ influx via transmembrane. Overall, within their cells, they have positive potential and thus pump out protons as seen in *Bacillus* and *Termoplasma* sp. Also, cell membrane biosynthetic enzymes constitute a rich diversity in certain bacterial genomes that help adapt them to unpredictable pH (Baker-Austin and Dopson, 2007). A further adaptation through the generation of a Donnan potential (in suitable 1:1 stoichiometries) inside the cytosol formed by high cation concentration (positive charge), preventing H^+ influx amidst favorable gradient concentration of anti-porters K^+/H^+. These limited anticarriers and many proton pumps together facilitate H^+ efflux and prevent cytoplasmic acidification (Baker-Austin and Dopson, 2007; Enami et al., 2010; Tiquia-Arashiro et al., 2016). Over and above a smaller genome is another remarkable feature of these microorganisms (Baker-Austin and Dopson, 2007).

4.2.7 ALKALIPHILES (ALKALI LOVING ORGANISMS)

Primarily alkaliphiles effectively and efficiently flourish at a pH of ≥ 9 (Krulwich et al., 2009), or sometimes gradually evolve even at a pH of 6.5. They could be obtained from garden soils under normal environmental conditions, even though viable counts are considerably greater from alkaline environments. Several alkaliphiles originating under such conditions predominantly appertain to *Natrialba* spp., *Natronolimnobius* spp., *Halorubrum* spp., *Bacillus* spp., *Pseuodomonas* spp., *Rhodobacterium* spp., *Methylobacter* spp., *Cyanospira* spp., *Chroococcus* spp., *Synechococcus* spp., *Spirulina* spp., *Natronomonas* spp., and *Bogoriella* spp. (Dhakar and Pandey, 2016; Grum-Grzhimaylo et al., 2015). Few well-known alkaliphilic regions are Turkey, India, and Russia, etc. (Grant, 2006; Groth et al., 1997; Rees et al., 2004; Sorokin et al., 2014; Antony et al., 2013)

4.2.8 ADAPTATION

In addition to other adaptations similar halophiles, alkaliphiles show high pH adaptability (Kanekar et al., 2012; Pikuta et al., 2008; Michels and Bakker, 1985). Since they thrive in a low H^+ environment, alkaliphiles change the cell's biochemical processes and influence the cell membrane's morphological configurations against the external environment's strong pH (Dhakar and Pandey, 2016) to combat the imbalance of H^+ ions (Elleuche et al., 2014). Also, significantly larger Na^+ extrusion produces a higher salt generated power which has been important to most of the bacterial ATP synthases being energized (Dhakar and Pandey, 2016; Tiquia-Arashiro et al., 2016). Perhaps the most prevalent adjustment is exerted in using antiporters (sodium/hydrogen and potassium/hydrogen type) driving H^+ relocation inside and outside the cell (Dhakar and Pandey, 2016), for example, *Natranaerobius thermophilus* and *Desulfovibrio vulgaris* are typical anaerobes that use Na^+/H^+ antiporters, whereas *Bacillus psuedofirmus* and also *Synchococcus elongatus* altogether represent $Na^+(K^+)/Na^+H^+$ channel transport (Krulwich et al., 2011; Mesbah et al., 2009, 2011). They further utilize teichurono/(ic) or polyglutamic acids (Aono et al., 1992, 1993), from acidic secondary walls, to attract H^+ and keep away OH^-, required to drive ATP synthesis. Thus alkaliphiles are able to hydrolyze amino acids, retain meager content of alkali amino acids, enriching negatively charged cell membrane constituents, and have greater unsaturated fats and channel proteins in membrane as possible methods to counteract raised pH (Preiss et al., 2015; Slonczewski et al., 2009; Mamo, 2019; DeCaen et al., 2014).

4.2.9 RADIOPHILES

In response to low doses, the effect of more than one particular class of radiations, frequently ionizing radiations and UV radiations, have acquired importance in microbial studies (Mesquita et al., 2013; Tomac and Yeannes, 2012; Paul et al., 2012; Xavier et al., 2014). Examples of bacterial microbes such as *Deinococcus radiodurans* and *Cryptococcus neoformans* (Cox et al., 2010), are costumed to extreme radioactivity and melanin secretory fungi are reportedly more capable of tolerating radiation levels (Dadachova and Casadevall, 2008). While it may appear at first glance that a conforming matching adaptive machinery is underlying against hazardous effects of both radiation types, moderate dissimilarities exist which are seemingly based on the type of harm sustained by each. Generally, ionizing radiations yield double-stranded interruptions, as a result, serious oxidative stress damage to proteins and lipids is caused (Slade and Radman, 2011). Good growth rates along with mixed gene switching abilities, microbes also have intensified adaptation to environmental provocations (DeLong, 2012). Auto-regulatory systems of microorganisms, therefore, provide a shield toward short intense variation limits in response to environmental triggers (Siasou et al., 2017). Some of the radiation-resistant strategies in organisms are:

1. Adaptive repair mechanisms (nucleotide excision repair pathway, excision repair pathway homologous recombination pathway)

2. Resistance/defense against antioxidants and enzymes
3. Compact nucleoid and single-stranded binding proteins (SSB)

These strategies help organisms to sustain mostly all radiations (Cox et al., 2010; Pavlopoulou et al., 2016; Lockhart and DeVeaux, 2013; Coker, 2019). It was further proven condensed nucleoid facilitates the effectiveness of DNA repair (Minsky et al., 2006; Daly et al., 2007), in fragmented DNA, created by radiation. Besides, UV radiation exposure involves the creation of photoproducts of thymine dimers and pyrimidine photoreactive products. About 80% of these DNA-borne lesions in microbes are induced by UV radiation at all levels of organization, evoking microbiome dynamics via crippled genetic processes (Matallana-Surget et al., 2008; Matallana-Surget and Wattiez, 2013; Santos et al., 2013; Ayala et al., 2014; de Oliveira et al., 2015; Jones and Baxter, 2017). Species, especially extremophilic bacteria, living in adverse conditions perpetually subjected to solar irradiance harm, have established a variety of photoprotection mechanisms to safeguard themselves from sustained UV stress taking care of vital processes like DNA repair, defensive UV-based oxidative stress, and regulated proteome (Kashefi and Lovley, 2003; Matallana-Surget et al., 2009; Lo´pez-Monde´jar et al., 2016; Gao and Garcia-Pichel, 2011; Matallana-Surget et al., 2014; Albarracín et al., 2012). In addition, altered carotenoids, antioxidant- dismutase, hydroperoxidases, duplicates of gene, and some reduced bi-pyrimidine numbers are also reported in radiophiles (Jones and Baxter, 2017). Nevertheless, ionizing radiations are typically a cause of cell death (Quintana-Cabrera et al., 2012).

4.2.10   PIEZOPHILIES-PRESSURE LOVING MICROBES

Among various adaptations, pressure adapted specified kinds are known as piezophillic organisms (also known as Barophiles). These organisms have above normal pressure needs (Michoud and Jebbar, 2016) and are found in waters of the Mariana Trench at a pressure of 110 MPa (Glud et al., 2013; Nunoura et al., 2015; Tarn et al., 2016). Additionally, some deep seas exerting a pressure of 38 MPa are known to shelter such living forms (Oger and Jebbar, 2010). A limited faction of the prokaryotic community exists under such high pressure and acts as a controlling force for diversity patterns (Kashefi and Lovley, 2003; Lo´pez-Monde´jar et al., 2016; Picard and Daniel, 2013; Cario et al., 2019). Certain physiological impacts of hydrostatic pressure are well documented for (piezosensitive bacteria) *E. coli* and some cold and pressure-adapted microbes like *Photobacterium profundum* (Michoud and Jebbar, 2016; Oger and Jebbar, 2010). Piezophiles are categorized into two types depending on pressure prerequisites:

1. Piezophilies—whose optimum growth is above atmospheric pressure
2. Piezotolerant—those that can survive at/above atmospheric pressure (Picard and Daniel, 2013; Abe, 2015).

While high pressure exerts a specific effect on microorganisms, detailed information on their cell functionalities, like motility of cells and differentiation is needed.

4.2.11 ADAPTATIONS

Just a handful of knowledge on adaptive forces/machinery of piezophiles exists. Their fine scalability to settle in very extreme pressure thus stipulates major improvements in physiology involving changeable cell structures through altered gene expressivity. At a cellular level they have shown week cell-divisions, accumulation of low weight organic compounds, incorporated oils {P(poly)U(unsaturated)F(fatty)A(acids)} and development of antioxidant proteins (Nath and Subbiah; Ichiye, 2018; Huang et al., 2017). Another major high-pressure adaptation in prokaryotes revealed from 16S rRNA prokaryotic gene quantification is their longer DNA helices (Lauro et al., 2007). They are also known to produce MUFA (monounsaturated fatty acids) oils (Siliakus et al., 2017; Michels and Bakker, 1985) and sometimes phosphatidyl-glycerol/choline in place of phosphatidyl-ethanolamine (Jebbar et al., 2015), to stabilize the plasma membrane as an upsurge in oils makes phospho-bilipid membranes resilient to high pressure(Jebbar et al., 2015). Knowledgeably, significant research work represents changes in a number of fatty acids in certain archaeols (Kaneshiro and Clark, 1995). Still, other pathways at the gene adaptation stage are usually not well understood. General improvements in the overall metabolism are known to hold proteins soluble and functioning in such microorganisms (Michoud and Jebbar, 2016; Hassan et al., 2016; Vannier et al., 2015; Amrani et al., 2014).

4.2.12 HALOPHILES

Halophiles are microorganisms that commonly flourish along a salty continuum, like salt waters, estuaries/inlets, salt lakes, ponds, brine springs, and brackish waters, etc., with a potential to equilibrate osmotic pressure and tolerate the destabilizing influence of salts (Tiquia-Arashiro et al., 2016). Despite these issues, such ecosystems are intense in the sense that they have restricted microbial diversity arising due to coupled environmental factors, one being hyper-salinity and the others being temperature, pH, low nutrients, or oxygen (Ventosa et al., 2015). Depending on the quantity of salt level present in a habitat, halophiles can be broadly categorized as slight, moderate, or highly halophilic. Moderate halophiles have optimal growth around 3%–5% sodium salt concentration, whereas organisms of high salty habitats luxuriantly grow in 15%–30% sodium salt concentration, and slight halophiles, flourish in 1%–3% sodium salt concentration (Ventosa et al., 2015; Enache et al., 2010; Yin et al., 2015). Halophilic bacteria include members from bacteroides, cyanobacterium, proteobacteria, lactobacillus and sulfur-green bacteria (Tiquia-Arashiro et al., 2016).

4.2.13 ADAPTATIONS

Comprehending the apparent adaptive mechanisms occurring in halophiles due to high salt concentrations is of greater interest at present. These salt-loving organisms follow the salt in/salt out procedure to withstand saline conditions (Yin et al., 2015). Explicitly, salt-in operation routinely maintain 4–5 M salinity/ion (K^+, Na^+, Cl^-) concentrations (Yamagishi et al., 2018). Either side of plasma membrane, (inside/outside), for example, haloarchaea, in a strategy collect KCl molecules using salts from their environment (Enache et al., 2010; Edbeib et al., 2016; DasSarma et al., 2012; DasSarma, 2009). The cytoplasm in these organisms consequently represents a significant challenging environment to standard biochemistry. Ultimate, hyper-saline pressures likely expel fluids from proteins resulting from protein coagulation directly attributable to exposed hydrophobic patches that bind together. To counteract this, in "salt in" strategy, appropriate adaptations occur in their proteins and enzymes. In particular, a rise in acidic amino acid residues, especially Aspartic and/or Glutamic acids occurs over protein's surface (Brininger et al., 2018; Ventosa, 2006;

Graziano and Merlino, 2014). Technically, acidic residues in proteins help form a "water-trap" that prevents the proteins from degradation and coagulates off from liquid (DasSarma and DasSarma, 2015; DasSarma et al., 2009). Note that archaea and some bacteria have demonstrated this adaptive approach (Oren, 2008). Halophiles usually maintain different concentration levels of salt/ions within and outside the cell wall pumping out salts from cytosol, alternatively can buildup osmolytes, also popularly known as compatible solutes, like those of Glycine, Betaine, Zwitterionic constituents, glycerol-trihydroxyl alcohol, and also some specific polyols (DasSarma et al., 2009; Roberts, 2005; Edbeib et al., 2016). While describing their adaptive mechanism, combinational adaptabilities may exist in certain halophiles.

"Salt out" strategy is a more active process (energy-dependent) regulating the turgor pressure of cells under stressful conditions (Math et al., 2012; Pikuta et al., 2008). The gained compatible solutes are consequently to be ejected via mechanistic-sensitive channels (Oren, 2008). Other than that, these halophiles are also thermally stable and have a wide range of pH tolerance. They consist of metabolically diverse genera such as anoxic phototrophic, methanogenic, sulfate reducer, denitrifying, and heterotrophic aerobic, microbes (Antunes et al., 2011). Evolutionary adaptations of the "salt-out" strategy have made it possible for these microorganisms to survive over different values of salinity.

4.3   MICROBIAL SEASONAL ADAPTATIONS TO STRESSORS: BACTERIAL AND FUNGAL LINEAGE ADAPTATIONS

Microorganisms are typically the earliest colonists and keystone players to produce a chain of activities that are fundamental to the advancement of high-level food webs in all fragile ecosystems (Bradley et al., 2016). Microorganisms' high adaptive ability and physiological versatility enable them to inhabit the biggest and extreme habitats. Microbial communities tend to change their composition, varying from hours to days when disturbed in their natural environment (Rubin et al., 2013). Microbial populations in soils respond to changes in the climate by adapting and connecting species functional aspect and genetic construct to physiochemical circumstances (Grayston et al., 2001; Wallenstein and Hall, 2012). Studies suggest that the assessment of certain environmental factors in determining the structure of microbial diversity can help to illuminate biogeochemical cycles in soil (Keith-Roach and Livens, 2002). Yet, even with their widespread habitability and their usefulness across land ecosystems and functioning, the heterogeneity and seasonal variation in microbiome, especially at small-regional and large-gobal study point of view are still understudied (Kazemi et al., 2016). In fact, microbes are sometimes expressed by physiochemical and biophysical mesh resulting in highly diverse and unpredictable ecological processes (Cray et al., 2013). Subsequent conceptual and empirical research has shown that either stochastic and/or deterministic forces control the distribution pattern of microbes at spatial and temporal scales (Caruso et al., 2011). Seasons induce dramatic versatile parameters, for example, temperature, precipitation, vegetation cover, and raptness of nutrients that remain indefinitely essential to microbial survivalism. Certainly, several species of microbes can escape or withstand adverse conditions by adopting dormant forms during growth, which can better tolerate high temperatures as compared to other active species. Distinct edaphic variables influence both bacterial and fungal community structures (Girvan et al., 2003; Lauber et al., 2009; Tedersoo et al., 2014; Frey et al., 2004). Habitat's nutritional and physicochemical characteristics, like pH, also determine spatiality and anthropocentric spread of microorganisms in medium (Bu´ee et al., 2009; Burke et al., 2009; Fierer and Jackson, 2006). Therefore, microbial populations are supposed to undergo pronounced shifts in conditions where these variations are very evident (McMahon et al., 2011; Monson et al., 2006). Often, some seasonal sway ascertained from soil microbiome behavior additionally is correlated to changes in the population structure of the microbe. This differential dynamics arises in response to microbial communities' developmental activities which in turn are susceptible to climate-controlling agents, viz., temperature and humidity. Rates of growth and expansion are unique characteristics of microbial populations which can differ independently, implying that certain climatic conditions supporting increased rate of microbial activity do not necessarily promote fast microbial growth and enhanced biomass. Elevated normal temperature is a general and likely cause of increased microbe activity. Therefore, owing to the decreased temperature, microbial development and activity usually decline in the winter season. For instance, bacteria clustering around the Lyman Glacier showed that the bacterial populations are assembled more deterministically than the fungal counterpart (Brown and Jumpponen, 2014). Also, bacteria have been shown to fluctuate in aquatic environments in response to parameters such as temperature, daytime, and nutrient concentrations. Moreover, these parameters are season to season reproducible even in marine and freshwater habitats (Staley et al., 2015). On the other hand, prior research on steppe bacterial population assembly, in the North China environment, was mainly influenced by stochastic processes. However, this observation was centered on only one species (Soininen et al., 2007; Zhang et al., 2011, 2013). Even though there is a relatively positive association through temperature and microbe development, thermal aptitude in microbiome is dependent upon microbial species and on spectrum of temperature experienced. In general, exceedingly high temperatures are deleterious to many microorganisms. It has been seen that rising temperatures manipulate microbial respiration, and persistent respiration of microbes under glacial ice concrete can trigger oxygen depletion and CO_2 and CH_4 accumulation (Crawford and Braendle, 1996). The amount of water and temperature reportedly influences microbe longevity and distribution (Whitford, 1996). Interestingly, there is still a great deal of confusion on how sensitive and diverse microbial classes are to a range of temperatures.

In contradistinction to temperature, moisture content, the timing of moisture-dryness and freezing–unfreezing cycle are known to change soil-stable structure along with long term profound influences, mostly on the abundance of organic material. Both aspects heavily influence microbiome arrangement and metabolomics. Microbial action persists all through winter under thick snow cover and is an integral element of yearly carbon budget allocations and nitrogen (N) mineralization (Fahnestock et al., 1999; Schimel et al., 2004), with profound impacts on nutrient availability at the beginning of spring. Additionally, increased freezing and severity of freeze-thaw events during low snow was correlated with disordered soil aggregates, plant, and microbial cells (Lauenroth and Bradford, 2012). Winter conditions reportedly are of necessary relevance to ecosystem processes with possible chances of rippling impacts over the growing season (Phoenix and Lee, 2004; Post et al., 2009). It is needless to say, just a few experimental investigations on the implications of irregular cold weather, particularly to soil microbial populations and functions in natural forests, have been managed to carry out so far. Field-based experiments on snow removal in one of the forests have shown higher soil frost causes root mortality (Groffman et al., 2001). These changes also affected the relationships between plants and microbes with overall decrease in micro-flora (Sorensen et al., 2016a, 2016b). Forest restricts decay and nitrogen mineralization, and therefore nitrogen remains attached to organic matter. Snow cover causes significant CO_2 release while erratic snow causes soil forest to decrease microbial respiration (Brooks et al., 2011). Though, it is not understood how well various eukaryotes especially fungi and bacteria react toward events in growing conditions during winter. However, these predicted seasonal dynamics will vary in different soil environments, for example, microbial biomass in treeless soils has been highest around winters, that is, when temperature stands minimal (Edwards et al., 2006).

Moore-Kucera and Dick (2008), affirmed that different seasons affect the forest's microbial populations. Seasonal changes in symbiotic mycor-rhizae, saprotrophs, and bacterial population are widely involving access to carbon (C) through exudates of plants or litter or at some point, in decom-position (Kaiser et al., 2010; Vŏřísková et al., 2014; López-Mondéjar et al., 2015; Santalahti et al., 2016; Žifčáková et al., 2016). Biotic activity and the abundance of substrates in forest areas greatly affect the ecological decomposer populations (Pioli et al., 2018). Together, fungal and bacterial populations are indispensable for reducing organic material decomposers in forest ecosystems (Baldrian et al., 2012; Baldrian, 2016; Llado et al., 2017). Both microbes are vital in the process of decomposition while fungi decompose recalcitrant substances the bacteria decompose readily biodegradable N-rich substrates (Rousk et al., 2016). In the case of forest soils, microbial functions, biomass, and population composition remain extremely diverse, enduring seasonal variations in temperature, nutrient, and carbon supply. Myriad microorganisms were shown to be functional in forest soils in both summer and winter (Kaiser et al., 2010; Vŏřísková a et al., 2014; Žifčáková et al., 2016; Žifčáková et al., 2017). It has just been speculated from the soil of the needle-bearing forest, the fungal actions have been high enough during summers as opposed by bacteria which substitute them in winter to carry out the break-down of dead decay plant/microbial matter (Žifčáková et al., 2017). In the degradation of dead fungal biomass, bacteria are also regarded as essential (Brabcová et al., 2016; Llado et al., 2017). In addition, cellulose polymer oxidation has also been associated with various individuals of bacterial species of forest soils (López-Mondéjar et al., 2015, 2016; Štursova et al., 2012).

In forest soils, tree species are among other strong drivers of soil microbial communities (Urbanová et al., 2015). The microbiome can change spectacularly in forests with the seasons with respect to quality of growth medium and biological plant phenology, for example, genus *Russula* is dominant in plant growth season, also *Mortierella* sp. dominates all through fall and mid-winters (Shigyo et al., 2019). Similarly, temperate forest soils reported *Actinobacteria* abundance more during winter followed by considerate weaker quantity in *Acidobacteria* sp. as well as *Proteobacteria* (Shigyo et al., 2019). A study that took into account the amount and quality of carbon and nutrients cyclization in forest soil by decomposition of plant, litter, and rhizodeposits (Pfeiffer et al., 2013; Haichar et al., 2014) and others showed mycorrhizal associations are regulated by the type of vegetation (van der Heijden et al., 2008). Ectomycorrhizal fungi and ericoid mycorrhizal fungi and N_2-fixing bacteria are deemed advantageous microbes for growing plants in forest soils (Read and Perez-Moreno, 2003; van der Heijden et al., 2008). Thus, intricate environmental systems mainly focus on nutrient uptake and microbial community. From a broader perspective, fungal seasonal dynamics have been illustrated little so far than the bacterial flora (Tedersoo et al., 2014). Siles's (Siles et al., 2017) studied seasonal dynamics of fungi through elevations in seasonally shedding and needle-bearing forests. This study showed that community composition is interlinked to seasonal modifications of organic constituents of soils. Though, elevations are less studied but a weighty aspect for controlling microbiome (Shigyo et al., 2019). However, heterogeneity in the nature of microbial populations still remains unclear.

The nature of microbial populations residing underground with their very diverse approaches in which microorganisms interact with environment mystify different microbial reactions. Among other factors soil respiration seemingly is said to play indispensable survive mainly because of: the gush of CO_2 and CH_4, modulating overall dependence of C-stock levels, and its own susceptibility to rising temperatures. Also, it helps to modulate overall organic composting procedure thus acts as a C-prevalence controller. Inefficient respiration means less decay and efficacious respirations mean more decay, that is, the more will be the release of carbon (C) and oxygen (O). However, extreme weather likely causes significant deleterious effects on soils that can affect all soil biodiversity services; estimation of such impacts is indeed needed. There is also a strong need to select proficient permutations of environmental indicators to study microbiota in ecosystems along with the index of soil biological quality. Metabolic activities of the soil-dependent member species of the microbe, which presently have been defined by the whole of litter abundance, soil quality itself, and other ecological external conditions (Samuel, 2010). The soil enzyme dehydrogenase tends to reflect soil metabolic profile as such an excellent soil quality measure (Arunkumar et al., 2013).

Recent developments in high-throughput technology have revealed the diversity and genomic competence of bacteria to decompose, comparing soil C and nutrient-cycling mechanisms with respective roles played by bacteria and fungi (Baldrian, 2016; Llado et al., 2017; Rousk et al., 2016; S'tursova et al., 2012). Few workers also studied total fungi and bacterial communities using both DNA and RNA sequencing. Environmental samples having active and inactive community components are often detected via PCR-based approach. Illumina HiSeq and Oxford Nanopore (ONT) are recent technological advancements that can help study genomes of community unveiling adaptive mechanisms of extremophiles and also document diversity from extreme environmental samples, even seasonally. Prospective seasonal microbe research is primarily fundamental to nurture biotechnological strategies that help improve on microbial understanding and accomplish unified soil microbial planning. This scientific research is undoubtedly very appealing and will support long-term environmental consequences and research development.

4.4 BIOTECHNOLOGICAL POTENTIALITIES

In recent years, microbial biotechnology has revolutionized the field of food manufacturing, sewage treatment, waste management, manufacturing of antibiotics, and organic acids, etc., and the technology is gaining its pace overcoming traditional barriers of production through the inception of extremozymes. Extremozymes are the biocatalysts that operate under deleterious conditions like those of excessive hot/cold temperatures, dangerous radiation, and pH (Sarmiento et al., 2015). Extremophilic enzymes are usually taken from most species of all typified classes of extremophillic organisms and are procured/generated from microorganisms such as bacteria, algae, fungi under unfavorable conditions (Anitori, 2012; Krishnaraj and Sani, 2017). Thus, extremozymes present the foundation/ amelioration of working conditions with regard to sustainable industrial developments. Next to stability, these enzymes do have a higher activity and catalytic rate compared to their normal mesophilic counterparts. Their extreme structural stability furhter assists in genetic engineering or site-directed enzyme engineering mutagenesis/approaches for protein development.

4.4.1   *BACTERIAL LINEAGE EXTREMOPHILE APPLICATIONS*

Recently serine-based protease obtained from *Glaciozyma antarctica* demonstrated towering activity around 20°C (Alias et al., 2014; Sarmiento et al., 2015). Some heat-labile Uracil-DNA *N*-glycosylase exploration in *Psychrobacter* species and its vector cloned/expressed form in *E. coli*, has optimal functional temperature is 20°C–25°C (Lee et al., 2009). Recombinant Cryonase, a cold-active nuclease, isolate of *Shewanella* sp., can digest all kinds of genomes (single/double/linear/circular) and can be inactive after half an hour incubation at 70°C (Awazu et al., 2011). Nowadays, recombinase polymerase amplification is carried out using cold-active recombinases facilitating the insertional oligonucleotide primer in dsDNA. Also, psychrophilles are suppliers of DNA-ligase combating cold and supporting significantly higher enzyme activities. The ligase extracted from *Pseudoalteromonas haloplanktis*, displays activity at 4°C (Georlette et al., 2000; Sarmiento et al., 2015). Furthermore, cold-adapted lipases have also been reported to target stains from triglycerides (Jiewei et al., 2014; Ji et al., 2015). More recently among lipases an efficient lipase enzyme, source strain *Pseudomonas stutzeri,* has displayed better enzyme activity at a temperature of 20°C and a pH need of $8^{1/2}$ (Li et al., 2014). Often it is useful for detergent/surfactant/oxidant, large scale making (Li et al., 2014). Use of alpha-amylase which converts starch into water soluble form through 1,4-α-glucosidic bond hydrolysis (Hmidet et al., 2009; Sarmiento et al., 2015) has been yielded from *Bacillus cereus* which is detergent stable (Roohi et al., 2013). Alpha-amylase so obtained has an optimum operation at 22°C, stable activity at low temperatures of 4°C–37°C, and alkaline pH >7–11. Yet, another amylase isolated from marine *Zunongwangia profunda* has potential applicability in detergents (Qin et al., 2014). Hydrolytic cellulase enzymes, derived from fungus *Humicola insolens* can withstand 15°C. This cellulase finds use in color protecting and brightening detergents, digests damaged fabric fibers, and eliminates (β)beta glucan dirt stains. The enzyme's optimum activity ranges around 30°C–60°C but is also active at fairly low temperatures (Sarmiento et al., 2015).

Other prominent representatives of cold-active enzymes, such as mannanases help degrade hydrates of carbon in food articles, etc., whereas pectinase is known to help degrade polysaccharides, widely a part of edible food products such as fruits, etc. In agri-foodstuff industry, cold-active α-amylases from *Microbacterium foliorum* (Kuddus et al., 2012) and

Zunongwangia profunda (Qin et al., 2014) are among fine isolates used in beverage fermentation, confectionary, and juice processing. Moreover, *Thalassospirafrigidphilos profundus* (Ghosh et al., 2012; Pulicherla et al., 2013) and *Pseudoalteromonas haloplanktis* (Van de Voorde et al., 2014) have wide applicability in the degradation of lactose from cold-stored dairy products using harnessed cold-active isolate, β-galactosidases (Asenjo et al., 2014). Del-Cid et al. (2014) also reported Xylanases source from *P. haloplanktis*, and *Flavobacterium* sp. An early study by Kitamoto et al. (1988), has identified *Thermoanaerobacterium thermosulfurigenes* and *Clostridium thermocellum* as the prospective aspirants in the underlying saccharification process. Sukumar et al. (2013), in recent times suggested the use of *Fervidobacterium gondwanense* and *Bacillus* sp. in industrial glucose isomerization. These strains have 70°C optimal temperature needs. Some reports of *Amylopullulanases* used in the starch liquefaction process are also there (Satyanarayana et al., 2004).

Currently, few hyper-thermophilic xylanases are commercially used in pulp and paper making which have turned out to be the derivatives of *Themus maritime, Streptomyces* sp., *Thermotoga thermarum*, and *Thermoascus aurantiacus* (Shi et al., 2013). Apart from xylanases, enzymes like laccases have huge applications in the textile industry, dye industry, wastewater treatment, and bioremediation (Virk et al., 2012). An excellent outlined procedure of microbial-bleaching using laccase isolate on wheat straw pulp from *Thermophilus* bacteria (a hyperthermophilic bacterium), efficient at 85°C, has been given by Zheng et al. (2012). Overall, the major reported use of extremophilic proteins is through xylanase, lipase, esterase, and cellulase; with hyperthermophiles sourcing for the pectinase, amylase, and cellulase enabling the pulp and paper industry to adopt more sustainable, effective and refined procedures (Reid and Ricard, 2000; de Souza and de Oliveira Magalhaes, 2010; Kuhad et al., 2011).

Further, metabolites and enzymes from them are useful in biodegradation and bioremediation of polluted habitats, source of biofuel or bioenergy, to thermostablize DNA polymerases in a polymerase chain reaction (Arora, 2019). To date, more than 3000 new biological metabolites/enzymes have been screened from microbes, and there is still more to be explored. All these do not need costly maintenance and have high reaction rates even under cold conditions that reduce the energy needs. In addition, the emancipated research on extremophilic metabolites can be useful to understand extreme ecological environments besides industrial benefits. Over the last few years, enthralling findings have been brought into the microbiome in extreme conditions which can improve the quality of our life. Also, recent research proposes to fully realize the spatio-temporal changes, potential risks, and opportunities presented by life in these habitats. Extreme ecosystems should, therefore, be primarily aimed at ecological research to explore the richness of genetic diversity for different eco-friendly resources.

4.4.2   *FUNGAL EXTREMOPHILE APPLICATIONS*

Enzyme phytases at a large scale are obtained from *Aspergillus ficuum, A. niger, A. fumigatus* (Zhang et al., 2013), and some *Penicillium* strains (Zhao et al., 2012). The enzyme is known to cleave the phosphomonoester bond of phytic acid to release inorganic phosphate and myoinositol (Singh et al., 2015; Gaind and Singh, 2015) and is added to animal feed to utilize phosphate and limit phosphorus pollution in surroundings (Özdemir and Uzel, 2020). Earlier, Xu (Xu et al., 1996) and Berka (Berka et al., 1997) reported thermostable laccase genesis in *Myceliophthora thermophila,* which was later cloned in *Aspergillus oryzae* functional activity upto 70°C. Often, fungal isolates such as laccase, protease, lipase, cellulase, chitinase, amylase, peroxidase, glucosidase, etc., are widely used for pharmaceutical and medical fields (Red Biotechnology), environmental fields (Grey Biotechnology) and in aqua-marine fields (Blue Biotechnology) (Barone et al., 2019; Frazzetto, 2003). A commercial source of amylases is reportedly a fungus, *Aspergillus oryzae.* (Gupta et al., 2003) Polyextremophilic amylases obtained from marine fungi are now being used in aqua-farming, biorefinery, food fermentatives, fabric industries, paper mills, medicines, and even in household applications (Suriya et al., 2016; Gopinath et al., 2017). The halophilic *Aspergillus penicillioides* produces amylase that exhibits high catalysis even in 300–400 g/L salinity, while in extreme halophilic marine prokaryotes its production gets reduced due to salt stress (Ali et al., 2015). Another fungal enzyme, lipases is substantially gaining interest as it catalyzes the break-down of a lipid, in particular, helps to take away with discoloration. It is also a source for the by-product oils (PUFA/MUFA), nutrition and even used as green fuel (Schreck and Grunden, 2014). Evidences show genus *Aspergillus, Peicillium, Rhodotortula,* and *Candida* produce lipases (Wang et al., 2009; De Almeida et al., 2013). These are chiefly present in feeds, detergents, biosensors, diagnostic tools, oleo-chemicals, and bioremediation agents (Verma et al., 2017; Guerrand, 2017).

Likewise, esterases enzyme is harnessed for pharmaceutical, cosmetic, and food industries (Kuddus, 2018). Fungal metabolites/enzymes are now known to hold antibiotics, anti-inflammatory, anticancer, and hypercholesterolemia properties (Barone et al., 2019). For instance, tetranorlabdane diterpenoid, asperolide E, from *Aspergillus wentii* showed cytotoxicity against breast, cervical, and pulmonary cancer cell lines (Li et al., 2016). *Aspergillus wentii* is also a source of five diterpenoids which showed cytotoxicity in human lung adenocarcinomas (Li et al., 2016). Likewise, *Penicillium commune* yields Xanthocillin-X chrysogine, and meleagrin all used as an anti-metastatic drug in treating liver, early prostates and lobular cancer lines (Shang et al., 2012; Lv et al., 2013). One of the similar metabolites is Wentilactone A, derived from *Aspergillus dimorphicus*, induce cell arrest before the G2/M cell phase in lung carcinomas (Lv et al., 2013). Another type of Wentilactone, Wentilactone B helps in blocking proliferation and migration of human hepatoma cells (Shang et al., 2012). Amazingly, there are some recent reports on enhanced taxol production by *Aspergillus aculeatinus* (Qiao et al., 2017). On the other hand, *Aspergillus versicolor* can produce antioxidant compounds. Thus extremophiles can act as potential therapeutics for neurological disorderness/illness including Alzheimer and Parkinson syndromes (Barone et al., 2019).

In gray biotechnology, fungi represent a promising eco-friendly and cost-effective biotechnological alternative that is employed to polluted environments to degrade and remove and/or reduce the toxicity of pollutants (Cerniglia, 1997; Marco-Urrea and García-Romera, 2015). Few fungi can also absorb, degrade, and accumulate recalcitrant dyes from the immediate environment (Prasad, 2018; Harms et al., 2011; Deshmukh et al., 2016). Moreover, halophilic fungal lignin destroying protein is succeeding to show immense bioremediation potential (Passarini et al., 2011), while few others are well-known for the degradation of toxic dyes and PHAs (Haritash and Kaushik, 2009). Other proteinecous substances like lignin peroxidase and manganese peroxidase mostly assist in decomposing aromatic hydrocarbons (Bonugli-Santos et al., 2010). Laccases help reduce the toxicity of chemicals containing polycyclic/nitrogen/chlorinated/aromatic compounds, pest-destroying agents and dyes (Whiteley and Lee, 2006). Fungi also show tolerance opposing high-level concentrations of metal stressors, for example, genus *Aspergillus* and *Penicillium* show stability in arsenic dominated sites, particularly, *Aspergillus sydowii,* exhibit incredible tolerance toward arsenic concentrations (Vala and Sutariya, 2012) and hence, relevant researches into fungal lineages that show evidence of such distinctive properties can result in new and unforeseen biotechnological discoveries.

4.5 CONCLUSIONS

In this chapter, we sought to characterize the microbiome associated with extreme habitats, their seasonal occurrence, and important biotechnological potential. Overall seasonal dynamics, multifariousness, and plentifulness in microbiome show heterogenity. Patterning in dynamism is asynchronous for fungi and bacteria but are influenced by seasonal abiotic factors. According to research findings, erratic dynamism in them is output of nonidentical seasonal performance, showing a sharp relationship out of seasonal abiotics and below ground microbiome. Reports also suggest that micro-biota exhibited the occurrence of different taxa during winter seasons and summer seasons and in relation to different latitudes. Periodical dynamics for adequate and inadequate represented groups are distinguishable among taxa, with these differences arise primarily due to climate. Also, the fertility of the soil is a major determinant in the case of fungi but soil bacteria are associated with soil water contents. In backdrop atmosphere, features of soil and attributes of nearby plant community speculating physical riffs can accordingly have echoing impacts on microbiome. This implies the endow-ments of abiotics made toward fungi and bacteria of soils for distinctive patterning in seasons. In conclusion, fungal and bacterial counteraction is deliberated through external agents and their internal environment to either resist or escape the external stressful conditions. Continuous and sporadic dynamics of bacteria and fungi can be studied through standard molecular techniques such as TRFLP, qPCR, and metagenomics.

Furthermore, in the light of scientific explorations, the microbiome has wide applications in terms of bio-economy. The bacterial and fungal communities, especially extremophiles, hold within themselves a whole new range of exotic and eco-friendly molecules aimed for bioprospec-tion and bioremediation by use of extremozymes yielding products like chemicals, biomaterials, pharmaceuticals, food, feed, and biofuels. Future research on a comprehensive understanding of cellular processes in these extremophiles is essential for sound biotechnological applications. New technological advances are aiming to extract these biomolecules to meet our growing needs. More and more innovations and improvements are being focused to use extremozymes for a sustainable future.

KEYWORDS

- seasonal dynamics
- bacteria
- fungi
- extreme environment
- bioremediation

REFERENCES

Abe, F. Effects of High Hydrostatic Pressure on Microbial Cell Membranes: Structural and Functional Perspectives. In *High Pressure Bioscience*; Springer: Dordrecht, 2015; pp. 371–381.

Albarracín, V. H.; Pathak, G. P.; Douki, T.; Cadet, J.; Borsarelli, C. D.; Gärtner, W., et al. Extremophilic Acinetobacter Strains from High-Altitude Lakes in Argentinean Puna: Remarkable UV-B Resistance and Efficient DNA Damage Repair. *Orig. Life Evol. Biosph.* **2012**, *42*, 201–221. doi: 10.1007/s11084-012-9276-3.

Ali, I.; Akbar, A.; Anwar, M.; Prasongsuk, S.; Lotrakul, P.; Punnapayak, H. Purification and Characterization of a Polyextremophilic α-Amylase from an Obligate Halophilic Aspergillus Penicillioides Isolate and Its Potential for Souse with Detergents. *BioMed Res. Int.* **2015**, *2015*, 1–8.

Alias, N.; Ahmad Mazian, M.; Salleh, A. B.; Basri, M.; Rahman, R. N. Molecular Cloning and Optimization for High Level Expression of Cold-Adapted Serine Protease from Antarctic Yeast *Glaciozyma Antarctica* PI12. *Enzyme Res.* **2014**, *2014*. 197938. doi: 10.1155/2014/197938.

Amaro, A. M.; Chamorro, D.; Seeger, M., et al. Effect of External pH Perturbations on in Vivo Protein Synthesis by the Acidophilic Bacterium Thiobacillus Ferrooxidans. *J. Bacteriol.* **1991**, *173* (2), 910–915.

Amrani, A.; Bergon, A.; Holota, H., et al. Transcriptomics Reveal Several Gene Expression Patterns in the Piezophile Desulfovibrio Hydrothermalis in Response to Hydrostatic Pressure. *PLoS One* **2014**, *9* (9), e106831.

Anitori, R. P., Ed. *Extremophiles: Microbiology and Biotechnology*; Caister Academic Press: Norfolk, 2012. isbn:978-1-904455-98-1.

Antony, C. P.; Kumaresan, D.; Hunger, S.; Drake, H. L.; Murrell, J. C.; Shouche, Y. S. Microbiology of Lonar Lake and Other Soda Lakes. *ISME J* **2013**, *7*, 468–476.

Antunes, A.; Ngugi, D. K.; Stingl, U. Microbiology of the Red Sea (and Other) Deep-Sea Anoxic Brine Lakes. *Environ. Microbiol.* **2011**, *3*, 416–433. [CrossRef].

Aono, R.; Ito, M.; Horikoshi, K. Regeneration of Protoplasts Prepared from Alkaliphilic Strains of Bacillus spp. *Biosci. Biotechnol. Biochem.* **1993**, *57*, 1597–1598.

Aono, R.; Ito, M.; Horikoshi, K. Instability of the Protoplast Membrane of Facultative Alkaliphilic Bacillus sp. C-125 at Alka Line pH Values Below the pH Optimum for Growth. *Biochem. J.* **1992**, *285*, 99–103.

Arora, N. K.; Panosyan, H. Extremophiles: Applications and Roles in Environmental Sustainability, **2019**, 217–218.

Arunkumar, K.; Singh, R. D.; Patra, A. K.; Sahu, S. K. Probing of Microbial Community Structure, Dehydrogenase and Soil Carbonin-Relation to Different Land Uses in Soils of Ranichauri (Garhwal Himalayas). *Int. J. Curr. Microbiol. Appl. Sci.* **2013**, *2*, 325–338.

Asenjo, J. A.; Andrews, B. A.; Acevedo, J. P.; Parra, L.; Burzio, L. O. Protein and DNA Sequence Encoding a Cold Adapted Xylanase. US Patent No 8679814 B2. United States Patent and Trademark Office, 2014.

Awazu, N.; Shodai, T.; Takakura, H.; Kitagawa, M.; Mukai, H.; Kato, I. Microorganism-Derived Psychrophilic Endonuclease. US Patent No 8034597 B2. United States Patent and Trademark Office, 2011.

Ayala, A.; Munoz, M. F.; Arguelles, S. Lipid Peroxidation: Production, Metabolism, and Signaling Mechanisms of Malondialdehyde and 4-hydroxy-2-nonenal. *Oxid. Med. Cell. Longev.* **2014**, *2014*, 360438. doi: 10.1155/2014/360438.

Baker-Austin, C.; Dopson, M. Life in Acid: pH Homeostasis in Acidophiles. *Trends Microbiol.* **2007**, *15* (4), 165–171.

Baldrian, P.; Kolarik, M.; Štursová, M., et al. Active and Total Microbial Communities in Forest Soil are Largely Different and Highly Stratified during Decomposition. *ISME J.* **2012**, *6*, 248–258.

Baldrian, P. Forest Microbiome: Diversity, Complexity and Dynamics. *FEMS Microbiol. Rev.* **2016**, *41*, 109–130.

Banciu, H. L.; Muntyan, M. S. Adaptive Strategies in the Double-Extremophilic Prokaryotes Inhabiting Soda Lakes. *Curr. Opin. Microbiol.* **2015**, *25*, 73–79.

Bardgett, R. D. *The Biology of Soil: A Community and Ecosystem Approach*; Oxford University Press: New York, 2005.

Barone, G.; Varrella, S.; Tangherlini, M.; Rastelli, E.; Dell'Anno, A.; Danovaro, R.; Corinaldesi, C. Marine Fungi: Biotechnological Perspectives from Deep-Hypersaline Anoxic Basins. *Diversity* **2019**, *11* (7), 113.

Bell, E. A.; Boehnke, P.; Harrison, T. M.; Mao, W. L. Potentially Biogenic Carbon Preserved in a 4.1 Billion-Year-Old Zircon. *Proc. Natl. Acad. Sci.* **2015**, *112*, 14518–14521.

Berka, R. M.; Schneider, P.; Golightly, E. J.; Brown, S. H.; Madden, M.; Brown, K. M., et al. Characterization of the Gene Encoding an Extracellular Laccase of Myceliophthora Thermophila and Analysis of the Recombinant Enzyme Expressed in Aspergillus Oryzae. *Appl. Environ. Microbiol.* **1997**, *63*, 3151–3157.

Bertrand, J. C.; Brochier-Armanet, C.; Gouy, M.; Westall, F. For Three Billion Years, Microorganisms Were the Only Inhabitants of the Earth. In *Environmental Microbiology: Fundamentals and Applications*; Bertrand, J.; Caumette, P.; Lebaron, P.; Matheron, R.; Normand, P.; Sime Ngando, T., Eds.; Springer: Dordrecht, 2015; pp 25–71. doi: 10.1007/978-94-017-9118-2_4.

Bonugli-Santos, R. C.; Durrant, L. R.; da Silva, M.; Sette, L. D. Production of Laccase, Manganese Peroxidase and Lignin Peroxidase by Brazilian Marine-Derived Fungi. *Enzyme Microb. Technol.* **2010**, *46*, 32–37.

Brabcová, V.; Nováková, M.; Davidová, A., et al. Dead Fungal Mycelium in Forest Soil Represents a Decomposition Hotspot and a Habitat for a Specific Microbial Community. *New Phytol.* **2016**, *210*, 1369–1381.

Bradley, J. A.; Arndt, S.; Šabacká, M.; Benning, L. G.; Barker, G. L.; Blacker, J. J.; Yallop, M. L.; Wright, K. E.; Bellas, C. M.; Telling, J.; Tranter, M.; Anesio, A. M. Microbial Dynamics in a High-Arctic Glacier Forefield: A Combined Field, Laboratory, and Modeling Approach. *Biogeosciences* **2016**, *13*, 5677–5696.

Brininger, C.; Spradlin, S.; Cobani, L., et al. The More Adaptive to Change, the More Likely You Are to Survive: Protein Adaptation in Extremophiles. *Semin. Cell Dev. Biol.* **2018**, *84*, 158–69. PubMed Abstract | Publisher Full Text | F1000 Recommendation.

Brooks, P. D.; Grogan, P.; Templer, P. H.; Groffman, P.; Öquist, M. G.; Schimel, J. Carbon and Nitrogen Cycling in Snow-Covered Environments. *Geogr. Compass* **2011**, *5* (9), 682–699.

Brown, S. P.; Jumpponen, A. Contrasting Primary Successional Trajectories of Fungi and Bacteria in Retreating Glacier Soils. *Mol. Ecol.* **2014**, *23* (2), 481–497.

Buée, M.; Reich, M.; Murat, C., et al. 454 Pyrosequencing Analyses of Forest Soils Reveal an Unexpectedly High Fungal Diversity. *New Phytol.* **2009**, *184*, 449–456.

Burke, D. J.; Lopez-Gutierrez, J. C.; Smemo, K. A., et al. Vegetation and Soil Environment Influence the Spatial Distribution of Root Associated Fungi in a Mature Beech-Maple Forest. *Appl. Environ. Microb.* **2009**, *75*, 7639–7648.

Cario, A.; Oliver, G. C.; Rogers, K. L. Exploring the Deep Marine Biosphere: Challenges, Innovations and Opportunities. *Front. Earth Sci* **2019**, *7*, 225.

Caruso, T.; Chan, Y.; Lacap, D. C.; Lau, M. C.; McKay, C. P.; Pointing, S. B. Stochastic and Deterministic Processes Interact in the Assembly of Desert Microbial Communities on a Global Scale. *ISME J.* **2011**, *5*, 1406–1413. (Brown and Jumpponen, 2014).

Casanueva, A.; Tu_n, M.; Cary, C.; Cowan, D. A. Molecular Adaptations to Psychrophily: The Impact of 'Omic' Technologies. *Trends Microbiol.* **2010**, *18*, 374–381. [CrossRef] [PubMed].

Cavicchioli, R. Cold-Adapted Archaea. *Nat. Rev. Microbiol.* **2006**, *4*, 331. [CrossRef] [PubMed].

Celik, Y.; Drori, R.; Pertaya-Braun, N., et al. Microfluidic Experiments Reveal That Antifreeze Proteins Bound to Ice Crystals Suffice to Prevent Their Growth. *Proc. Natl. Acad. Sci. USA* **2013**, *110* (4), 1309–1314. PubMed Abstract | Publisher Full Text | Free Full Text.

Cerniglia, C. E. Fungal Metabolism of Polycyclic Aromatic Hydrocarbons: Past, Present and Future Applications in Bioremediation. *J. Ind. Microbiol. Biotechnol.* **1997**, *19*, 324–333. [CrossRef] [PubMed].

Chang, S. S.; Kang, D. H. Alicyclobacillus spp. in the Fruit Juice Industry: History, Characteristics, and Current Isolation/Detection Procedures. *Crit. Rev. Microbiol.* **2004**, *30*, 55–74.

Coker, J. A. Recent Advances in Understanding Extremophiles. *F1000Research* **2019**, *8*.

Cooney, D.G.; Emerson, R. *Thermophilic Fungi: An Account of Their Biology Activities and Classification*; W.H. Freeman & Co.: San Francisco, CA, 1964.

Corliss, J. B.; Baross, J. A.; Hoffman, S. E. An Hypothesis Concerning the Relationships Between Submarine Hot Springs and the Origin of Life on Earth. *Oceanologica Acta*, Special issue, 1981.

Cox, M. M.; Keck, J. L.; Battista, J. R. Rising from the Ashes: DNA Repair in Deinococcus Radiodurans. *PLoS Genet.* **2010,** *6* (1), e1000815. PubMed Abstract | Publisher Full Text | Free Full Text.

Crawford, R. M. M.; Braendle, R. Oxygen Deprivation Stress in a Changing Environment. *J. Exp. Bot.* **1996,** *47*, 145–159. (Henry 2007).

Cray, J. A.; Bell, ANW, Bhaganna, P.; Mswaka, A. Y.; Timson, D. J.; Hallsworth, J. E. The Biology of Habitat Dominance; Can Microbes Behave as Weeds? *Microb. Biotechnol.* **2013,** *6*, 453–492.

Cregger, M. A.; Schadt, C. W.; McDowell, N. G.; Pockman, W. T.; Classen, A. T. Response of the Soil Microbial Community to Changes in Precipitation in a Semiarid Ecosystem. *Appl. Environ. Microbiol.* **2012,** *78* (24), 8587–8594.

Dadachova, E.; Casadevall, A. Ionizing Radiation: How Fungi Cope, Adapt, and Exploit with the Help of Melanin. *Curr. Opin. Microbiol.* **2008,** *11* (6), 525–531.

Daly, M. J.; Gaidamakova, E. K.; Matrosova, V. Y., et al. Protein Oxidation Implicated as the Primary Determinant of Bacterial Radioresistance. *PLoS Biol.* **2007,** *5* (4), e92. PubMed Abstract | Publisher Full Text | Free Full Text | F1000 Recommendation.

Daniel, I.; Oger, P.; Winter, R. Origins of Life and Biochemistry under High-Pressure Conditions. *Chem. Soc. Rev.* **2006,** *35*, 858–875. [CrossRef].

DasSarma, S.; DasSarma, P Halophiles. In *eLs*. Wiley: Chichester, 2012. doi: 10.1002/9780470015902. a0000394.pub3.

DasSarma, S. Halophiles. *Encycl. Life Sci.* **2009,** 1–9.

DasSarma, S.; DasSarma, P. Halophiles and Their Enzymes: Negativity Put to Good Use. *Curr. Opin. Microbiol.* **2015,** *25*, 120–126. PubMed Abstract | Publisher Full Text | Free Full Text 118–139.

DasSarma, S.; Coker, J. A.; DasSarma, P. Archaea. In *Encyclopedia of Microbiology*. Oxford Academic Press: Oxford, 2009; pp 1–23. Publisher Full Text.

De Almeida, A. F.; Tauk-Tornisielo, S. M.; Carmona, E. C. Acid Lipase from Candida Viswanathii: Production, Biochemical Properties, and Potential Application. *BioMed Res. Int.* **2013,** *2013*, 1–10.

De Maayer, P.; Anderson, D.; Cary, C.; Cowan, D. A. Some Like It Cold: Understanding the Survival Strategies of Psychrophiles. *EMBO Rep.* **2014,** *15*, 508–517. [CrossRef] [PubMed].

de Oliveira, T. B.; Gomes, E.; Rodrigues, A Thermophilic Fungi in the New Age of Fungal Taxonomy. *Extremophiles* **2015.**

de Souza, P. M.; de Oliveira Magalhaes, P. Application of Microbial Alpha-Amylase in Industry—A Review. *Braz. J. Microbiol.* **2010,** *41*, 850–861. doi: 10.1590/S1517-83822010000400004 [PMC free article] [PubMed] [CrossRef] [Google Scholar].

DeCaen, P. G.; Takahashi, Y.; Krulwich, T. A.; Ito, M.; Clapham, D. E. Ionic Selectivity and Thermal Adaptations within the Voltage-Gated Sodium Channel Family of Alkaliphilic Bacillus. *Elife* **2014,** *3*, e04387.

Del-Cid, A.; Ubilla, P.; Ravanal, M. C.; Medina, E.; Vaca, I.; Levican, G., et al. Cold-Active Xylanase Produced by Fungi Associated with Antarctic Marine Sponges. *Appl. Biochem. Biotechnol.* **2014**

DeLong, E. F. Microbial Microevolution in the Wild. *Science* **2012**, *336*, 422–424. doi: 10.1126/science.1221822.

Deshmukh, R.; Khardenavis, A. A.; Purohit, H. J. Diverse Metabolic Capacities of Fungi for Bioremediation. *Indian J. Microbiol.* **2016**, *56*, 247–264.

Dhakar, K.; Pandey, A. Wide pH Range Tolerance in Extremophiles: Towards Understanding an Important Phenomenon for Future Biotechnology. *Appl. Microbiol. Biotechnol.* **2016**, *100* (6), 2499–2510.

Dohm, J. M.; Maruyama, S. Habitable Trinity. *Geosci. Front.* **2015**, *6*, 95–101.

Druschel, G. K.; Baker, B. J.; Gihring, T. M.; Banfield, J. F. Acid Mine Drainage Biogeochemistry at Iron Mountain, California. *Geochem Trans.* **2004**, *5*, 13–32.

Edbeib, M. F.; Wahab, R. A.; Huyop, F. Halophiles: Biology, Adaptation, and Their Role in Decontamination of Hypersaline Environments. *World J. Microbiol. Biotechnol.* **2016**, *32* (8), 135.

Edwards, K. A.; McCulloch, J.; Kershaw, G. P.; Jefferies, R. L. Soil Microbial and Nutrient Dynamics in a Wet Arctic Sedge Meadow in Late Winter and Early Spring. *Soil Biol. Biochem.* **2006**, *38* (9), 2843–2851.

Elleuche, S.; Schroeder, C.; Sahm, K.; Antranikian, G. Extremozymes—Biocatalysts with Unique Properties from Extremophilic Microorganisms. *Curr. Opin. Biotechnol.* **2014**, *29*, 116–123.

Enache, M. Ã. D. Ã. L. I. N.; Kamekura, M. A. S. A. H. I. R. O. Hydrolytic Enzymes of Halophilic Microorganisms and Their Economic Values. *Rom. J. Biochem.* **2010**, *47* (1), 46–59.

Enami, I.; Adachi, H.; Shen, J. R. Mechanisms of Acido-Tolerance and Characteristics of Photosystems in an Acidophilic and Thermophilic Red Alga, Cyanidium Caldarium. In: Red Algae in the Genomic Age; Seckbach, J.; Chapman, D. J. Eds.; Springer: Dordrecht, 2010; pp 373–389.

Fahnestock, J. T.; Jones, M. H.; Welker, J. M. Wintertime CO2 Efflux from Arctic Soils: Implications for Annual Carbon Budgets. *Global Biogeochem. Cycl.* **1999**, *13*, 775–779.

Fierer, N.; Jackson, R. B. The Diversity and Biogeography of Soil Bacterial Communities. *Proc. Natl. Acad. Sci. USA* **2006**, *103*, 626–631.

Frazzetto, G. White Biotechnology. *EMBO Rep.* **2003**, *4*, 835–837.

Frey, S. D.; Knorr, M.; Parrent, J. L., et al. Chronic Nitrogen Enrichment Affects the Structure and Function of the Soil Microbial Community in Temperate Hardwood and Pine Forests for. *Ecol. Manage.* **2004**, *196*, 159–171.

Gaind, S.; Singh, S. Production, Purification and Characterization of Neutral Phytase from Thermotolerant Aspergillus Flavus ITCC 6720. *Int. Biodeterior. Biodegrad.* **2015**, *99*, 15–22.

Gao, Q.; Garcia-Pichel, F. Microbial Ultraviolet Sunscreens. *Nat. Rev. Microbiol.* **2011**, *9*, 791–802. doi: 10.1038/nrmicro2649 [PubMed] [CrossRef] [Google Scholar].

Georlette, D.; Jonsson, Z. O.; Van Petegem, F.; Chessa, J.; Van Beeumen, J.; Hubscher, U., et al. A DNA Ligase from the Psychrophile *Pseudoalteromonas haloplanktis* Gives Insights into the Adaptation of Proteins to Low Temperatures. *Eur. J. Biochem.* **2000**, *267*, 3502–3512. doi: 10.1046/j.1432-1327.2000.01377.x.

Ghosh, M.; Pulicherla, K. K.; Rekha, V. P.; Raja, P. K.; Sambasiva Rao, K. R. Cold Active Beta-Galactosidase from Thalassospira sp. 3SC-21 to Use in Milk Lactose Hydrolysis:

A Novel Source from Deep Waters of Bay-of-Bengal. *World J. Microbiol. Biotechnol.* **2012**, *28*, 2859–2869. doi: 10.1007/s11274-012-1097-z.

Girvan, M. S.; Bullimore, J.; Pretty, J. N., et al. Soil Type Is the Primary Determinant of the Composition of the Total and Active Bacterial Communities in Arable Soils. *Appl. Environ. Microbiol.* **2003**, *69*, 1800–1809.

Glud, R. N.; Wenzhöfer, F.; Middelboe, M.; Oguri, K.; Turnewitsch, R.; Canfield, D. E., et al. High Rates of Microbial Carbon Turnover in Sediments in the Deepest Oceanic Trench on Earth. *Nat. Geosci.* **2013**, *6*, 284–288. doi: 10.1038/ngeo1773.

Goordial, J.; Davila, A.; Lacelle, D.; Pollard, W.; Marinova, M. M.; Greer, C. W., et al. Nearing the Cold-Arid Limits of Microbial Life in Permafrost of an Upper Dry Valley, Antarctica. *ISME J.* **2016**, *10*, 1613–1624. doi: 10.1038/ismej.2015.239.

Gopinath, S. C. B.; Anbu, P.; Arshad, M. K. M.; Lakshmipriya, T.; Voon, C. H.; Hashim, U.; Chinni, S. V. Biotechnological Processes in Microbial Amylase Production. *BioMed Res. Int.* **2017**, *2017*, 1–9. [CrossRef].

Grant, W. D. Alkaline Environments and Biodiversity. In *Extremophiles*; Gerday, E. C.; Glansdorff, N., Eds.; UNESCO, Eolss Publishers: Oxford, UK, 2006. http://www.eolss. net.

Grayston, S. J.; Griffith, G. S.; Mawdsley, J. L.; Campbell, C. D.; Bardgett, R. D. Accounting for Variability in Soil Microbial Communities of Temperate Upland Grassland Ecosystems. *Soil Biol. Biochem.* **2001**, *33*, 533–551. doi: 10.1016/S0038-0717(00)00194-2.

Graziano, G.; Merlino, A. Molecular Bases of Protein Halotolerance. *Biochim. Biophys. Acta* **2014**, *1844*, 850–858.

Groffman, P. M.; Driscoll, C. T.; Fahey, T. J., et al. Colder Soils in a Warmer World: A Snow Manipulation Study in a Northern Hardwood Forest Ecosystem. *Biogeochemistry* **2001**, *56*, 135–150. (Sorensen et al. 2016a,b, 2018).

Groth, I.; Schuma, P.; Rainey, F. A.; Martin, K.; Schuetze, B.; Augsten, K. Bogoriella Caseilytica gen. nov., sp. nov., a New Alkaliphilic Actinomycete from a Soda Lake in Africa. *Int. J. Syst. Bacteriol.* **1997**, *47* (3), 788–794.

Grum-Grzhimaylo, A. A.; Georgieva, M. L.; Bondarenko, S. A.; Debets, A. J. M.; Bilanenko, E. N. On the Diversity of Fungi from Soda Soils. *Fungal Divers.* **2015**. doi: 10.1007/ s13225-015-0320-2

Guerrand, D. Lipases Industrial Applications: Focus on Food and Agroindustries. *OCL* **2017**, *24*, D403.

Gupta, G. N.; Srivastava, S.; Khare, S. K.; Prakash, V. Extremophiles: An Overview of Microorganism from Extreme Environment. *Int. J. Agric., Environ. Biotechnol.* **2014**, *7* (2), 371–380.

Gupta, G. N.; Srivastava, S.; Khare, S. K.; Prakash, V. Extremophiles: An Overview of Microorganism from Extreme Environment. *Int. J. Agric. Environ. Biotechnol.* **2014**, *7* (2), 371–380.

Gupta, R.; Gigras, P.; Mohapatra, H.; Goswami, V. K.; Chauhan, B. Microbial α-Amylases: A Biotechnological Perspective. *Process Biochem.* **2003**, *38*, 1599–1616.

Haichar, F. E. Z.; Santaella, C.; Heulin, T., et al. Root Exudates Mediated Interactions Belowground. *Soil Biol. Biochem.* **2014**, *77*, 69–80.

Haritash, A. K.; Kaushik, C. P. Biodegradation Aspects of Polycyclic Aromatic Hydrocarbons (PAHs): A Review. *J. Hazard. Mater.* **2009**, *169*, 1–15.

Harms, H.; Schlosser, D.; Wick, L. Y. Untapped Potential: Exploiting Fungi in Bioreme-diation of Hazardous Chemicals. *Nat. Rev. Microbiol.* **2011,** *9,* 177–192. [CrossRef] [PubMed].

Hassan, N.; Rafiq, M.; Hayat, M.; Shah, A. A.; Hasan, F. (2016). Psychrophilic and Psychrotrophic Fungi: A Comprehensive Review. *Rev. Environ. Sci. Bio/Technol.* **2016,** *15* (2), 147–172.

Hassan, N.*;* Rafiq, M.*;* Hayat, M.*;* Shah, A. A.*;* Hasan, F. Psychrophilic and Psychrotrophic Fungi*:* A Comprehensive Review. *Rev. Environ. Sci. Bio/Technol.* **2016,** *15 (2),* 147–172.

Haubensak, K. A.; Hart, S. C.; Stark, J. M. Influences of Chloroform Exposure Time and Soil Water Content on C and N Release in Forest Soils. *Soil Biol. Biochem.* **2002,** *34,* 1549–1562.

Hmidet, N.; Ali, N. E. H.; Haddar, A.; Kanoun, S.; Alya, S. K.; Nasri, M. Alkaline Proteases and Thermostable α-Amylase Co-produced by *Bacillus licheniformis* NH1: Characterization and Potential Application as Detergent Additive. *Biochem. Eng. J.* **2009,** *47,* 71–79. doi: 10.1016/j.bej.2009.07.005.

Huang, Q.; Rodgers, J. M.; Hemley, R. J.; Ichiye, T. Extreme Biophysics: Enzymes under Pressure. *J. Comput. Chem.* **2017,** 38, 1174–1182. [CrossRef] [PubMed].

Hullar, M. A.; Kaplan, L. A.; Stahl, D. A. Recurring Seasonal Dynamics of Microbial Communities in Stream Habitats. *Appl. Environ. Microbiol.* **2006,** *72,* 713–722. Abstract/ FREE Full TextGoogle Scholar.

Huston, A. L. Biotechnological Aspects of Cold-Adapted Enzymes. In *Psychrophiles: From Biodiversity to Biotechnology*; Margesin, R.; Schinner, F.; Marx, J. C.; Gerday, C., Eds.; Springer: Berlin; Heidelberg, 2008; pp 347–363.

Ichiye, T. Enzymes from Piezophiles. In *Seminars in Cell & Developmental Biology*; Elsevier: Amsterdam, The Netherlands, 2018; pp 138–146.

Jackson, C. R.; Millar, J. J.; Payne, J. T.; Ochs, C. A. Free-Living and Particle-Associated Bacterioplankton in Large Rivers of the Mississippi River Basin Demonstrate Biogeo-graphic Patterns. *Appl. Environ. Microbiol.* **2014,** *80,* 7186. doi: 10.1128/AEM.01844-14.

Jamroze, A.; Perugino, G.; Valenti, A., et al. The Reverse Gyrase from Pyrobaculum Calidifontis, a Novel Extremely Thermophilic DNA Topoisomerase Endowed with DNA Unwinding and Annealing Activities. *J. Biol. Chem.* **2014,** *289* (6), 3231–3243.

Jebbar, M.; Franzetti, B.; Girard, E., et al. Microbial Diversity and Adaptation to High Hydrostatic Pressure in Deep-Sea Hydrothermal Vents Prokaryotes. *Extremophiles* **2015,** *19* (4), 721–740.

Ji, X.; Chen, G.; Zhang, Q.; Lin, L.; Wei, Y. Purification and Characterization of an Extracellular Cold-Adapted Alkaline Lipase Produced by Psychrotrophic Bacterium *Yersinia enterocolitica* Strain KM1. *J. Basic Microbiol.* **2015,** *55,* 718–728. doi: 10.1002/ jobm.201400730 [PubMed] [CrossRef] [Google Scholar].

Jiewei, T.; Zuchao, L.; Peng, Q.; Lei, W.; Yongqiang, T. Purification and Characterization of a Cold-Adapted Lipase from Oceanobacillus Strain PT-11. *PLoS One* **2014,** *9,* e101343. doi: 10.1371/journal.pone.0101343 [PMC free article] [PubMed] [CrossRef] [Google Scholar].

Jones, D. L.; Baxter, B. K. DNA Repair and Photoprotection: Mechanisms of Overcoming Environmental Ultraviolet Radiation Exposure in Halophilic Archaea. *Front Microbiol.* **2017,** *8,* 1882. PubMed Abstract | Publisher Full Text | Free Full Text | F1000 Recommendation.

Kaiser, C.; Koranda, M.; Kitzler, B., et al. Belowground Carbon Allocation by Trees Drives Seasonal Patterns of Extracellular Enzyme Activities by Altering Microbial Community Composition in a Beech Forest Soil. *New Phytol.* **2010**, *187*, 843–858.

Kanekar, P. P.; Kanekar, S. P.; Kelkar, A. S.; Dhakephalkar, P. K. Halophiles-Taxonomy, Diversity, Physiology, and Applications. In *Microorganisms in Environmental Management: Microbes and Environment*; Satyanarayana, T.; Johri, B. N.; Prakash, A., Eds.; Springer: Dordrecht, 2012; pp 1–34.

Kaneshiro, S. M.; Clark, D. S. Pressure Effects on the Composition and Thermal Behavior of Lipids from the Deep-Sea Thermophile Methanococcus Jannaschii. *J. Bacteriol.* **1995**, *177* (13), 3668–3672.

Karan, R.; Capes, M. D.; DasSarma, S. Function and Biotechnology of Extremophilic Enzymes in Low Water Activity. *Aquat. Biosyst.* **2012**, *8*, 4. [CrossRef].

Kashefi, K.; Lovley, D. R. Extending the Upper Temperature Limit for Life. *Science* **2003**, *301*, 934. doi: 10.1126/science.1086823.

Kawahara, H. The Structures and Functions of Ice Crystal-Controlling Proteins from Bacteria. *J. Biosci. Bioeng.* **2002**, *94* (6), 492–496. PubMed Abstract | Publisher Full Text.

Kazemi, S.; Hatam, I.; Lanoil, B., Bacterial Community Succession in a High-Altitude Subarctic Glacier Foreland Is a Three-Stage Process. *Mol. Ecol.* **2016**, *25*, 5557–5567.

Keith-Roach, M. J.; Livens, F. R. Microbial Interactions with Radionuclides—Summary and Future Perspectives. In *Radioactivity in the Environment* (Vol. 2); Elsevier, 2002; pp 383–390.

Kelley, D. S.; Karson, J. A.; Blackman, D. K.; Frueh-Green, G. L.; Butterfield, D. A.; Lilley, M. D.; … Rivizzigno, P. An Off-Axis Hydrothermal Vent Field Near the Mid-Atlantic Ridge at 30 N. *Nature* **2001**, *412* (6843), 145–149.

Kitamoto, N.; Yamagata, H.; Kato, T.; Tsukagoshi, N.; Udaka, S. Cloning and Sequencing of the Gene Encoding Thermophilic Beta-Amylase of *Clostridium* Thermosulfurogenes. *J. Bacteriol.* **1988**, *170*, 5848–5854.

Konings, W. N.; Albers, S. V.; Konings, S., et al. The Cell Membrane Plays a Crucial Role in Survival of Bacteria and Archaea in Extreme Environments. *Antonie Van Leeuwenhoek* **2002**, *81* (1–4), 61–72.

Krishnaraj, R. N.; Sani, R. K. Introduction to Extremozymes. Extremophilic Enzymatic Processing of Lignocellulosic Feedstocks to Bioenergy. Springer, Cham, 2017. 1–4.

Krulwich, T. A.; Hicks, D. B.; Ito, M. Cation/Proton Antiporter Complements of Bacteria: Why So Large and Diverse? Mol Microbiol **2009**, *74* (2), 257–260.

Krulwich, T. A.; Sachs, G.; Padan, E. Molecular Aspects of Bacterial pH Sensing and Homeostasis. *Nat. Rev. Microbiol.* **2011**, *9* (5): 330–343. PubMed Abstract | Publisher Full Text | Free Full Text.

Kuddus, M.; Roohi, S.; Ahmad, I. Z. Cold-Active Extracellular α-Amylase Production from Novel Bacteria *Microbacterium foliorum* GA2 and *Bacillus cereus* GA6 Isolated from Gangotri Glacier, Western Himalaya. *J. Genet. Eng. Biotechnol.* **2012**, *10*, 151–159. doi: 10.1016/j.jgeb.2012.03.002.

Kuddus, M. Cold-Active Enzymes in Food Biotechnology: An Updated Mini Review. *J. Appl. Biol. Biotechnol.* **2018**, *6*, 58–63.

Kuhad, R. C.; Gupta, R.; Singh, A. Microbial Cellulases and Their Industrial Applications. *Enzyme Res.* **2011**, *2011*, 280696. doi: 10.4061/2011/280696 [PMC free article] [PubMed] [CrossRef] [Google Scholar].

López-Mondéjar, R.; Vŏřísková, J.; Větrovský, T., et al. The Bacterial Community Inhabiting Temperate Deciduous Forests is Vertically Stratified and Undergoes Seasonal Dynamics. *Soil Biol. Biochem.* **2015,** *87,* 43–50.

Lauber, C. L.; Hamady, M.; Knight, R., et al. Pyrosequencing-Based Assessment of Soil pH as a Predictor of Soil Bacterial Community Structure at the Continental Scale. *Appl. Environ. Microbiol.* **2009,**75, 5111–5120.

Lauenroth, W. K.; Bradford, J. B. Ecohydrology of Dry Regions of the United States: Water Balance Consequences of Small Precipitation Events. *Ecohydrology* **2012,** *5,* 46–53.

Lauro, F. M.; Chastain, R. A.; Blankenship, L. E.; Yayanos, A. A.; Bartlett, D. H. The Unique 16S rRNA Genes of Piezophiles Reflect Both Phylogeny and Adaptation. *Appl. Environ. Microbiol.* **2007,** *73,* 838–845.

Lazzaro, A.; Hilfiker, D.; Zeyer, J. Structures of Microbial Communities in Alpine Soils: Seasonal and Elevational Effects. *Front. Microbiol.* **2015,** *6,* 1330.

Lee, M. S.; Kim, G. A.; Seo, M. S.; Lee, J. H.; Kwon, S. T. Characterization of Heat-Labile Uracil-DNA Glycosylase from Psychrobacter sp. HJ147 and Its Application to the Polymerase Chain Reaction. *Biotechnol. Appl. Biochem.* **2009,** *52*(Pt 2), 167–175. doi: 10.1042/BA20080028 [PubMed] [CrossRef] [Google Scholar].

Li, X. L.; Zhang, W. H.; Wang, Y. D.; Dai, Y. J.; Zhang, H. T.; Wang, Y., et al. A High-Detergent-Performance, Cold-Adapted Lipase from *Pseudomonas stutzeri* PS59 Suitable for Detergent Formulation. *J. Mol. Catal. B. Enzym.* **2014,** *102,* 16–24. doi: 10.1016/j.molcatb.2014.01.006.

Li, X.; Li, X. M.; Li, X. D.; Xu, G. M.; Liu, Y.; Wang, B. G. 20-Nor-Isopimarane Cycloethers from the Deep-Sea Sediment-Derived Fungus: Aspergillus Wentii SD-310. *RSC Adv.* **2016,** *6,* 75981–75987.

Lipson, D. A.; Schmidt, S. K. Seasonal Changes in an Alpine Soil Bacterial Community in the Colorado Rocky Mountains. *Appl. Environ. Microbiol.* **2004,** *70,* 2867–2879.

Llado, S.; Lopez-Mondejar, R.; Baldrian, P. Forest Soil Bacteria: Diversity, Involvement in Ecosystem Processes, and Response to Global Change. *Microbiol. Mol. Biol. Rev.* **2017,** *81,* e00063–16.

López-Mondéjar, R.; Zühlke, D.; Becher, D., et al. Cellulose and Hemicelluloses Decomposition by Forest Soil Bacteria Proceeds by the Action of Structurally Variable Enzymatic Systems. *Sci. Rep.* **2016,** *6,* 25279.

Lockhart, J. S.; DeVeaux, L. C. The Essential Role of the Deinococcus Radiodurans ssb Gene in Cell Survival and Radiation Tolerance. *PLoS One* **2013,** *8* (8), e71651. PubMed Abstract | Publisher Full Text | Free Full Text.

Luo, X.; Wang, M. K.; Hu, G.; Weng, B. Seasonal Change in Microbial Diversity and Its Relationship with Soil Chemical Properties in an Orchard. *PLoS One* **2019,** *14* (12).

Lv, C.; Hong, Y.; Miao, L.; Li, C.; Xu, G.; Wei, S.; Wang, B.; Huang, C.; Jiao, B. Wentilactone A as a Novel Potential Antitumor Agent Induces Apoptosis and G2/M Arrest of Human Lung Carcinoma Cells, and Is Mediated by HRas-GTP Accumulation to Excessively Activate the Ras/Raf/ERK/p53-p21 Pathway. *Cell Death Dis.* **2013,** *4,* e952–e963.

Macalady, J.; Banfield, J. F. Molecular Geomicrobiology: Genes and Geochemical Cycling. *Earth Planet Sci. Lett.* **2003,** *209,* 1–17.

Maheshwari, R.; Bharadwaj, G.; Bhat, M. K. Thermophilic Fungi: Their Physiology and Enzymes. *Microbiol. Mol. Biol. Rev.* **2000,** *64,* 461–488.

Mamo, G. Challenges and Adaptations of Life in Alkaline Habitats. *Adv. Biochem. Eng. Biotechnol.* **2019**. PubMed Abstract | Publisher Full Text | F1000 Recommendation.

Marco-Urrea, E.; García-Romera, I.; Aranda, E. Potential of Non-Ligninolytic Fungi in Bioremediation of Chlorinated and Polycyclic Aromatic Hydrocarbons. *New Biotechnol.* **2015**, *32*, 620–628. [CrossRef] [PubMed].

Matallana-Surget, S.; Meador, J. A.; Joux, F.; Douki, T. Effect of the GC Content of DNA on the Distribution of UVB-Induced Bipyrimidine Photoproducts. *Photochem. Photobiol. Sci.* **2008**, *7*, 794–801. doi: 10.1039/b719929e [PubMed] [CrossRef] [Google Scholar].

Matallana-Surget, S.; Wattiez, R. Impact of Solar Radiation on Gene Expression in Bacteria. *Proteomes* **2013**, *1*, 70–86. doi: 10.3390/proteomes1020070 [PMC free article] [PubMed] [CrossRef] [Google Scholar].

Matallana-Surget, S.; Joux, F.; Raftery, M. J.; Cavicchioli, R. The Response of the Marine Bacterium Sphingopyxis Alaskensis to Solar Radiation Assessed by Quantitative Proteomics. *Environ. Microbiol.* **2009**, *11*, 2660–2675. doi: 10.1111/j.1462-2920.2009.01992.x [PubMed] [CrossRef].

Matallana-Surget, S.; Derock, J.; Leroy, B.; Badri, H.; Deschoenmaeker, F.; Wattiez, R. Proteome-Wide Analysis and Diel Proteomic Profiling of the Cyanobacterium Arthrospira Platensis PCC 8005. *PLoS One* **2014**, *9*, e99076. doi: 10.1371/journal.pone.0099076 [PMC free article] [PubMed] [CrossRef] [Google Scholar].

Math, R. K.; Jin, H. M.; Kim, J. M., et al. Comparative Genomics Reveals Adaptation by Alteromonas sp. SN2 to Marine Tidal-Flat Conditions: Cold Tolerance and Aromatic Hydrocarbon Metabolism. *PLoS One* **2012**, *7*.

McDermott, J. M.; Sylva, S. P.; Ono, S.; German, C. R.; Seewald, J. S. Geochemistry of Fluids from Earth's Deepest Ridge-Crest Hot-Springs: Piccard Hydrothermal Field, Mid-Cayman Rise. *Geochim. Cosmochim. Acta* **2018**, *228*, 95–118. doi: 10.1016/j.gca.2018.01.021.

McMahon, S. K.; Wallenstein, M. D.; Schimel, J. P. Across-Seasonal Comparison of Active and Total Bacterial Community Composition in Arctic Tundra Soil Using Bromodeoxyuridine Labeling. *Soil Biol. Biochem.* **2011**, *43*, 287–295. doi: 10.1016/j.soilbio.2010.10.013.

Merlino, A.; Krauss, I. R.; Castellano, I.; De Vendittis, E.; Rossi, B.; Conte, M.; Vergara, A.; Sica, F. Structure and Flexibility in Cold-Adapted Iron Superoxide Dismutases: The Case of the Enzyme Isolated from Pseudoalteromonas Haloplanktis. *J. Struct. Biol.* **2010**, *172*, 343–352. [CrossRef].

Mesbah, N. M.; Cook, G. M.; Wiegel, J. The Halophilic Alkalithermophile Natranaerobius Thermophiles Adapts to Multiple Environmental Extremes Using a Large Repertoire of Na+ (K+)/H+ Antiporters. *Mol. Microbiol.* **2009**, *74*, 270–281.

Mesbah, N. M.; Wiegel, J. The Na+-Translocating F1FO-ATPase from the Halophilic, Alkalithermophile Natranaerobius Thermophiles. *Biochim. Biophys Acta (BBA)— Bioenerget.* **2011**, *1807*, 1133–1142.

Mesquita, N.; Portugal, A.; Pinar, G.; Loureiro, J.; Coutinho, A. P.; Trovao, J., et al. Flow Cytometry as a Tool to Assess the Effects of Gamma Radiation on the Viability, Growth and Metabolic Activity of Fungal Spores. *Int. Biodeter. Biodegr.* **2013**, *84*, 250–257. doi: 10.1016/j.ibiod.2012.05.008.

Michels, M.; Bakker, E. P. Generation of a Large, Protonophore-Sensitive Proton Motive Force and pH Difference in the Acidophilic Bacteria Thermoplasma Acidophilum and Bacillus Acidocaldarius. *J. Bacteriol.* **1985**, *161* (1), 231–237. PubMed Abstract | Free Full Text.

Michels, M.; Bakker, E. P. Generation of a Large, Protonophore-Sensitive Proton Motive Force and pH Difference in the Acidophilic Bacteria Thermoplasma Acidophilum and Bacillus Acidocaldarius. *J. Bacteriol.* **1985,** *161* (1), 231–237.

Michoud, G.; Jebbar, M. High Hydrostatic Pressure Adaptive Strategies in an Obligate Piezophile Pyrococcus Yayanosii. *Sci. Rep.* **2016,** *6* (1), 1–10.

Minsky, A.; Shimoni, E.; Englander, J. Ring-Like Nucleoids and DNA Repair through Error-Free Nonhomologous End Joining in Deinococcus Radiodurans. *J. Bacteriol.* **2006,** *188* (17), 6047–6051; discussion 6052. PubMed Abstract | Publisher Full Text | Free Full Text.

Monson, R. K.; Lipson, D. L.; Burns, S. P.; Turnipseed, A. A.; Delany, A. C.; Williams, M. W., et al. Winter Forest Soil Respiration Controlled by Climate and Microbial Community Composition. *Nature* **2006,** *439,* 711–714. doi: 10.1038/nature04555.

Moore-Kucera, J. and Dick, R. P. PLFA Profiling of Microbial Community Structure and Seasonal Shifts in Soils of a Douglas-Fir Chronosequence. *Microb. Ecol.* **2008,** *55,* 500–511. (Keith- Roach et al. 2002).

Morgenstern, I.; Powlowski, J.; Ishmael, N.; Darmond, C.; Marqueteau, S.; Moisan, M. C.; Quenneville, G.; Tsang, A. A Molecular Phylogeny of Thermophilic Fungi. *Fungal Biol.* **2012,** *116,* 489–502.

Mulkidjanian, A. Y.; Bychkov, A. Y.; Dibrova, D. V.; Galperin, M. Y.; Koonin, E. V. Open Questions on the Origin of Life at Anoxic Geothermal Fields. *Origins Life Evol. Biospheres* **2012,** *42* (5), 507–516.

Nath, A.; Subbiah, K. Insights into the Molecular Basis of Piezophilic Adaptation: Extraction of Piezophilic.

Natvig, D. O.; Taylor, J. W.; Tsang, A.; Hutchinson, M. I.; Powell, A. J. Mycothermus Thermophilus gen. et Comb. nov., a New Home for the Itinerant Thermophile Scytalidium Thermophilum (Torula Thermophila). *Mycologia* **2015,** *107,* 319–327.

Nunoura, T.; Takaki, Y.; Hirai, M.; Shimamura, S.; Makabe, A.; Koide, O., et al. Hadal Biosphere: Insight Into the Microbial Ecosystem in the Deepest Ocean on Earth. *Proc. Natl. Acad. Sci.* **2015,** *112,* E1230–E1236. doi: 10.1073/pnas. 1421816112.

Oberson, J.; Rawyler, A.; Brändle, R.; Canevascini, G. Analysis of the Heat-Shock Response Displayed by Two Chaetomium Species Originating from Different Thermal Environments. *Fungal Genet. Biol.* **1999,** *26* (3), 178–189.

Oger, P. M.; Jebbar, M. The Many Ways of Coping with Pressure. *Res. Microbiol.* **2010,** *161,* 799–809.

Oren, A. Microbial Life at High Salt Concentrations: Phylogenetic and Metabolic Diversity. *Saline Syst.* **2008,** *4,* 2. PubMed Abstract | Publisher Full Text | Free Full Text.

Özdemir, S. C.; Uzel, A. Bioprospecting of Hot Springs and Compost in West Anatolia Regarding Phytase Producing Thermophilic Fungi. *Sydowia* **2020,** *72,* 1–11.

Özdemir, S. C.; Uzel, A. Bioprospecting of Hot Springs and Compost in West Anatolia Regarding Phytase Producing Thermophilic Fungi. *Sydowia* **2020,** *72,* 1–11.

Passarini, M. R. Z.; Rodrigues, M. V. N.; da Silva, M.; Sette, L. D. Marine-Derived Filamentous Fungi and Their Potential Application for Polycyclic Aromatic Hydrocarbon Bioremediation. *Mar. Pollut. Bull.* **2011,** *62,* 364–370.

Paul, A.; Dziallas, C.; Zwirnmann, E.; Gjessing, E. T.; Grossart, H.-P. UV Irradiation of Natural Organic Matter (NOM): Impact on Organic Carbon and Bacteria. *Aquat. Sci.* **2012,** *74,* 443–454. doi: 10.1007/s00027-011-0239-y.

Pavlopoulou, A.; Savva, G. D.; Louka, M.; Bagos, P. G.; Vorgias, C. E.; Michalopoulos, I., et al. Unraveling the Mechanisms of Extreme Radioresistance in Prokaryotes: Lessons from Nature. *Mutat. Res. Rev. Mutat. Res.* **2016**, *767*, 92–107. doi: 10.1016/j.mrrev.2015.10.001.

Pérez, V.; Hengst, M.; Kurte, L.; Dorador, C.; Jeffrey, W. H.; Wattiez, R.; ... Matallana-Surget, S. Bacterial Survival under Extreme UV Radiation: A Comparative Proteomics Study of Rhodobacter sp., Isolated from High Altitude Wetlands in Chile. *Front. Microbiol.* **2017**, *8*, 1173.

Pfeiffer, B.; Fender, A.; Lasota, S., et al. Leaf Litter Is the Main Driver for Changes in Bacterial Community Structures in the Rhizosphere of Ash and Beech. *Appl. Soil Ecol.* **2013**, *72*, 150–160.

Phoenix, G. K.; Lee, J. A. Predicting Impacts of Arctic Climate Change: Past Lessons and Future Challenges. *Ecol. Res.* **2004**, *19*, 65–74.

Picard, A.; Daniel, I. Pressure as an Environmental Parameter for Microbial Life—A Review. *Biophys. Chem.* **2013**, *183*, 30–41. doi: 10.1016/j.bpc.2013. 06.019.

Pikuta, E. V.; Hoover, R. B.; Tang, J. Microbial Extremophiles at the Limits of Life. *Crit. Rev. Microbiol.* **2008**, *33* (3), 183–209. PubMed Abstract | Publisher Full Text 33, 183–209.

Pioli, S.; Antonucci, S.; Giovannelli, A.; Traversi, M. L.; Borruso, L.; Bani, A.; ... Tognetti, R. Community Fingerprinting Reveals Increasing Wood-Inhabiting Fungal Diversity in Unmanaged Mediterranean Forests. *Forest Ecol. Manage.* **2018**, *408*, 202–210.

Post, E.; Forchhammer, M. C.; Bret-Harte, M. S., et al. Ecological Dynamics across the Arctic Associated with Recent Climate Change. *Science* **2009**, *325*, 1355–1358.

Prasad, R. *Mycoremediation and Environmental Sustainability*; Springer: Berlin, Germany, 2018; p 2.

Prasad, R., Ed. *Advances and Applications through Fungal Nanobiotechnology*; Springer: Dordrecht, 2016.

Preiss, L.; Hicks, D. B.; Suzuki, S.; Meier, T.; Krulwich, T. A. Alkaliphilic Bacteria with Impact on Industrial Applications, Concepts of Early Life Forms, and Bioenergetics of ATP Synthesis. *Front. Bioeng. Biotechnol.* **2015**, *3*, 75.

Pulicherla, K. K.; Kumar, P. S.; Manideep, K.; Rekha, V. P.; Ghosh, M.; Sambasiva Rao, K. R. Statistical Approach for the Enhanced Production of Cold-Active Beta-Galactosidase from *Thalassospira frigidphilosprofundus*: A Novel Marine Psychrophile from Deep Waters of Bay of Bengal. *Prep. Biochem. Biotechnol.* **2013**, *43*, 766–780. doi: 10.1080/10826068.2013.773341.

Qiao, W.; Ling, F.; Yu, L.; Huang, Y.; Wang, T. Enhancing Taxol Production in a Novel Endophytic Fungus, Aspergillus Aculeatinus Tax-6, Isolated from Taxus Chinensis var. Mairei. *Fungal Biol.* **2017**, *121*, 1037–1044.

Qin, Y.; Huang, Z.; Liu, Z. A Novel Cold-Active and Salt-Tolerant Alpha-Amylase from Marine Bacterium *Zunongwangia profunda*: Molecular Cloning, Heterologous Expression and Biochemical Characterization. *Extremophiles* **2014**, *18*, 271–281. doi: 10.1007/s00792-013-0614-9.

Quintana-Cabrera, R.; Fernandez-Fernandez, S.; Bobo-Jimenez, V., et al. γ-Glutamylcysteine Detoxifies Reactive Oxygen Species by Acting as Glutathione Peroxidase-1 Cofactor. *Nat. Commun.* **2012**, *3*, 718. PubMed Abstract | Publisher Full Text | Free Full Text.

Rampelotto, P. H. Extremophiles and Extreme Environments. *Life* **2013**, *3*, 482–485.

Raymond-Bouchard, I.; Tremblay, J.; Altshuler, I., et al. Comparative Transcriptomics of Cold Growth and Adaptive Features of a Eury- and Steno-Psychrophile. *Front. Microbiol.* **2018,** *9,* 1565. PubMed Abstract | Publisher Full Text | Free Full Text | F1000 Recommendation.

Read, D. J.; Perez-Moreno, J. Mycorrhizas and Nutrient Cycling in Ecosystems—A Journey Towards Relevance? *New Phytol.* **2003,** *157,* 475–492.

Reed, C. J.; Lewis, H.; Trejo, E.; Winston, V.; Evilia, C. Protein Adaptations in Archaeal Extremophiles. *Archaea* **2013,** 2013. [CrossRef].

Rees, H. C.; Grant, W. D.; Jones, B. E.; Heaphy, S. Diversity of Kenyan Soda Lake Alkaliphiles Assessed by Molecular Methods. *Extremophiles* **2004,** *8* (1), 63–71.

Reid, I. I.; Ricard, M. Pectinase in Papermaking: Solving Retention Problems in Mechanical Pulps Bleached with Hydrogen Peroxide. *Enzyme Microb. Technol.* **2000,** *26,* 115–123. doi: 10.1016/S0141-0229(99)00131-3 [PubMed] [CrossRef] [Google Scholar].

Roberts, M. F. Organic Compatible Solutes of Halotolerant and Halophilic Microorganisms. *Saline Syst.* **2005,** *1,* 5. PubMed Abstract | Publisher Full Text | Free Full Text.

Roohi, R.; Kuddus, M.; Saima, S. Cold-Active Detergent-Stable Extracellular α-Amylase from *Bacillus cereus* GA6: Biochemical Characteristics and Its Perspectives in Laundry Detergent Formulation. *J. Biochem. Technol.* **2013,** *4,* 636–644.

Rousk, K.; Michelsen, A.; Rousk, J. Microbial Control of Soil Organic Matter Mineralization Responses to Labile Carbon in Subarctic Climate Change Treatments. *Global Change Biol.* **2016,** *22,* 4150–4161.

Rubin, B. E.; Gibbons, S. M.; Kennedy, S.; Hampton-Marcell, J.; Owens, S.; Gilbert, J. A. Investigating the Impact of Storage Conditions on Microbial Community Composition in Soil Samples. *PLoS One* **2013,** *8* (7).

Štursova, M.; Žifčáková, L.; Leigh, M. B., et al. Cellulose Utilization in Forest Litter and Soil: Identification of Bacterial and Fungal Decomposers. *FEMS Microbiol. Ecol.* **2012,** *80,* 735–746.

Samuel, A. D. Dehydrogenase: An Indicator of Biological Activities in a Preluvo Soil. *Res. J. Agr. Sci.* **2010,** *42,* 306–310. (Arunkumar et al. 2013).

Santalahti, M.; Sun, H.; Jumpponen, A., et al. Vertical and Seasonal Dynamics of Fungal Communities in Boreal Scots Pine Forest Soil. *FEMS Microbiol. Ecol.* **2016,** *92,* fiw170.

Santiago, M.; Ramírez-Sarmiento, C. A.; Zamora, R. A., et al. Discovery, Molecular Mechanisms, and Industrial Applications of Cold-Active Enzymes. *Front. Microbiol.* **2016,** *7,* 1408.

Santos, A. L.; Oliveira, V.; Baptista, I.; Henriques, I.; Gomes, N. C.; Almeida, A., et al. Wavelength Dependence of Biological Damage Induced by UV Radiation on Bacteria. *Arch. Microbiol.* **2013,** *195,* 63–74. doi: 10.1007/s00203-012-0847-5 [PubMed] [CrossRef] [Google Scholar].

Sarmiento, F.; Peralta, R.; Blamey, J. M. Cold and Hot Extremozymes: Industrial Relevance and Current Trends. *Front. Bioeng. Biotechnol.* **2015,** *3,* 148.

Sarmiento, F.; Peralta, R.; Blamey, J. M. Cold and Hot Extremozymes: Industrial Relevance and Current Trends. *Front. Bioeng. Biotechnol.* **2015,** *3,* 148.

Satyanarayana, T.; Noorwez, S. M.; Kumar, S.; Rao, J. L.; Ezhilvannan, M.; Kaur, P. Development of an Ideal Starch Saccharification Process Using Amylolytic Enzymes from Thermophiles. *Biochem. Soc. Trans.* **2004,** *32* (Pt 2), 276–278. doi: 10.1042/bst0320276 [PubMed] [CrossRef] [Google Scholar].

Schadt, C. W.; Martin, A. P.; Lipson, D. A.; Schmidt, S. K. Seasonal Dynamics of Previously Unknown Fungal Lineages in Tundra Soils. *Science* **2003**, *301* (5638), 1359–1361.

Schimel, J. P.; Bilbrough, C.; Welker, J. M. Increased Snowdepth Affects Microbial Activity and Nitrogen Mineralization in Two Arctic Tundra Communities. *Soil Biol. Biochem.* **2004**, *36*, 217–227.

Schreck, S. D.; Grunden, A. M. Biotechnological Applications of Halophilic Lipases and Thioesterases. *Appl. Microbiol. Biotechnol.* **2014**, *98*, 1011–1021. [CrossRef].

Schröder, C.; Burkhardt, C.; Antranikian, G. What We Learn from Extremophiles. *Chem Texts* **2020**, *6* (1), 1–6.

Schulze-Makuch, D.; Guinan, E. Another Earth 2.0? Not So Fast. *Astrobiology* **2016**, *16*, 817–821. doi: 10.1089/ast.2016.1584.

Schulze-Makuch, D.; Schulze-Makuch, A.; Houtkooper, J. M. The Physical, Chemical and Physiological Limits of Life. *Life* **2015**, *5*(3), 1472–1486.

Selbmann, L.; Egidi, E.; Isola, D.; Onofri, S.; Zucconi, L.; de Hoog, G. S.; … Lantieri, A. Biodiversity, Evolution and Adaptation of Fungi in Extreme Environments. *Plant Biosyst. Int. J. Dealing Aspects Plant Biol.* **2013**, *147* (1), 237–246.

Shang, Z.; Li, X.; Meng, L.; Li, C.; Gao, S.; Huang, C.; Wang, B. Chemical Profile of the Secondary Metabolites Produced by a Deep-Sea Sediment-Derived Fungus Penicillium Commune SD-118. *Chin. J. Oceanol. Limnol.* **2012**, *30*, 305–314. [CrossRef].

Sharp, C. E.; Brady, A. L.; Sharp, G. H.; Grasby, S. E.; Stott, M. B.; Dunfield, P. F. Humboldt's Spa: Microbial Diversity Is Controlled by Temperature in Geothermal Environments. *ISME J.* **2014**, *8*, 1166–1174. doi: 10.1038/ismej.2013.237. Shock, E. L.; Boyd, E. S. Principles of Geobiochemistry. *Elements* **2015**, *11*.

Shi, H.; Zhang, Y.; Li, X.; Huang, Y.; Wang, L.; Wang, Y., et al. A Novel Highly Thermostable Xylanase Stimulated by Ca^{2+} from *Thermotoga thermarum*: Cloning, Expression and Characterization. *Biotechnol. Biofuels* **2013**, *6*, 26. doi: 10.1186/1754-6834-6-26 [PMC free article] [PubMed] [CrossRef] [Google Scholar].

Shigyo, N.; Umeki, K.; Hirao, T. Plant Functional Diversity and Soil Properties Control Elevational Diversity Gradients of Soil Bacteria. *FEMS Microbiol. Ecol.* **2019**, *95* (4), fiz025.

Shimada, H.; Nemoto, N.; Shida, Y.; Oshima, T.; Yamagishi, A. Complete Polar Lipid Composition of Thermoplasma Acidophilum HO-62 Determined by High-Performance Liquid Chromatography with Evaporative Light-Scattering Detection. *J. Bacteriol.* **2002**, *184*, 556–563.

Siasou, E.; Johnson, D.; Willey, N. J. An Extended Dose–Response Model for Microbial Responses to Ionizing Radiation. *Front. Environ. Sci.* **2017**, *5*, 6.

Siddiqui, K. S.; Cavicchioli, R. Cold-Adapted Enzymes. *Annu. Rev. Biochem.* **2006**, *75*, 403–433. [CrossRef][PubMed].

Siles, J. A.; Cajthaml, T.; Filipova, A.; Minerbi, S.; Margesin, R. Altitudinal, Seasonal and Interannual Shifts in Microbial Communities and Chemical Composition of Soil Organic Matter in Alpine Forest Soils. *Soil Biol. Biochem.* **2017**, *112*, 1–13.

Siliakus, M. F.; van der Oost, J.; Kengen, S. W. M. Adaptations of Archaeal and Bacterial Membranes to Variations in Temperature, pH and Pressure. *Extremophiles.* **2017**, *21* (4), 651–670. PubMed Abstract | Publisher Full Text | Free Full Text |.

Singh, B.; Satyanarayana, T. Fungal Phytases: Characteristics and Amelioration of Nutritional Quality and Growth of Non-Ruminants. *J. Anim. Physiol. Anim. Nutr.* **2015,** *99,* 646–660.

Slade, D.; Radman, M. Oxidative Stress Resistance in Deinococcus Radiodurans. *Microbiol. Mol. Biol. Rev.* **2011,** *75* (1), 133–191. PubMed Abstract | Publisher Full Text | Free Full Text.

Slonczewski, J. L.; Fujisawa, M.; Dopson, M., et al. Cytoplasmic pH Measurement and Homeostasis in Bacteria and Archaea. *Adv. Microb. Physiol.* **2009,** *55,* 1–79, 317. PubMed Abstract|Publisher Full Text.

Soininen, J.; Lennon, J. J.; Hillebrand, H. A Multivariate Analysis of Beta Diversity across Organisms and Environments. *Ecology* **2007,** *88,* 2830–2838.

Sorensen, P. O.; Templer, P. H.; Finzi, A. C. Contrasting Effects of Winter Snowpack and Soil Frost on Growing Season Microbial Biomass and Enzyme Activity in Two Mixed-Hardwood Forests. *Biogeochemistry* **2016a,** *128* (1–2), 141–154.

Sorensen, P. O.; Templer, P. H.; Christenson, L.; Duran, J.; Fahey, T.; Fisk, M. C.; … Finzi, A. C. Reduced Snow Cover Alters Root–Microbe Interactions and Decreases Nitrification Rates in a Northern Hardwood Forest. *Ecology* **2016b,** *97* (12), 3359–3368.

Sorokin, D. Y.; Berben, T.; Melton, E. D.; Overmars, L.; Vavourakis, C. D.; Muyzer, G. Microbial Diversity and Biogeochemical Cycling in Soda Lakes. *Extremophiles* **2014,** *18,* 791–809.

Staley, C.; Gould, T. J.; Wang, P.; Phillips, J.; Cotner, J. B.; Sadowsky, M. J. Species Sorting and Seasonal Dynamics Primarily Shape Bacterial Communities in the Upper Mississippi River. *Sci. Total Environ.* **2015,** *505,* 435–445.

Stetter, K. O. Extremophiles and Their Adaptation to Hot Environments. *FEBS Lett.* **1999,** *452* (1–2): 22–25. PubMed Abstract | Publisher Full Text.

Stierle, A. A.; Stierle, D. B.; Goldstein, E.; Parker, K.; Bugni, T.; Baarson, C.; Gress, J.; Blake, D. A Novel 5-HT Receptor Ligand and Related Cytotoxic Compounds from an Acid Mine Waste Extremophile. *J. Nat. Prod.* **2003,** *66* (8), 1097–1100.

Sukumar, M. S.; Jeyaseelan, A.; Sivasankaran, T.; Mohanraj, P.; Mani, P.; Sudhakar, G., et al. Production and Partial Characterization of Extracellular Glucose Isomerase Using Thermophilic Bacillus sp. Isolated from Agricultural Land. *Biocatal. Agric. Biotechnol.* **2013,** *2,* 45–49. doi: 10.1016/j.bcab.2012.10.003 [CrossRef] [Google Scholar].

Sun, W.; Xia, C.; Xu, M.; Guo, J.; Sun, G. Seasonality Affects the Diversity and Composition of Bacterioplankton Communities in Dongjiang River, a Drinking Water Source of Hong Kong. *Front. Microbiol.*<journal name/italics?> **2017.** https://doi.org/10.3389/fmicb.2017.01644

Suriya, J.; Bharathiraja, S.; Krishnan, M.; Manivasagan, P.; Kim, S. K. Marine Microbial Amylases. In *Advances in Food and Nutrition Research*; Elsevier: Amsterdam, The Netherlands, 2016; Volume 79, pp 161–177.

Takai, K.; Nakamura, K.; Toki, T.; Tsunogai, U.; Miyazaki, M.; Miyazaki, J., et al. Cell Proliferation at 122 C and Isotopically Heavy CH4 Production by a Hyperthermophilic Methanogen Under High-Pressure Cultivation. *Proc. Natl. Acad. Sci. USA* **2008,** *105,* 10949–10954. doi: 10.1073/pnas.0712334105.

Tarn, J.; Peoples, L. M.; Hardy, K.; Cameron, J.; Bartlett, D. H. Identification of Free-Living and Particle-Associated Microbial Communities Present in Hadal Regions of the Mariana Trench. *Front. Microbiol.* **2016,** *7,* 665. doi: 10.3389/fmicb.2016.00665.

Tedersoo, L.; Bahram, M.; P˜olme, S., et al. Global Diversity and Geography of Soil Fungi. *Science* **2014**, *346*, 1256688.

Tiquia-Arashiro, S.; Rodrigues, D. F. *Extremophiles: Applications in Nanotechnology*; Springer International Publishing: Cham, Switzerland, 2016; p 193.

Tomac, A; Yeannes, M. I. Gamma Radiation Effect on Quality Changes in Vacuum-Packed Squid (Illex Argentinus) Mantle Rings during Refrigerated (4–5°C) Storage. *Food Sci. Technol.* **2012**, *47*, 1550–1557. doi: 10.1111/j.1365-2621.2012.03005.x.

Urbanová, M.; Šnajdr, J.; Baldrian, P. Composition of Fungal and Bacterial Communities in Forest Litter and Soil Is Largely Determined by Dominant Trees. *Soil Biol. Biochem.* **2015**, *84*, 53–64.

Vala, A. K.; Sutariya, V. Trivalent Arsenic Tolerance and Accumulation in Two Facultative Marine Fungi. *Jundishapur J. Microbiol.* **2012**, *5*, 542–545.

Van de Voorde, I.; Goiris, K.; Syryn, E.; Van den Bussche, C.; Aerts, G. Evaluation of the Cold-Active *Pseudoalteromonas haloplanktis* β-Galactosidase Enzyme for Lactose Hydrolysis in Whey Permeate as Primary Step of d-Tagatose Production. *Process Biochem.* **2014**, *49*, 2134–2140. doi: 10.1016/j.procbio.2014.09.010.

van de Vossenberg, J. L. C. M.; Driessen, A. J.; Konings, W. N. The Essence of Being Extremophilic: The Role of the Unique Archaeal Membrane Lipids. *Extremophiles* **1998a,** *2,* 163–170.

van de Vossenberg, J. L. C. M.; Driessen, A. J. M.; Zillig, W.; Konings, W. N. (1998a) Bioenergetics and Cytoplasmic Membrane Stability of the Extremely Acidophilic, Thermophilic Archaeon Picrophilus Oshimae. *Extremophiles* **1998b,** *2,* 67–74.

van der Heijden, M. G.; Bardgett, R. D.; van Straalen, N. M. The Unseen Majority: Soil Microbes as Drivers of Plant Diversity and Productivity in Terrestrial Ecosystems. *Ecol. Lett.* **2008**, *11,* 296–310.

Vannier, P.; Michoud, G.; Oger, P., et al. Genome Expression of Thermococcus Barophilus and Thermococcus Kodakarensis in Response to Different Hydrostatic Pressure Conditions. *Res. Microbiol.* **2015**, *166* (9), 717–725.

Ventosa, A.; Haba, R.; Sanchez-Porro, C.; Papke, R. T., Microbial Diversity of Hypersaline Environments: A Metagenomic Approach. *Curr. Opin. Microbiol.* **2015**, *25*, 80–87.

Ventosa, A. Unusual Microorganisms from Unusual Habitats: Hypersaline Environments. In *Prokaryotic Diversity: Mechanisms and Significance*; Logan, N. A.; Lappin-Scott, H. M.; Oyston, P. F. C., Eds.; Cambridge University Press: Cambridge, 2006; pp 223–253.

Verma, S.; Prasanna, R.; Saxena, J.; Sharma, V.; Nain, L. Deciphering the Metabolic Capabilities of a Lipase Producing Pseudomonas Aeruginosa SL-72 Strain. *Folia Microbiol.* (Praha) **2012**, *57*, 525–531. [CrossRef] 137. Guerrand, D. Lipases Industrial Applications: Focus on Food and Agroindustries. *OCL* **2017**, *24*, D403.

Virk, A. P.; Sharma, P.; Capalash, N. Use of Laccase in Pulp and Paper Industry. *Biotechnol. Prog.* **2012**, *28*, 21–32. doi: 10.1002/btpr.727 [PubMed] [CrossRef] [Google Scholar].

Vǒřísková J, Brabcová V, Cajthaml T et al. Seasonal Dynamics of Fungal Communities in a Temperate Oak Forest Soil. *New Phytol.* **2014**, 201, 269–278.

Waldrop, M. P.; Firestone, M. K. Seasonal Dynamics of Microbial Community Composition and Function in Oak Canopy and Open Grassland Soils. *Microb. Ecol.* **2006**, *52*, 470–479.

Wallenstein, M. D.; Hall, E. K. A Trait-Based Framework for Predicting When and Where Microbial Adaptation to Climate Change Will Affect Ecosystem Functioning. *Biogeochemistry* **2012**, *109*, 35–47. doi: 10.1007/s10533-011-9641-8.

Wang, M.; Jiang, X.; Wu, W.; Hao, Y.; Su, Y.; Cai, L.; … Liu, X. Psychrophilic Fungi from the World's Roof. *Persoonia: Mol. Phylogeny Evol. Fungi* **2015**, *34*, 100.

Wang, L.; Chi, Z.; Wang, X.; Liu, Z.; Li, J. Diversity of Lipase-Producing Yeasts from Marine Environments and Oil Hydrolysis by Their Crude Enzymes. *Ann. Microbiol.* **2009**, *57*, 495–501.

Whiteley, C. G.; Lee, D. J. Enzyme Technology and Biological Remediation. *Enzyme Microb. Technol.* **2006**, *38*, 291–316.

Whitford, W. G. The Importance of the Biodiversity of Soil Biota in Arid Ecosystems. *Biodivers. Conserv.* **1996**, *5*, 185–195.

Xavier, M. L.; Dauber, C.; Mussio, P.; Delgado, E.; Maquieira, A.; Soria, A., et al. Use of Mild Irradiation Doses to Control Pathogenic Bacteria on Meat Trimmings for Production of Patties Aiming at Provoking Minimal Changes in Quality Attributes. *Meat Sci.* **2014**, *98*, 383–391. doi: 10.1016/j.meatsci.2014.06.037.

Xu, F.; Shin, W.; Brown, S. H.; Wahleithner, J. A.; Sundaram, U. M.; Solomon, E. I. A Study of a Series of Recombinant Fungal Laccases and Bilirubin Oxidase That Exhibit Significant Differences in Redox Potential, Substrate Specificity, and Stability. *Biochim. Biophys. Acta.* **1996** Feb 8, *1292 (2), 303–311.*

Yadav, A. N.; Sachan, S. G.; Verma, P.; Saxena, A. K. Bioprospecting of Plant Growth Promoting Psychrotrophic Bacilli from Cold Desert of North Western Indian Himalayas. *Indian J. Exp. Biol.* **2016**, *52*, 142–150.

Yamagishi, A.; Kawaguchi, Y.; Hashimoto, H.; Yano, H.; Imai, E.; Kodaira, S., et al. Environmental Data and Survival Data of Deinococcus Aetherius from the Exposure Facility of the Japan Experimental Module of the International Space Station Obtained by the Tanpopo Mission. *Astrobiology* **2018**, *18*, 1369–1374. doi: 10.1089/ast.2017.1751.

Yin, J.; Chen, J. C.; Wu, Q., et al. Halophiles, Coming Stars for Industrial Biotechnology. *Biotechnol. Adv.* **2015**, *33* (7), 1433–1442. PubMed Abstract | Publisher Full Text.

Žifčáková, L.; Větrovský, T.; Howe, A., et al. Microbial Activity in Forest Soil Reflects the Changes in Ecosystem Properties between Summer and Winter. *Environ. Microbiol.* **2016**, *18*, 288–301.

Žifčáková, L.; Větrovský, T.; Lombard, V., et al. Feed in Summer, Rest in Winter: Microbial Carbon Utilization in Forest Topsoil. *Microbiome* **2017**, *5*, 122.

Zhang, X.; Liu, W.; Bai, Y.; Zhang, G.; Han, X. Nitrogen Deposition Mediates the Effects and Importance of Chance in Changing Biodiversity. *Mol. Ecol.* **2011**, *20*, 429–438.

Zhang, Z.; Miao, L.; Lv, C.; Sun, H.; Wei, S.; Wang, B.; Huang, C.; Jiao, B. Wentilactone B Induces G2/M Phase Arrest and Apoptosis via the Ras/Raf/MAPK Signaling Pathway in Human Hepatoma SMMC-7721 Cells. *Cell Death Dis.* **2013**, *4*, e657–e669.

Zhang, Z.; Miao, L.; Lv, C.; Sun, H.; Wei, S.; Wang, B.; Huang, C.; Jiao, B. Wentilactone B Induces G2/M Phase Arrest and Apoptosis via the Ras/Raf/MAPK Signaling Pathway in Human Hepatoma SMMC-7721 Cells. *Cell Death Dis.* **2013**, *4*, e657–e669.

Zhao, Y.; Chen, H.; Shang, Z.; Jiao, B.; Yuan, B.; Sun, W.; Wang, B.; Miao, M.; Huang, C. SD118-Xanthocillin X (1), a Novel Marine Agent Extracted from Penicillium Commune, Induces Autophagy through the Inhibition of the MEK/ERK Pathway. *Mar. Drugs* **2012**, *10*, 1345–1359.

Zheng, Z.; Li, H.; Li, L.; Shao, W. Biobleaching of Wheat Straw Pulp with Recombinant Laccase from the Hyperthermophilic *Thermus thermophilus*. *Biotechnol. Lett.* **2012**, *34*, 541–547. doi: 10.1007/s10529-011-0796-0 [PubMed] [CrossRef] [Google Scholar].

CHAPTER 5

Antimicrobial Resistance, Climate Change, and Public Health

KHURSHID AHMAD TARIQ*

Department of Zoology, Islamia College of Science & Commerce (UGC Autonomous), Srinagar, Jammu & Kashmir, India

*E-mail: drkatariq@gmail.com.

ABSTRACT

Nine decades earlier, in 1928, when Alexander Fleming invented the first antibiotic—penicillin, no one could have imagined the time when antibiotic usage will be one of the great threats itself to humans and other lifeforms on this planet. Undoubtedly, the invention of antimicrobials chiefly the antibiotics is the greatest achievement of modern medical science because they revolutionized human life on this planet by increasing the life expectancy rates, decreasing the incidence of diversity of infectious diseases, simplifying the various operative procedures, and even curing us from many skin cuts that may otherwise prove fatal to us. But now we have such a dependence on antibiotics that it is equally tough to convince ourselves to imagine our life without antibiotics. However, at the same time, we should be quite aware with the harmful and challenging aspects of antibiotics in our life and in the environment because they are now one of the recognized and greatest emerging threats to public and animal health. In fact, on one hand, we are fighting a silent and diplomatic war with pathogens and on the other hand, the side effects of antimicrobials are threatening our environment too, meaning we are caught in a double-sword effect of antibiotics, that is, the emergence of superbugs (special resistant microbes with decreased susceptibility to antibiotic action) and millions of deaths secondarily due to antimicrobial resistance. Considering the evidence, it is clear that the antimicrobial resistance is not a localized phenomenon, neither it is a regional problem, nor it is restricted largely to any area or research institute, or agricultural farm, but it is a problem of major international concern and significance. It will not be wrong to label it as a pandemic proportionally as valid as any pandemic due to infectious disease/s. Therefore, rather than searching and researching for next-generation antibiotics to play tricks with drug-resistant microbes, time is enough ripe to look for minimizing this antimicrobial resistance and to search for reliable alternatives to antimicrobials taking cues from the principles of nature itself. The other greatest trouble starts when we know that the climate of earth is evolving with consequences for agriculture, water resources, public health, ecosystem health, etc. The two are a real threat to man and the whole biosphere. In fact, recent reports from Europe indicate that a novel association exists between climate change and antibiotic resistance with a trend toward increased antimicrobial resistance under warmer weather conditions. Therefore, it is quite interesting to prove if the same situation exists across the world differing in geography, socio-economy, health care, and climate change scenario. This chapter aims to look very deep into all these aspects of antibiotic use ranging from antibiotic resistance, emerging diseases, climate change, and alternative to conventional antibiotics to safeguard public and animal health.

5.1 INTRODUCTION

An in-depth discussion of all antimicrobials and antimicrobial resistance is not possible in a single book. Rather this chapter highlights a general description of antimicrobial resistance and its environmental and public health implications in the presence of emerging and re-emerging infectious diseases, the changing weather and climatic patterns, and the resulting effects on food security and economy. Overall, this chapter provides a general information on antimicrobial resistance and public health in relation to changing climatic conditions worldwide. It does not refer to any particular class of antimicrobial drugs, neither it provides the details on the mechanism of drug action nor describes any methodology to assess the antimicrobial resistance. It does not specifically stress on any special climate change event or public health implications. The information and knowledge presented in this chapter is intended as a source of awareness for students and researchers in the field of biology and climate change. Therefore, this perspective is not encompassing and the reader is directed to such manuscripts, articles, chapters, reviews, and books on individual publications on an antimicrobial class, specific mechanism of action, development of specific antimicrobial resistance against a particular drug or antimicrobial, specific diseases and its target antibiotic, climate change scenario vis-à-vis a particular microbial disease, emerging and re-emerging infectious diseases to obtain a detailed information, and knowledge enrichment in the field of antimicrobials, antimicrobial resistance, and climate change.

5.2   DEFINITION AND ROOT CAUSE OF THE PROBLEM

Generally speaking, microbes or microorganism is a collective term used to indicate viruses, bacteria, mycoplasma, fungi, protozoa, and other microparasitic agents found everywhere in humans, animals, within other microbes, and widely distributed in nature under varied environmental conditions from the Arctic to Antarctic. The microbes are either harmful to humans, animals, and agriculture or they are directly useful as important components of our microbiota (useful microbial flora of our body) and as potential agents of industrial importance. Microbes have been responsible for the onset of various infectious diseases in humans and animals leading to some devastating epidemics and pandemics taking a heavy toll on life. The weapons to fight these microbes are the antimicrobial compounds generally called antibiotics, antifungals, and antiparasitic agents. Infectious diseases being a major cause of morbidity and mortality, in humans and animals, have worsened due to the ongoing antimicrobial resistance crisis and in order to tackle this crisis, more studies analyzing the causes, routes, and reservoirs are needed throughout the world for a universal sustenance (Teresa et al., 2019).

The discovery of antibiotics with their amazing life-curing properties is a wonderful phase in the history of human life. Nathans and Cars (2014) state that the two major ways that modern medicine saves lives are through antibiotic treatment of severe infections and the performance of medico-surgical procedures under the shield of antibiotics. They facilitate the surgeries and other operative procedures to safeguard human health. They decrease the incidence of childbirth deaths (death of both mothers and babies), protect us from otherwise incurable microbial infections, and thereby decrease unwarranted human deaths. However, right from the beginning of antibiotic therapy, the evolution of microbial resistance gave rise to the emergence of superbugs or rogue bacteria (resistant microbes), which can fight any of the broad-spectrum antibiotics. Their infections known as superinfections are not curable by these last-resort drugs. This evolutionary process alone results in the death of at least 700,000 global deaths per year. Therefore, multidrug-resistant infections, superbugs, and antimicrobial resistance are urgent health-care issues and are not less than a crisis. Similarly, it is one of the biggest global threats to public health due to changing climate conditions in the world with a trend toward warmer and drier weather patterns. A recent report in April 2019 released by UNO showed that if no action is taken, drug resistance infections and diseases could be responsible for 10 million human deaths each year by the end of the next three decades, that is, 2050 (WHO, 2019). Despite the substantial worldwide efforts, superbugs and superinfections are on escalation and are spreading quite fast. Considering the continued rise in antimicrobial resistance and the emergence of superinfections and multidrug-resistant diseases in man, it is obvious that the current efforts to manage and combat this health crisis are in a struggling phase.

No doubt, antimicrobials protect us from untimely diseases and pre-age deaths but at the same time, thousands die alone due to such infections which emerge due to antimicrobial resistance where the clinically available antibiotics fail to show any promise. People are indiscriminately consuming antibiotics as per their own choices whether or not they actually need them and are thus inviting untimely death to themselves because of their side effects. We are gravely addicted to the antibiotics and other related drugs to protect ourselves against the minutest of pathogenic infections. We have developed such a great dependency on these wonder drugs that we keep a reserve of antibiotics in our houses just like other daily use items. Moreover, we can easily procure them from the market like other daily consumables and there is no check for this menace. So we are caught in a double sword effect of antibiotics, that is, emergence of superbugs (decreased susceptibility to antibiotic action) and millions of deaths secondarily due to microbial resistance.

Overall, antibiotics are one of the greatest achievements of modern medical science in terms of their revolutionary role by increasing the life expectancy, decreasing the incidence of diversity of infectious diseases, simplifying different operative procedures, maintaining the public health quality, and increasing the livestock and agricultural productivity. However, they are also quite harmful or dangerous in terms of their recognized environmental side effects (antibiotic pollution), antibiotic or antimicrobial resistance, destruction of useful microbiota of our body, and other toxic effects.

5.3 ANTIMICROBIAL AGENTS AND CHALLENGE OF ALTERNATIVE ANTIMICROBIALS

5.3.1 WHAT AN ANTIMICROBIAL AGENT IS?

Any chemical or therapeutic agent either of synthetic or natural origin whether in liquid, cream, powder, spray, ointment, gel, tablet, or any other form of preparation like bioactive coatings which has the capability of killing or disabling or arresting the microbial growth, inhibiting cell wall synthesis, disrupting enzymatic action, altering general microbial metabolism, blocking or disrupting the synthesis of folic acid like essential nutrients, stopping development, division or preventing any other biological activity like protein biosynthesis, membrane functioning, DNA or RNA synthesis, disruption of structural and functional aspects of biomolecules is generally labeled as an antimicrobial. The antimicrobial agent can be an antiviral, antibacterial, antifungal, antiprotozoal, antiparasitic, or any other such agent capable of action against a pathogen or a microorganism at any stage of its cellular functioning, therefore, is simply called an anti-infective agent. Overall an antimicrobial agent is a chemical warfare molecule that has helped man to launch a survival battle against the dangerous pathogens both in human and livestock sector by either killing a microbe (bactericidal) or inhibiting the microbial growth (bacteriostatic) and the battle goes on against an emerging infection by the discovery of new and novel antimicrobials. Man has developed a variety of drugs with a narrow-spectrum to broad-spectrum action against the pathogens derived from bacteria, fungi, protozoa, mycoplasma, and other parasitic biological agents including helminth parasites.

5.3.2 GOLDEN ERA OF ANTIBIOTICS

Perhaps the full bloom of antimicrobial drug discovery happened during the 1950s–1980s spanning between 1955 and 1985. During these times we witnessed the discovery, testing, and marketing of over 100 antimicrobial/ antibacterial agents. Hence, the discovery of new antimicrobials never stopped but progressed despite reports of the development of antibiotic resistance coming against the previously and commonly used antibiotics. Similarly, the growing threat of antibiotic-resistant bacteria has never stopped since the dawn of the discovery of antibiotics. It is not only a global issue but a global disease itself considering its rising threats and increased mortalities in humans and animals. Keeping the other causes constant, the excessive and indiscriminate use and consumption of any antimicrobial class in humans and animals have been the greatest of the causes behind the development of antimicrobial resistance supported by the microbial genetics and natural selection on resistance genes within the bacteria and other microbial species. We know that the bacterial plasmids commonly designated as R plasmids carry the resistance genes, mutate them and transmit them from generation to generation besides the role of some chromosomal gene mutations in certain cases. The evolutionary biology of drug resistance is a fascinating field in itself because most determinants of antimicrobial resistance are not based on only simple gene mutations but on interlinked genes and their sequences work as per the environment they are exposed to. On one hand, the infectious agents are evolving thanks to their fast rate of genetic mutations, old infectious diseases are getting difficult to treat due to multidrug resistance, for example, in case of tuberculosis, emerging new viral infectious diseases like Ebola, Nipah, Zikah, coronaviruses and on the other hand, antibiotic resistance is becoming a permanent fixture in the microbial world. Therefore, people who are advocating the concept of antibiotic apocalypse as the world is now heading toward post-antibiotic era full of new challenges and health issues because many infections including some old and new ones in absence of a cure against them will prove a new threat as it is now quite hard to discover new antibiotics and antimicrobials. Although new research and innovation to search for alternatives to antimicrobials is growing at a very fast rate worldwide taking advantage of new technology including nanobiotechnology (combining recombinant DNA technology and nanotechnology) that is expected to bring a paradigm shift in antimicrobial discovery. Further discussion in this regard follows in the next section.

5.3.3 *TECHNOLOGY AND ANTIBIOTIC DISCOVERY*

Can we imagine a world without antibiotics? Everyone's answer must definitely be no because we cannot afford going back to the 1920s and 1930s as even a minor prick or infection can kill us. The field of antibiotics and drug resistance is becoming more and more complex due to the tremendous growth of biotechnology and medical science. The antimicrobial resistance is a continuous problem in presence of evolutionary selection and transfer of resistant genes to succeeding generations of microbes or pathogens. The emergence of multidrug-resistant pathogens is in fact a driving force behind the cutting-edge research to find alternative antimicrobials. Technology including nanobiotechnology has revolutionized the field of biology including medical science and pharmaceutical sciences. In the coming decades, the artificial intelligence and nanobiotechnology are thought to dominate and revolutionize life more on earth. A new powerful broad-spectrum antibiotic called Halicin has been discovered using artificial intelligence (a deep learning model in mice) by looking at a novel structure and has been reported to kill the world's nastiest bacteria—the *Mycobacterium tuberculosis* in murine/ mice models (Jonathan et al., 2020). It is being considered to emerge as a strong alternative to antimicrobials besides the discovery of potent antimicrobial peptides, probiotics, immunotherapy, homeopathy, herbal drugs, bacteriophage therapy, predatory fungi and bacteria, bacteriocins, etc. Besides, preventing or reducing the transfer and dissemination of antimicrobial-resistant genes is part of a high priority research involving modern technology, for example, the antibiotic nanoparticles to fight drug-resistant microbes is a high priority research. The research on antimicrobial resistance shall assume a more complex nature in future also despite technological advancements because even though novel or newly discovered antibiotics might exert a number of novel effects on the bacterial cell and its systems even at low antibiotic concentrations, a number of bacterial cells within a population will be again unaffected due to inherent nature of adaptations of resistance genes in microbes and the phenotypic tolerance to drug concentrations. As we will be able to understand and know more about the genome and genomics of mammals or humans alone, more we will be able to understand the link between resistance and prospects of resistance less action of new antimicrobials and to understand the genetics of target action of drugs. From both biological and pharmaceutical point of view, the field of technology and drug discovery is expected to provide more opportunities for marketing and investment in this sector.

5.4 ANTIMICROBIAL RESISTANCE AND ILL EFFECTS OF ANTIMICROBIALS

5.4.1 *ANTIMICROBIAL RESISTANCE*

Resistance to antimicrobial agents or antimicrobial resistance can be defined in a number of ways and lots of definitions are available in the literature, but here we will take into account two definitions. One definition of resistance is based on the idea that a specific pathogen or microbial agent or an infectious agent in a host due to some natural or inherent genetic mechanism present in the pathogen, will depict a low or least probability of responding to a normal or standard antimicrobial drug administered in the host. Another definition is based on the principle of acquiring special or specific characteristics in the genetic material (DNA) of a microbe which allows it to change the minimum inhibitory concentration of the drug through specialized modes of drug destruction mechanisms to increase the minimum inhibitory concentration of the drug over that of its wild genotype. Both the definitions satisfy the antimicrobial resistance concept that microbes change when exposed to an antimicrobial drug and emerge as new or novel drug-resistant strains.

It is interesting to note that many bacterial species including some plants living in varied ecosystems naturally produce certain antimicrobial agents or compounds that have been found to result in the development of antimicrobial resistance. Here man or industry has no role at all. Therefore, what can we guess or interpret out of this fact is, even without anthropogenic activities or intervention of any kind, the natural environment too contributes to the development of antimicrobial resistance and selection of resistant microorganisms and their populations operating under the principles of natural selection (Hermine, 2019).

The different mechanisms by which microorganism shows resistance to antimicrobial agents are the production of a variety of enzymes, synthesis of modified drug or pro drug targets, alteration of permeability of cell wall, alteration of metabolism and metabolic pathway, and alteration of membrane pumps, etc. Overall the basis of the development of resistance can be either genetic (chromosome, plasmid, and transposon-mediated) or nongenetic (loss of cell wall, development of membrane coverings, nonreplicating status, etc.). Recent studies have indicated that antimicrobial resistance in bacteria does not develop only under the selective pressures of indiscriminate use of antibiotics but the increased temperature too has been found as a significant factor in enhancing the development of antibiotics or other forms of resistance (MacFadden et al., 2018). According to this study, an increase in temperature of 10°C resulted in the increase of antibiotic resistance by 4.2% in *Escherichia coli* indicating that the changing climate or global temperature could result in devastating human infections during warmer months of the year (Tina, 2019) due to antibiotic-resistant bacteria. Therefore, to put things in perspective from a different angle, on one side global life is threatening due to emerging and re-emerging infectious diseases and food insecurity, on the other side climate change is a new threat and it seems practically impossible for the man to overcome any of these novel disasters. In fact, a report from the European Congress of Clinical Microbiology and Infectious Diseases (ECCMID, 2019) has indicated that a novel association exists between climate change and antibiotic resistance. So, it is quite interesting to prove if the same situation exists across the world buffering in geography, socio-economy, and health care. It might differ in those nations which prefer Ayurveda and homeopathy system of medicine to allopathic drugs to treat the cases of infectious diseases. However, cross-resistance will be a challenge.

5.4.2   DEVELOPMENT OF RESISTANCE

Right from the beginning of antibiotic usage, naturally antibiotic resis-tance evolved because it is a natural adaptation due to the evolution and mutation of resistance genes in a pathogen against which the chemicals are aimed to either kill or arrest the microbial growth and development. In fact, a particular antibiotic chooses itself a specific population or genotype of microbes to depict resistance against it. The microbes have evolved a variety of mechanisms and methods to develop antimicrobial resistance which subsequently results in the generation of superbugs or rogue bacteria that dominate the microbial world. Resistance to drugs or antibiotics is mediated by a variety of microbial enzymes coded majorly by plasmid genes which become active due to antimicrobial chemotherapy.

   The development of antimicrobial resistance against antibiotics or antivirals or antifungals or any other antimicrobial agent is a natural phenomenon and whatever the drugs or chemicals are used against microbial infections, the same chemical agent provides the selection pressure to the microbial populations to select their resistant genes and every time a new chemical or drug is used the more likely resistance develops against the succeeding drugs too. This is due to the inherent genetic mechanism available within the microbes that results in the drug treatment leading to the selection, survival, and propagation of resistant-pathogen populations (biologically fit populations to either tolerate or withstand the drug effects), thereby transferring resistance genes to the next generation, hence its propagation to upcoming generations. Therefore, once the resistance genes are selected and become fixed in the microbial genome then resistance becomes an established and permanent phenomenon. Therefore, it is a normal process for microorganisms. However, we accelerate this process by adopting some undue measures. For example, the main risk factors for the development of antibiotic or antiviral resistance and the emergence of superbugs are self-medication without any knowledge of the infection along with untimely, frequent, and indiscriminate usage, over and under dosage and improper administration of the particular drug or its ingredient. We are promoting the building up of antibiotic concentration in the environment, animals, and human beings. In either of the ways, we are building up the heavy load of antibiotics in our body and are hampering the normal physiological processes of our body.

The superbugs or rogue problem is not only of the bacterial origin but also it gets further complicated with the emergence of human viral superinfections. Using antiviral agents and antibiotics indiscriminately to treat viral diseases has resulted in the evolution of viral superinfections, which are subsequently resistant to antiviral drugs (a human being which has been previously infected by one virus or a viral strain gets co-infected by a different virus or a new resistant strain of it after antiviral therapy). Therefore, taking antimicrobials unnecessarily and frequently puts us at a greater risk of these superbugs. They must be our last choice and not the first response medicine for infectious diseases that can be treated or controlled by following proper disease management programs or standard procedures of disease prevention like hygiene.

5.4.3 ANTIMICROBIAL RESISTANCE AND BACTERIAL BIOFILMS

It is one of the most significant factors contributing to the development of microbial resistance and incidence of diversity of new microbial infections. It is a sessile association of various species of microorganisms particularly, bacteria characterized by adhering of microbial cells to each other on some living and nonliving surfaces due to extracellular polymeric substances secreted by bacterial cells themselves (Muhsin et al., 2015). The biofilms are not easily accessible to the components of our immune system and they show extreme antibiotic resistance because of low penetration and slow diffusion of antibiotics through their exopolysaccharide film matrix. Further, the presence of neutralizing enzymes in them render the antibiotics ineffective. A bacterial biofilm is highly infectious in nature and is a great public health threat because of its less susceptibility and accessibility to disinfectants and antibiotics. They are formed both within living systems as well outside the living systems in residential houses and external environments at almost every available place. The multiplication of resistant bacteria in the biofilms results in the formation of micro-colonies and the individual member bacteria can detach as per their will and undergo rapid multiplications and avail the chances of infecting the fresh populations of individuals. Some common multidrug-resistant bacteria which form biofilms are *E. coli*, *Streptococcus epidermidis*, *Staphylococcus aureus*, *Haemophilus influenza,* and *Pseudomonas aeruginosa*, etc. Generally, hospital infections or nosocomial infections are caused due to biofilms contaminating the hospital buildings and equipment or people's skin, respiratory discharge and gastrointestinal discharge, therefore, are a source of great infections worldwide (Okshevsky and Louise, 2016). Therefore, nosocomial infections are a serious concern and have attracted the attention of hospital administrator's right from the beginning of the development of microbial resistance. Strict handwashing and hand hygiene, elimination of overcrowding, increased breastfeeding rates in neonatal and children hospitals are some of the important measures practiced to reduce or lower the hospital infections by bacterial biofilms. We do not perhaps know that most of hospital admissions worldwide are either due to side effects of antimicrobials or infections by drug-resistant microbes coming from these bacterial biofilms (Saima et al., 2014).

5.4.4 ENVIRONMENTAL SIDE EFFECTS OF ANTIMICROBIALS OR ANTIBIOTIC POLLUTION

Anthropogenic activities have been on a rise since the time man learnt to use different ecosystems and life forms, which have resulted in the development of such a diverse category of stressful environmental situations of which antimicrobial compounds and their accumulation or discharge into our surroundings make a significant share. This microbial drug product or by-product-based enrichment of our environment after their use in human and livestock sector has threatened life in various forms on this planet with aquatic life as more prone to its effects. One of the major effects is on the useful microbial flora of soil and its concentration across the food chains that finally prove disastrous to birds and mammals, for example, use of various categories of antimicrobial compounds and pesticides in agriculture and horticulture sector gets leached into our environment and unnecessarily either kill the nontarget animals and plants or disturb their biochemical and metabolic pathways. Further, resistance of microbes becomes one of the greatest and emerging threats to our environment due to their direct and indirect effects on various forms of life.

At the same time, accumulation of solid and liquid wastes of antimicrobial compounds released from the pharmaceutical companies and manufacturing industries or discharge of packing material of antimicrobial compounds, waste accumulation of expired or unused drug products from hospitals, medical institutes, healthcare centers, and even households is another grave issue. In fact, the environmental waste management of accumulated medicinal products is quite challenging.

Another problem is the release or discharge or excretion of metabolized and partially metabolized antimicrobial compounds and other anti-infective drugs via human and animal excreta into the environment where they prove a havoc due to their toxicity in soil and aquatic ecosystems and their nonbiodegradable nature is a complicated problem that even the most advanced and modern technology-based treatment plants are not able to treat or degrade them. However, if anything can degrade them it is again some good soil microbial flora but unfortunately, they get disturbed due to the accumulation and bioaccumulation of such products due to antibiotic pollution.

The excessive effects of antimicrobial pollutants on the free living useful soil and water microbial populations and even on some useful soil nematodes have therefore disturbed the various ecological functions like the nutrient cycles, ecological relationships between various living organisms including both plants and animals, the ecosystem productivity, the process of natural decomposition and scavenging, and food web dynamics or trophic connections between different food chains in different ecosystems.

However, we must know that antimicrobial pollution and accumulation of resistant genes of these microbes or superbugs in our environment is a natural process rather than an evolutionary aspect of normal biological activities for certain forms of life although it has proved a novel type of environmental pollution albeit more of a anthropogenic nature. In this perspective, every person or an individual has a role to perform within his/her self, within houses, within workplaces, within markets, within the agricultural field, within industries and factories to minimize and discourage the discharge of expired, or excessive or unused or waste antibiotics and their packing's into our immediate surroundings, and ultimately remote water bodies. For example, during the COVID-19 pandemic times what has basically happened is the emergence of a new form of environmental pollution due to the accumulation of excessive medicinal products used to manage coronavirus infection, personal protection equipment, used test kits, syringes, gloves, masks, and their packings, etc. Overall, our protocol and behavior are not that much sufficient and standard to relieve our environment. Man has always enforced stress at the cost of his own benefits and greed. In the next section, readers will come to know the antimicrobial resistance and other harmful effects of antimicrobials besides antimicrobial pollution and environmental contamination.

5.4.5   OTHER ASSOCIATED PROBLEMS OF ANTIMICROBIALS

The biggest problem of the present day humans is the practice of self-medication with the use of antibiotics as no exception. Although antibiotic prescription and usage is not advisable in an age group of less than two years but it is hardly obeyed anywhere and wherever, it is being used in infants and children it results in abnormal development of various sensory structures like the ear sensory system. Similarly, issues of blindness and autism have also been reported in children with the indiscriminate use of broad-spectrum antibiotics like Vancomycin. With antimicrobial usage, resistance is not the only problem but the toxic side effect of the antibiotics is a burning issue. Although it seems that the use of antimicrobial agents relieves us from harmful infections, the excessive and indiscriminate use of certain broad-spectrum antibiotics deprives us from the useful microbial flora (also called microbiome or normal microbial flora) of our body, which being a vital component of our anatomical and physiological systems perform various indispensable roles in human health. The majority of the microbiota, which occurs in the gut besides other locations in the body, are quite essential to maintain our health and influence the anatomy, physiology, susceptibility to pathogens, and our overall wellbeing to fight with pathogens. Therefore, their suppression and subsequent replacement by drug-resistant microbes is a grave issue of the excessive use of antibiotics. Another recent dimension that adds fuel to this already existing fire of antimicrobial ill effects is climate change and global warming. Worldwide climate change has been predicted to result in the emergence and re-emergence of various categories of microbial infections whose control will be a real challenge in the presence of antibiotic resistance and immunologically weak hosts. Therefore, we need some comprehensive studies and researches on the possible impacts of climate change on the human infectious diseases and the subsequent effect on public health keeping in view the emerging and re-emerging potential of some infections and difficulties in managing those infections in the presence of antimicrobial resistance.

5.5 ANTIPARASITIC OR ANTHELMINTIC RESISTANCE

The development of resistance in macroparasites against various anti-parasitic drugs including anthelmintics is another dimension of antimicrobial resistance. Just like bacterial antibiotic resistance, the evolution of anthelmintic resistance is also a great challenge before the biologists and pharmaceutical industry. Macroparasites including nematodes, cestodes, and trematodes besides microbes are also responsible for causing a variety of parasitic diseases in plants, humans, and animals. These diseases are collectively known as parasitoses or helminthoses and are more a problem in developing countries due to their weak economies and traditional practices of personal protection and hygiene. Controlling antimicrobial resistance will require approaches to develop, share, and preserve antimicrobials that are scaled to the scientific, economic, and ethical dimensions of the crisis (Nathan, 2020). The principal mode for parasitic control involves the broad spectrum or narrow spectrum chemical drugs called anthelmintics, compound/s which destroy, kill and remove helminth parasites from the host body. But due to the emergence of anthelmintic resistant parasites, there is a need to evaluate other alternative strategies to develop a sustainable, effective, and safe anthelmintics—the concept of herbal anthelmintics (Tariq, 2018).

For each chemical class of anthelmintics, resistance to one member usually confers resistance to the other members. It is possible, and increasingly common, to have multiple resistances where parasites develop resistance sequentially and independently to several anthelmintic classes (Tariq, 2017a).

The development of antiparasitic and anthelmintic resistance is again a very simple natural phenomenon and whatever the better way and drugs man chooses to treat and control parasitic diseases, the more likely resistance develops due to the presence of already resistant genes in the target populations and a selective drug proves only as a stimulus to differentiate a susceptible and a resistant genotype in the parasite populations and communities. Once the resistance genes become fixed in the parasite genome the resistance becomes a prolonged phenomenon. Besides the parasite genotype, the untimely drug usage, over/under dosage, adopting of the improper route of administration in the host are the other risk factors for the development of antiparasitic and anthelmintic resistance. Further, there are so many important debatable issues surrounding the drug resistance both at the farmer and industrial level that there is a need for the evaluation of other alternative drugs and methods/strategies to control the infections.

But believing the natural selection principles, even if new chemical drugs or other novel antimicrobial products are developed against resistant parasites and their different stages, finally they will also be counterbalanced by the problem of drug resistance. Medicinal plant products and nanoparticles are being investigated worldwide and are referred to as potential alternative candidates for managing the menace of drug resistance. Therefore, the phenomenon of selection process of resistance in microbes cannot be stopped or eliminated permanently but can be delayed or temporarily stopped by introducing novel products of natural origin with standardized doses in replacement to conventional drugs for a sustainable management of antimicrobial resistance.

5.6   CLIMATE CHANGE, ENVIRONMENTAL HEALTH, AND PARASITISM

5.6.1   CLIMATE CHANGE: HAS IT HAPPENED

The 21st century's biggest issue for all countries has been the climate change. The biggest concern of climate change is its everlasting effect on the environment and the association of life with our environment in terms of its effect on different ecosystems. Ever since the evidences have accumulated that climate change alone is causing tens of thousands of deaths every year the situation has become quite complex when combined with antimicrobial resistance preponderance. So the problem is whether due to antimicrobial resistance or due to climate change or due to emerging and re-emerging infectious diseases influenced by climate change. Not to believe that the climate of the world has changed over the past several decades or not to believe that climate change will continue to hurt us is rather a blunder although people still carry different thoughts regarding climate change and even conspirational theories are also circulating. Let us accept that climate change is a reality, it has happened due to some natural causes and also due to mounting anthropogenic pressures in terms of agricultural and industrial practices.

The time I was writing this text I came across an article that, "sun has entered into a lockdown and it too is going to affect the life on earth in several ways including unexpected weather changes." It is at the same time viewed that antimicrobial resistance is not less than a pandemic equal in intensity and proportions to any infectious disease of pandemic nature because at least 700,000 people die annually in the world due to antimicrobial resistance alone. At the same time, when almost all the world had entered into a social distancing and enforced shutdown of major activities, nature started acting its own way and so many reports came across the world that the pollution level on earth has declined while purification and natural sanitization has inclined. Now what is all this about, it is the principle of natural selection that decides the final course of action on this earth.

According to Skuce et al. (2013), the direct effects of the predicted changes in climatic conditions, weather, frequency of extreme events, or altered thermal environments will potentially have an impact on animal production efficiency and through the imposition of stress, resulting in consequences ranging from increased mortality to stress-induced pathologies, altered disease resistance, and poor welfare.

Overall, the current scenario indicates that the climate has become warmer throughout the major portions of the world. The predictable effects of parasitic or pathogenic infections on animal health and productivity in relation to climate change have attracted the attention of parasitologists, environmentalists, veterinarians, climatologists, economists, and politicians worldwide. The emphasis vis-à-vis climate change and infectious diseases for long-term predictions are based on some important aspects like, how will climate change affect the epidemiology of parasitic infections and host–parasite dynamics?, can we predict the impact of climate change on parasitic and other infectious diseases of animals and the possible effect on productivity?, how can we ensure the sustainable management of our livestock in the presence of changing climate trends and the increasing incidence of infectious diseases?. Further, it is not only about the negative aspects of climate change and the infectious disease relationship, but the field has become more interesting that parasites are the best bio-monitoring tools/indicators of climate change itself.

At the same time, livestock contributes 18% of global greenhouse gases and is one of the major contributors to global climate change. A recent study has indicated that climate-change effects have turned Antarctic snow green due to growth of green algae (Woelders et al., 2018). Does it mean that the white continent is turning green to its real name due to climate crisis thereby paving the way to the emergence of new or novel species in erstwhile nongreen area—the Antarctica dominated by mosses and lichens? Finally, it means that it will be capable of capturing the atmospheric carbon dioxide.

5.6.2 HOST SWITCHING AND BIODIVERSITY SHIFTS IN THE CONTEXT OF CHANGING CLIMATIC TRENDS

At present, the climate change is the greatest threat not only to the economy and food security but also to the human and animal health and welfare as the living systems are already affected by climate change. The changes in climate and weather patterns in relevance to pathogenic viruses, bacteria, fungi, protozoa, and other macroparasites at field and farm level have become an interesting aspect. Long-term epidemiological surveys with improved methodologies and modeling procedures to study the ensuing relationship between the epidemiology of parasitic diseases and climate change will help to better understand the parasite–host dynamics.

The effects of climate change on the environment are no doubt enormous but hilarious too. What is expected is that global climate change might either make it easier for some parasites to infect their hosts or find new hosts which are better adapted to changed climatic conditions. Or changed climatic conditions may cause alteration in the dynamics of parasite transmission due to changed environmental conditions thereby the effects of some parasitic diseases may be exacerbated, or parasitic diseases may spread into naïve populations which would have been earlier free from these kinds of parasitic infections, ultimately these conditions may result in increasing the potential for host switching with ultimate consequences for biodiversity including those of parasites themselves. These predicted changes in the nature of host–parasite relationships are known as climate change-induced ecological perturbations. Climate change can also impact three main aspects that affects the equilibrium of parasite–host relationships; *biodiversity* (faster development of parasite free-living stages, faster reproduction of primary and more often the invertebrate secondary hosts or vectors), *population density* (increasing population of parasites and young susceptible hosts in the population), and *immune-competence*. Thus in consonance with the pattern changes in macro- and micro-climatic factors, parallel changes in epidemiology and pathogenesis of some major parasitic diseases directly influenced by environmental and other associated risk factors is expected. Infectious vector-borne diseases like *fascioliasis* and *schistosomiasis* having a stage in the environment are likely to be more affected by climate change. For example, vectors such as molluscs and insects will be able to complete more generations/year and the parasites they carry will expand their reproductive potential. To understand and predict the outcome of climate and host–parasite interactions will be more understood from the next sections of this manuscript.

5.6.3 *INTERCONNECTION OF CLIMATE CHANGE, BIODIVERSITY, AND PARASITIC DISEASES*

The effects of parasites and parasitism on ecological communities are diverse and range from negative to positive relationships because parasites not only act as pathogens but also act as predators and preys. They have a significant role in driving trophic communities and food web dynamics. Parasites are a very important component of life on earth in terms of their diversity and beneficial roles they play in the form of being a part of food chains, biomass, ecosystem detoxification, host evolution, indicators of environmental health, bio-monitoring, etc. About 50% of life is in the form of parasites, therefore, they have a great evolutionary and ecological importance. Despite being the cause of a variety of animal and plant diseases, parasites are not a cause of high mortality but their infections are responsible for severe morbidity due to less advances and inadequacy in the management of parasitic diseases in the presence of antimicrobial and anthelmintic resistance. For example, climatic factors, such as the mean warm-season temperature, significantly contribute to the prediction of antimicrobial resistance in different healthcare systems and societies, and climate change may increase the transmission of antimicrobial resistance, particularly the transmission of *P. aeruginosa* (Kaba et al., 2019).

The climate of the world is changing with a general trend toward average warmer and drier temperatures. There is enough of the evidence and research worldwide showing that climate change is altering the spread of parasitic diseases and it will have a devastating effect on health, food security, and economy. The climate change will have an impact directly and indirectly on livestock farming systems, the animals, and the helminth pathogens they contain (Tariq, 2017b). Another challenge will be the increasing human migration from endemic/high infected areas to nonendemic regions of the world which will create global human unrest and a new type of conflict called the climate conflict. It is, therefore, very crucial and imperative to recognize the possible impacts of global climate warming on the parasites of humans and other forms of life.

The Himalayas has been recognized as the high-impact region of the possible effects of changing global climate. Rather, it has already shown its effect on various species of birds like Black-necked Crane, Himalayan Vulture, and Himalayan Snow Cock. Regions that will receive very little precipitation or untimely arrival of rain will experience droughts, will have inadequate water supply to even maintain personal hygiene, which in turn will exacerbate the prevalence of various parasitic infections. However, we must keep it in mind that global climate change will not lead to increasingly favorable environmental conditions for all of the known parasites, but can also result in the elimination or extinction of many parasite species which will fail to adapt to the changing environmental and host conditions of climate change. Because it is equally possible that host animals might perish while resisting the unwanted climatic changes. Some parasites might increase their geographical range by using new animals as hosts. Antibiotic resistance, similar to climate change, is a shared global problem (Gelband and Laxminarayan, 2015).

As per the current climate change scenario worldwide, it is expected that patterns of parasitic diseases may change and life-cycle dynamics of a parasitic organism is also going to change. With this transmission potential of parasites is also going to change resulting in the introduction of diseases into new areas (emergence) and or cause a dramatic increase of the disease incidence in already endemic areas (re-emergence). So various parasitic diseases have a risk of spreading with a changing climate and negatively affect the public health. For example, the recently emerged Nipah Virus disease in some Asian countries from fruit-eating bats has also been linked to the changing climatic conditions.

The breeding potential of various species of insects that act as vectors of various parasitic diseases such as malaria, dengue, etc., is dependent on the level of environmental precipitation, which ultimately affects the potential of these vectors in transferring infections to animals and humans. Now, keeping in view the altered moisture conditions due to changing climatic conditions, therefore, the disease cycles are expected to vary on a larger scale and pose greater threats to human survival.

The invertebrate animals have the capability to adjust to the changing temperature conditions of the environment, therefore, invertebrates will pose novel challenges by playing their role as new hosts in influencing the transfer of new or novel parasitic infections to vertebrate animals. There are two possibilities vis-à-vis the biology of vector animals. Either we may see a decline in vector populations in coming years, coinciding with the absence of certain parasites or their populations may increase beyond their threshold levels, coinciding with the increase in the disease incidence of schistosomiasis, fascioliasis, and also re-emergence of some diseases.

Fortunately, as of now, the temperate climate of Kashmir valley has not proved conducive for different disastrous infections prevalent in tropical and subtropical regions of the world. However, the change in climatic conditions which is heading toward warmer climates might favor the emergence of new tropical infections like malaria, leishmaniasis, babesiasis, etc. under the changing climatic conditions. This might prove again very challenging due to less or complete absence of acquired immunity toward such infections. Further, less adaptability and experience to handle and manage the new parasitic diseases with more clinical issues will pose novel threats.

Soil-transmitted helminths like hookworms and roundworms which directly interact with the environmental conditions may also be altered by changing climatic conditions. This might happen because higher temperatures can cause faster development of their intermediate stages and greater survival rates with a significant increase in the prevalence of their infections. Similarly, the alterations in insect and bird distributions due to effects posed by climate change may result in the spread of parasitic diseases to new or novel areas.

Overall, it is expected that changing climate might cause water stress and heat shock that will influence the parasitic disease incidence to a larger extent. Therefore, we need studies on multiple aspects of parasites, their hosts, and disease ecology in response to changing climatic conditions to better understand the outcome of climate-pathogen/parasite interactions.

5.7 CLIMATE CHANGE AND BIODIVERSITY THREAT

5.7.1 CLIMATE CHANGE AND BIODIVERSITY

Around 2.4% of the earth's land surface supports about half of the world's plant and terrestrial vertebrate species. This much of the land surface witnesses around 36 areas as qualified hotspots of biodiversity-biologically rich but deeply threatened areas. Biodiversity has enormous ecological and economic significance and above all the life-sustaining building blocks of every form of life on earth. But above all, do we have that much of the good sense prevailing upon us that a common man understands what biodiversity means to us. Are we that much conscious and aware that every living organism beyond human matters and is quite vital in sustaining the complex food webs and trophic levels, which ultimately drive our life ahead? The politics within the human existence itself, and in social setup and national existence has not perhaps allowed man to take care or protect the life on earth while harvesting benefits for generations together since human inception. May it be the environmental pollution, plastic threat, chemical rampage, wildlife decline, misuse of gene technology, habitat destruction, and climate change all are equally responsible for their role in destroying the biological diversity at multiple levels. May it be a plant or an animal in wild or domestic form, they have fell victim or prey to over-exploitation or overharvesting by man due to human overpopulation and resulting materialism. This overpopulation of a single biological species alone is responsible for a decline or destruction of an innumerable variety of other life forms due to unsustainable or destructible use of biological diversity and various natural resources to fulfill human need and greed.

Man somehow tried to maintain a balance with a focus on access to genetic resources and other benefits by maintaining a fair and equitable sharing arising from their utilization or consumption. However, the concept of sustainable use has not progressed well since its emergence as a special ecological or developmental concept. We failed to capitalize this concept and it is now being felt that in view of the tremendous threats and extinction of life at its doorsteps, a convention on biological diversity will be adopted labeled as post-2020 global biodiversity framework as a stepping stone toward the 2050 vision of "living in harmony with nature" because biodiversity including wildlife and wild genotypes are declining from Africa to Asia at an alarming and unprecedented rate. For example, according to a recent grim report (Intergovernmental Science-Policy Platform on Biodiversity and Ecosystem Services at Paris) based on thousands of research publications, 85% of the world wetlands have been lost due to industrialization and urbanization alone, three-quarters of the world land and two-thirds of its marine environments have been significantly altered to such an extent by the anthropogenic actions that a million species of living organisms are threatened with extinction which is not less than a biological annihilation. Perhaps nature must be in action to revive its lost glory of biological diversity to maintain a balance between extinction and speciation, but human actions overcome the natural speciation, thereby extinction has taken a lead to wipe away the biological diversity once again-paying ways to man enforced sixth mass extinction event on earth.

Biodiversity is determined and influenced by so many environmental factors that are prone to alteration by changing climatic factors and variables due to global trends in climate change. Climate change is an additional stress on biodiversity besides other anthropogenic pressures on life and the environment due to some paradigm shifts in humans living on this planet. The predictions and models on possible impacts of climate change on biodiversity are actively pursued in the world and are part of the active worldwide research. Although significant initiatives have been taken at all levels within the economic threshold of all world countries to protect and safeguard biodiversity from climate change alterations and impacts with utmost thrust on sustainable management and development initiatives while using our environment and ecosystem for human benefits. However, nothing will work unless and until there is a greater participation of society and community groups that are directly affected by any action plan or policy in this regard. However, we should not forget that adaptation is the key to living on this planet and nature must be devising its own ways vis-à-vis natural selection to effectively manage the biological implications of climate change on the diversity of life.

5.7.2 IS INDISCRIMINATE ANTIPARASITIC USE JUSTIFIED CONSIDERING ECOLOGICAL RELEVANCE OF PARASITES

As parasites function as predators, preys, and hosts in an ecosystem, a considerable proportion of energy flows through parasites connecting various food chains and food webs. In simpler words, it is in the evolutionary interest of the parasite that its host shall *thrive* not *perish*, but the host always tries to get rid of the presence of a parasite by a variety of physiological and immunological mechanisms. However, as this relationship progresses, either the parasite evolves to become less harmful for its host or the host evolves to cope with the unavoidable presence of a parasite or both adjust to a mutual living. Second important thing is that the distribution and abundance of parasites in physical space (both in host as well as outside the host) and time and the factors (host, parasite, and environmental) ultimately regulate the host–parasite interaction and dynamics. Thus parasite ecology assumes a greater significance vis-à-vis parasite study. In fact, environmental parasitology has received a significant attention by different workers in recent times worldwide. This has been a debatable issue and has increased two-way curiosity in the minds of biologists, first to understand the environmental impact on parasites, and second the utility of parasites to assess environmental health and pollution.

Preston and Johnson in their paper titled "Ecological Consequences of Parasitism" published in *Nature* in 2010 emphasize that parasites can shape community structure through their multiple effects on trophic interactions, food webs, competition, biodiversity, and keystone/dominant species. This should incite our mind to think and know, as why parasites are ecologically important too if they are dangerous and frightening as pathogens. For time being let us forget the parasitic nuisance and associated problems and recall that parasites with varied host spectrum are such an important ecological and evolutionary adaptation of animal world that it drives the existence of diversified land and aquatic ecosystems. Parasites are not always dangerous as we perceive them but are ecologically very important for the healthy growth of our terrestrial and aquatic ecosystems. They are as good as monitors of ecosystem health and enable us to understand some of the basic ecological principles. Evidences are accumulating on a daily basis about the beneficial environmental relevance of parasites, because parasites are important components of life on earth in terms of their diversity and beneficial roles in the ecosystem, being integral parts of food chains, biomass, ecosystem detoxification, host evolution, indicators of environmental health, biomonitoring, etc. Therefore, environmental parasitology that lays stress on the beneficial roles of parasites and the ecological relevance of parasites is now given more consideration than their pathogenesis. Their significance can be understood by the fact that about 50% of life is in the form of parasites with great evolutionary and ecological importance rather than pathogenic significance. If latter would have been the main focus of parasites, then the other 50% of nonparasitic life would have seized to exist, but that actually does not happen.

The diversity and composition of helminths can be used to assess the ecosystem health particularly in an aquatic ecosystem, which has initially emerged as an ideal system to learn aspects of host, parasite, and environment in understanding parasite role to assess environmental health. Throughout the world, different workers have correlated parasitic infections to different environmental stresses. There are two ways by which parasites can be used as bio-indicators: either they can be used as *effect indicators*, or they can be used as *accumulation indicators*. Another aspect is whether to analyze an individual organisms or to focus on parasite populations and communities with respect to when assessing environmental pollution. In fact, environmental deterioration may have both positive and negative impacts on the parasitism. Pollutants can have detrimental effects on the host by weakening the immune system of the host, thereby making them more susceptible to parasite infections. Therefore, under such conditions, the immuno-compromised hosts are likely to be the safe havens for the parasites. At the same time, pollutants may also decrease the abundance of parasites as some parasites are more affected by contaminants due to their direct effects on the parasites and their intermediate stages. Although environmental parasitology is an old field of biology, it has now emerged as a novel area due to the ecological suitability of parasites in unraveling the hidden environmental enigmas of the parasite world in the coming times. Despite the tremendous ecological significance of parasite species, man has initiated the process of elimination and eradication for the elimination of parasites of domestic animals and man himself.

5.7.3 CAN WE AFFORD TO LOSE A BIOLOGICAL SPECIES BY ERADICATING A PARASITE?

At times when we talk of antimicrobials and antimicrobial resistance, we equally talk of parasites, if we can afford to lose a species by eradicating a pathogen or a parasite? Obeying the laws of nature and principles of existence of life in every form on this planet, our answer should be no, we cannot afford to lose a parasite at any cost, and to validate the answer, the following paragraphs will provide us a sufficient explanation.

What is inside us and eating us fearlessly while we lavishly enjoy a diversified and luxurious food? Well, it is a strategic, smart, and prudent natural agent, the parasite primarily created by nature to balance the energy dynamics of life in varied ecosystems on this earth. The evolutionary history of parasites enables us to understand that nature has created every form of life on this planet with a reason and purpose to enrich the earth with a diversity of ecologically interlinked species. The ecological linkage that balances the ecosystem's functioning. Basically, the parasite (macroparasite/microparasite) which is the focus of this section is a biological organism that lives in or on the body of another organism (different types of hosts) for shelter and to fulfill its metabolic and reproductive needs. As a student of parasitology one must know that co-evolution of a host and its parasite/s is a stable and specific relation guided by natural selection, however, this relationship is not complete for all parasites but is in evolutionary transition for many parasites that is visible as host pathogenesis in those particular cases. As laws of ecology guide us to understand that it can never be the objective of a parasite to harm or destroy its host and the resulting pathogenic aspects (if any) of parasitic infections in the host are actually the outcome of the parasite-host relationship manifested through morphological, physiological, biochemical and immunological mechanisms that determine the host pathogenesis (the damages and the destruction of host or parts of its body actually are the result of the byproducts of the host responses toward the presence of a parasite or its metabolic products). Thus it keeps the problem open for man to consider whether a parasite is a pathogen or a nature's connector or balancer of food web dynamics. Going further deep into the text of this section will help us to understand it.

Although parasites are integral components of an ecosystem, but pathogenesis duly happens in the host when infected by a parasitic species, accordingly the host is designated as a single infection host or mixed infection host. The control and treatment strategies against parasitic infections are not as much effective as they should be due to varieties of evasion strategies employed by parasites ranging from hiding in masked body sites, cryptobiosis (as of tardigrades), or hypobiosis (as of *Haemonchus*), or immune-evasion (as of *Plasmodium* and *Leishmania*), or immune-suppression (as of schistosomes) or antigenic variation (as of trypanosomes) or even mimicking the body components (as of nematodes) to avoid efficacy of antiparasitic drugs. We rather focus on parasite treatments but lack advances and adequacy in the management of parasitic infections which otherwise never want to disturb or destroy their hosts except for few calories of energy obeying the laws of thermodynamics. Although, we know that parasitic infections can cause major economic losses worldwide as a consequence of mortality of infected host animals, reduced weight gains, and other kinds of associated morbidity, however, the case is not generalized but differs across the parasite species across the various parts of the world. The general observation is that the developing world is relatively more affected than the developed world due to better cleanliness and hygiene, advanced husbandry conditions, and lifestyle to ward off parasitic infections.

As a rule, there is a density-dependent ecological principle governing the host–parasite relationship which naturally stops the destruction or extinction process in the host. The host is generally spread to innumerable populations within the species (monogeneans) or across the species (digenian parasites) in different ecosystems. As a matter of fact, when the density of a host reduces, parasite transmission rates should also drop, thereby when parasite abundance drops, it potentially prevents the host destruction, however, and it varies across the populations actually relying on the dynamics of the host–parasite relationship and the genetic composition of the host populations. But what actually prompts man to think of parasite extermination or eradication is in fact the human greed and self-interests. The phenomenon of parasite species eradication is, therefore, an unnatural offshoot of human selfishness and this is also one of the reasons that man is behind the sixth mass extinction of life on earth. In the next section, let us try to understand the useful aspects of parasites in the ecosystems, thereby their ecological relevance, and then seek justification for the parasite elimination or their eradication from the earth.

5.7.4 PARASITE ELIMINATION: IS IT JUSTIFIED

Now let us try to understand the relationship between species elimination or extinction of parasites due to drug action and its effects on biological diversity on earth. As human beings became conscious and as many parasites were discovered as the causative agent of some plant, animal, and human diseases, the man started developing the means and ways to destroy the parasite by introducing so many antiparasitic agents such as antibiotics, fungicides, anthelmintics, antiprotozoans, and other antimicrobial compounds and programs like vaccination and immunization. But their use is now being challenged at the ecological level because the existence of living organisms on this earth is based on the chain of relationships between the different species and its organisms in the multiple ecosystems. This certainly challenges the human intentions, which is in the race of eradicating various parasites from the planet that will finally result in the disturbance and extinction of so many parasite species. While certain eradication programs can be justified keeping in view the importance of human life or its domesticated animals.

Now the question is, why should man be after the parasite extinctions? Before trying to answer this question we should first answer this question that, is there any example of a parasite in the whole biological world that has caused the extinction of its host species ever? In true sense, a parasite does only limited or less damage to the host population before the latter learns and adapts to the ill effects of the former. In other words, it means that the goal of a parasite is never to kill or permanently harm its host, because the moment a parasite will result in the destruction of its host, it means the end to its own life means the parasite could get extinct. Therefore, the presence of a parasite temporarily can cause certain disease conditions in the host which in due course of time allows the host to overcome the ill effects and live a harmonious life in the presence of the parasite. In other words, sometimes the presence of a parasite is very essential to the well-being of the host, for example, the presence of one parasite in the host does not allow the establishment of other parasites in the same host (the concept of concomitant immunity).

Another interesting thing about the parasitic world is parasite itself is the host of other parasites, a phenomenon called as hyper-parasitism. In that instance, if one parasite is altered or irritated or eliminated it means the destruction of so many other biological objects. The latter can be very beneficial at various levels in the food chain and food web because, para-sites have very high diversity in the world. For example, as a special case of hyper-parasitism, the oligochaete (*Annelida*) parasitic worms that feed on their trematode parasites (hosts), that in turn actually emerge from their infected freshwater snail (Mollusca) hosts. Regarding the special litera-ture in most cases, the population dynamics of parasites and hosts show a relationship described by the Lotka–Volterra equation (or predator–prey equation) which describes the dynamics of the interaction of two different species (parasitism in the present case). Under natural conditions, a host species extinction is not caused by the parasite itself, most of the time these are the events like population bottleneck in the first place that lead to the extinction of the infected host species characterized with unhealthy populations, which are ecologically and genetically constrained due to low genetic diversity and problems with ecological adaptability and dimin-ished biotic potential. This generally happens with the founder populations following bottlenecks which become targets of highly virulent parasites leading to their extinction. Further, the parasite-driven extinction of host species (vertebrate host species) seems to be highly improbable due to the elaborate immune system of the host. There are highly susceptible and resistant individuals in a population of any species presenting a spectrum of defense at the herd and species level.

Species extermination by natural selection is a normal process. But to do so merely for human benefit seems to be an extraordinary sin. Some of the most disastrous human and animal diseases have been eradicated and are no longer a threat but instead, new infections have taken their place. Smallpox of human and rinderpest, also known as cattle plague, is the first animal disease that has been eradicated. Some of the infectious diseases are on the way of eradication. It may be possible to eradicate some of the world's deadliest parasitic diseases, but should we? The battle to wipe out the debilitating guinea worm parasite (*Dracunculus medinensis*) is ongoing and the target for its eradication is 2030, although the earlier target was 2020 to wipe this disease off the face of the earth. The word extinction can be of ordinary meaning for a layman but it sends red waves across the minds of a parasitologist and equally an ecologist. However, it seems justified for those parasitic cases in humans, against which no drug or vaccine is available. On the other hand, eradicating any organism has serious consequences for ecosystems. For example, mosquitos are natural pollinators of numerous crop species on which we humans depend, so they are not simply the annoying and painful disease-causing vectors but their niche extends beyond that. Now think of their elimination from this planet from specific ecosystems. Can their role as pollinators be taken over quickly and easily by other species? We have to understand and believe that parasites are not disgusting but are very indispensable in regulating the populations of their hosts and the overall ecosystem they are a part of.

5.8 PUBLIC HEALTH IMPLICATIONS OF CLIMATE CHANGE AND ANTIMICROBIAL RESISTANCE

The environmental, social, and other determinants like air, water, food, and shelter of public health are affected and influenced by climatic factors. Therefore, global climate change is expected to increase the rate of occurrence of various infectious diseases particularly in developing nations of the world. Malnutrition, desiccation, and heat shock are the other major determinants of public health and will be an additional burden. From the daily accumulating evidences about the changing climatic trends and shifting of climate toward warmer temperatures, its effects on world weather patterns, colder areas either becoming colder or hot areas becoming hotter or even cold climatic conditions of certain regions drifting toward hotter weather conditions. The problems of emerging and re-emerging infectious diseases, more occurrence of food-borne zoonotic infections, the sensitivity of certain pathogens to increasing temperatures, and certain pathogens mainly bacteria increasing their populations at higher temperatures, it is evident that man and the environment is going to witness more tough times ahead. The greater challenges of food insecurity are going to drastically drift and shit the ecosystem food chains toward more unnatural paths that chances of an eruption of new infections otherwise hidden in nature may be a new threat to human survival. That means new issues of public health directly and indirectly influenced by climate change will come into existence. The emerging microbial infections will pay new challenges due to the nonavailability of antimicrobial compounds to treat or manage them.

Further, due to already existing problems of the development of antimicrobial resistance, the resistant microbial infections are quite difficult to treat at present. The multidrug-resistant problem with certain pathogens is a burning issue. The measures to prevent it are being searched throughout the world. The diseases which make us most susceptible to the superbugs or super-infections are the common viral infections like cold and flu. Because we practice excessive and indiscriminate consumption of antibiotics to overcome these viral infections and it accelerates the development of drug resistance. Maintaining a very good hygiene (particularly hand hygiene), safe cooking practices, and timely vaccinations help in minimizing the superbug problem. However, considering the emergence of infectious diseases and the emerging need of antimicrobial therapy in this regard, at present we cannot even imagine or to think of doing away with the treasure of antibiotics dominating medical science. Similarly, it is equally challenging to look for other antimicrobials as an alternative to replace the existing ones against which resistance is present in the microbial world. Therefore, the best way at present is to minimize the use of antimicrobials, stop self-medication, and stop taking antibiotics for self-curable infections like common cold and flu, minimum use of antimicrobial creams and oint-ments. They should only be our last or final choice of treating infectious diseases that too under a strict medical supervision. Further, an appropriate and specific prescription by the clinicians, pharmacists, and medicos will greatly help in minimizing, or reducing the antimicrobial resistance. An alternative to antibiotics called phage therapy (using bacteriophages) was developed many years ago (in 1920s–1940s) but was discarded for use in a quick period of time due to many disadvantages like their unpredict-ability and maintenance problems. In the recent times, there has been a resurgence in the use of phages as a potent alternative to kills the resistant communities and populations of microbes due to the tremendous develop-ment and propagation of antibiotic-resistant genes between the microbial generations. So should we expect a time when a doctor or a pharmacist will prescribe a phage instead of a synthetic antibiotic to treat infectious diseases. Let us hope that new research reveals the best therapeutic poten-tial of phages to treat human and animal diseases. It will purely be a time to decide and tell us, whether it remains a dream or a reality.

5.9   CONCLUSIONS

No doubt, antibiotics ease our life, heal our wounds, and protect us from some of the terrific and incurable microbial infections thereby decrease the unwarranted human deaths, however, right from the beginning of the antibiotic therapy, the evolution of antimicrobial resistance and the emer-gence of superbugs or rogue bacteria that can fight off any broad-spectrum antibiotic have proved quite challenging to modern medicine and public health sector. These superinfections or superbug diseases do not get treated with the currently available antiparasitic or antipathogenic drugs and is a big challenge despite the discovery of new antibiotics, modifications of existing antimicrobials, and use of modern technology to avert drug resistance.

Another issue is that the superbugs or rogue microbe problem is not only of the bacterial origin but it also gets further complicated with the emergence of human superinfections of viral origin. Using antiviral agents and antibiotics indiscriminately to treat viral diseases has resulted in the evolution of viral superinfections, which are subsequently resistant to antiviral drugs (a human being which has been previously infected by one virus or a viral strain gets co-infected by a different virus or a new resistant strain of it after antiviral therapy). Its major cause has been the menace of self-medication without actually knowing the target and the symptoms. Therefore, taking antimicrobials unnecessarily and frequently puts us at an increased risk of these superbugs. However, for nature, every organism has a right to live, therefore, drug resistance is a natural phenomenon and an evolutionary means of biological fitness of an infectious agent.

Worldwide surveys and studies show us that the global temperature has risen during the past few decades or so, although differing evidences are accumulating from different corners of the world, some assuming to be not more than a conspiracy or a hoax. However, more evidences have accumulated that the increased local temperature has been shown to increase the antimicrobial resistance, therefore, climate change may produce more drug-resistant pathogens. As various infectious diseases called as zoonotic diseases are shared between animals and humans due to their close associations. Climate change in terms of rising temperatures and more extreme weathers is expected or has rather enhanced the emergence and re-emergence of various zoonotic infections in man, thereby pose a continuous threat to public health irrespective of region or religion. Therefore, developing novel strategies to first manage the ongoing threat of antimicrobial resistance in both man and in the livestock sector and then to plan to overcome the menace of microbial resistance and strategically deal with the superinfections will be of utmost importance and significance for not only the scientists and the medical community but the world political powers under the looming threat of climate change.

Man is making the best use of science and technology to manage the public health issues posed by antimicrobial resistance and global climate change and to be really successful in this endeavor, understanding and looking for sustainable remedies and alternatives to antimicrobial resistance and managing our resources and food security due to climate change shall ensure and safeguard the public and animal health during the crisis times. Finally, we cannot deny a civil conflict and mental health crisis if the triangular relationship between antimicrobial resistance, climate change, and public health is not brought under sustainable control within the ensuing times by making the best use of our resources and technology.

ACKNOWLEDGMENTS

Preparing this manuscript was not possible without consistent support and providing free time at home by my spouse (*Shaista*), Son (*Azzam*), and daughter (*Ayesha*). They made it easy for me to accomplish this task among the COVID-19 fear times. The challenging aspect during its writing phase was the ill-health of my father whose hospitalization for some days put a break to write it and later he did not prove any sort of hindrance to complete it well in time, thanks to Allah for bestowing him a safe recovery. I feel it equally important to thank the principal of our college—*Prof. Sheikh Ajaz Bashir* for his trust, confidence, and encouragement in my ways of doing in academia. Finally, not to acknowledge the continuing support of my mentor—*Prof. MZ Chishti* is beyond my thought, he is a constant inspiration.

KEYWORDS

- **antimicrobials**
- **resistance**
- **climate change**
- **herbal antibiotics**
- **emerging diseases**

REFERENCES

Bold Steps to Tackle Resistance. *Nat. Rev. Microbiol.* **2020**, *18*, 257. https://doi.org/10.1038/s41579-020-0356-5.

Gelband, H.; Ramanan, L. Antibiotic Resistance, Similar to Climate Change, Is a Shared Global Problem. *Trends Microbiol.* **2015** Sep, *23* (9), 524–526. doi: 10.1016/j.tim.2015.06.005.

Hermine, M. **2019**. https://www.sciencefocus.com/the-human-body/antibiotic-resistance-is-it-really-as-bad-as-climate-change/ScienceFocus. (accessed May 25, 2020).

Jonathan, M. S.; Yang, K.; Swanson, K., et al. A Deep Learning Approach to Antibiotic Discovery. *Cell* **2020** Feb, *180* (4). (accessed online June 14, 2020).

Kaba, H.; Kuhlmann, E.; Scheithauer, S. Novel Association between Antimicrobial Resistance and Climatic Factors in Europe: A 30-Country Observational Study. Presented at: The 29th European Congress of Clinical Microbiology & Infectious Diseases; April 13–16, 2019; Amsterdam, NL. Abstract 3045.

MacFadden, D. R.; McGough, S. F.; Fisman, D.; Santillana, M.; Brownstein, J. Antibiotic Resistance Increases with Temperature. *Nat. Climate Change* **2018**, *8*, 510–514.

Muhsin, J.; Ufaq, T.; Tahir, H.; Saadia, A. Bacterial Biofilm: Its Composition, Formation, and Role in Human Infections. *Res. Rev.: J. Microbiol. Biotechnol.* **2015**, *4* (3), 01–14.

Nathan, C. Resisting Antimicrobial Resistance. *Nat. Rev. Microbiol.* **2020,** *18*, 259–260. https://doi.org/10.1038/s41579-020-0348-5.

Nathan, C.; Cars, O. Antibiotic Resistance—Problems, Progress, and Prospects. *New Eng. J. Med.* **2014,** *371*, 1761–1763. doi: 10.1056/NEJMp1408040.

Okshevsky, M.; Louise, M. R. Big Bad Biofilms: How Communities of Bacteria Cause Long Term Infections. *Front. Young Minds* **2016,** *4*, 14. doi: 10.3389/frym.2016.00014.

Saima, A.; Jacquelyn, R. E. *NICU Environment and Principles of Infection Control. Book Chapter: Fetal and Neonatal Secrets*; Polin, R. A.; Spitzer, A. R., 3rd ed.; ISBN: 9780323091398; Elsevier, 2014; pp. 67–93 (accessed 21 May 2020).

Skuce, P. J.; Morgan, E. R.; van Dijk, J.; Mitchell, M. Animal Health Aspects of Adaptation to Climate Change: Beating the Heat and Parasites in a Warming Europe. *Animal* **2013,** *7* (2), 333–345.

Tariq, K. A. Anthelmintics and Emergence of Anthelmintic Resistant Nematodes in Sheep: Need of an Integrated Nematode Management. *Int. J. Veterinary Sci. Anim. Husbandry* **2017a,** *2* (1), 13–19.

Tariq, K. A. Helminth Infections of Livestock in Response to Changing Climatic Trends: A Review. *RRJZS,* **2017b,** *5* (1), 52–57.

Tariq, K. A. Use of Plant Anthelmintics as an Alternative Control of Helminthic Infections in Sheep. *Res. J. Zool.* **2018,** *1*, 1.

Teresa, G.; Pablo, L.; Fernando, S. G.; Sara, H. A.; Paula, B.; José, L. M. Antimicrobial Resistance: A Multifaceted Problem with Multipronged Solutions. *Microbiol. Open* **2019** Nov, *8* (11), e945. doi: 10.1002/mbo3.945.

The convention on biological diversity. https://www.cbd.int/convention/.

Tina, J. 2019. https://www.plymouth.ac.uk/news/pr-opinion/ global-challenges-intertwined-how-climate-change-is-linked-to-antimicrobial-resistance Magazine (accessed June 13, 2020).

WHO. 2019. https://www.who.int/news-room/detail/29-04-2019-new-report-calls-for-urgent-action-to-avert-antimicrobial-resistance-crisis (accessed June 1, 2020).

Woelders, L.; Lenaerts, J. T. M.; Hagemans, K., et al. Recent Climate Warming Drives Ecological Change in a Remote High-Arctic Lake. *Sci. Rep.* **2018**, 8, 6858. https://doi.org/10.1038/s41598-018-25148-7.

CHAPTER 6

Factors Responsible for Spatial Distribution of Enzyme Activity in Soil

TANIYA SENGUPTA RATHORE

Faculty of Life Science, Mandsaur University, Mandsaur, Madhya Pradesh, India

ABSTRACT

Soil is an integral part of our ecosystem. The organic compounds in soil are primarily delivered by the plant in the form of organic acids, amino acids, and biopolymers, such as cellulose, hemicelluloses, starch, and lignin. Some other organic components, such as chitin and peptidoglycan are received from microbial and microfauna present in the soil. Decomposition of these organic matters is carried out by a set of numbers of extracellular enzymes, which are produced by soil microbes which include bacteria and fungi. Measurement of these enzymes' activity represents the soil quality in terms of nutrient turnover. Many studies at different scales indicated enzyme distribution in soil. The scales considered ranges from a square meter to hectares. These studies concluded that enzyme distribution vary unevenly even in some square centimeters. This chapter summarizes the current research knowledge regarding spatial variability of soil enzymes at different scales and the factors of each scale responsible for the development of such establishment. The most common factors for all the scales include nutrient content, vegetation, microbial biomass, pH, and moisture content.

6.1 INTRODUCTION

Soil is a natural resource stands second after water in our mother earth playing important role in nutrient matter cycling, decomposing organic maters, water retention, etc. (Ritz et al., 2009). Soil quality needs to be monitored and measures should be taken to preserve in regular manner to improve food security and environmental protection. Quantitative measures, such as physiochemical and biological properties are the indicators for soil quality. Between these two properties, physiochemical property indicates the fundamental context of soil functions (Nosrati, 2013) whereas biological property certainly responds within a minute change in the soil environment (Paz-Ferreiro and Fu, 2014).

Due to high sensitivity to external factors, soil enzymes and their activities have become the potential indicator of soil quality. Soil enzymes are the biological catalysts required for the organic matter decomposition and breakdown, nutrient cycling, and also influences energy transformation processes, environmental soil quality, and agronomic productivity. Soil enzymes provide early detection of soil health differences because as compared to other soil quality parameters they respond faster to soil management changes and environmental factors (Srinivasarao, 2017).

Change in soil enzymes qualitatively and quantitatively determines the availability of nutrients and crop productivity. Cropping system and nutrient management are the different agricultural practices that influence the soil enzyme activities (Srinivasarao, 2014). Living or dead microbes, plant litters, and soil animals and insects are the sources of soil enzymes. Enzymes complexes with organic matter in the soil matrix form enzyme stabilization. Approximately, about 40%–60% of total enzyme activity comes from these stabilized enzymes. Therefore, the enzyme activities do not directly correlate with microbial activity. Enzyme activity generally increases with the rise in the concentration of organic matter content within the soil. High enzyme activity is the indicator for huge microbial communities and greater enzyme activity adsorbed through the humic materials. The activities of extracellular soil enzymes vary significantly with change in seasons, geographical locations, and soil depth (Dick and Gupta, 1994). The two kinds of enzymes are mainly found in the soil, that is, (1) Constitutive Enzyme: These enzymes are always present in more or less in constant amount and are generally not affected by the external factors or the level of substrate in the soil. (2) Inducible Enzyme: It may be present in a trace amount or may not be present, but with elevation in substrate level, the concentration of these enzymes get increased. Constitutive enzyme level firmly depends on the type of microbial community of soil, which directly or indirectly changes with the soil quality.

6.2 MAJOR SOIL ENZYMES

Some important major soil enzymes are presented in Table 6.1. This table also describes the soil functions for which these enzymes work as indicator and the major soil reaction catalyzed by these enzymes.

6.2.1 AMYLASE

Amylases are starch-digesting enzymes to form oligosaccharides. The main sources of amylase in soil are microbes, plants, and animals. Out of which microbes are the most important source producing about 19 types of hydrolase enzymes to hydrolyze starch, namely, α-amylase, β-amylase, glucoamylase, α-glucosidase, debranching enzymes, and transferases (E.C. 2), such as CGTase, 4-α-glucanotransferase, and a branching enzyme (Taniguchi and Honda, 2009). The species of filamentous fungus *Aspergillus* and *Rhizopus* are highly important for the amylases production (Pandey et al., 2006). The genus *Bacillus* from prokaryotes produces many extracellular enzymes, of which amylases are of the most significant. Najafi and Deobagkar (2005) reported that the optimum temperature and pH of the *B. subtilis* α-amylase were 55°C and 6.0 respectively. Soluble starch on hydrolysis by amylase produces glucose (70%–75%) and maltose (20%–25%) as end products. Amylase enzyme is generally classified into two forms based on the bond on which they act. Those are α- amylase and β-amylase. The *α*-amylase enzyme acts on the starch by hydrolyzing 1,4-*α*-glucoside bonds to produce glucose molecule, whereas β-amylase liberates β-maltose mainly from starch on hydrolysis. The functions of both α-amylase and β-amylase enzymes are greatly influenced by different factors, which include cultural practices, dominate vegetation, environmental, and soil types (Fig. 6.1). Plants directly or indirectly but significantly influence the amylase enzyme activities of the soil. In direct manner, plants supply the enzyme from their residual litters or excreted products, whereas indirectly they produce and provide the substrate called starch for enzymatic activity of microorganisms present in the soil. The significance of these amylase enzymes in the soil could be greatly understood by studying the influence of physical, chemical, biological, and agronomic factors. Furthermore, it would help in proper soil management to maximize the benefit of amylase (Joachim et al., 2008).

TABLE 6.1   Major Enzymes of Soil (Rao et al., 2017).

Enzymes	Sources	Soil function	Reactions catalyzed	End product/activity
Amylase		C-cycling	Starch hydrolysis	
α Amylase	Plants, animal and microbial community			Glucose
β Amylase	Plants			Maltose
Dehydrogenase	Microorganisms	C-cycling	Oxidation of organic matters	Addition of H to NAD or NADP
Phenol Oxidase	Plants and microbes	C-cycling	Lignin hydrolysis	Small carbon compounds
Urease	Microbes, plants and invertebrates	N-cycling	Urea hydrolysis	Ammonia and CO_2
Phosphatase		P-cycling	Hydrolysis of esters and anhydrides of phosphoric acid	Phosphate
Alkaline phosphatase	Majorly bacteria			
Acid phosphatase	Plant, fungi and bacteria			
Protease	Microbes and plants	N-cycling	Nitrogen mineralization	Nitrogen compounds for plant
Chitinase	Microbes, pants and insects	N-cycling	Chitin breakdown	Carbohydrate and inorganic nitrogen
		C-cycling		
Cellulase	Plants	C-cycling	Cellulose breakdown	Maltose

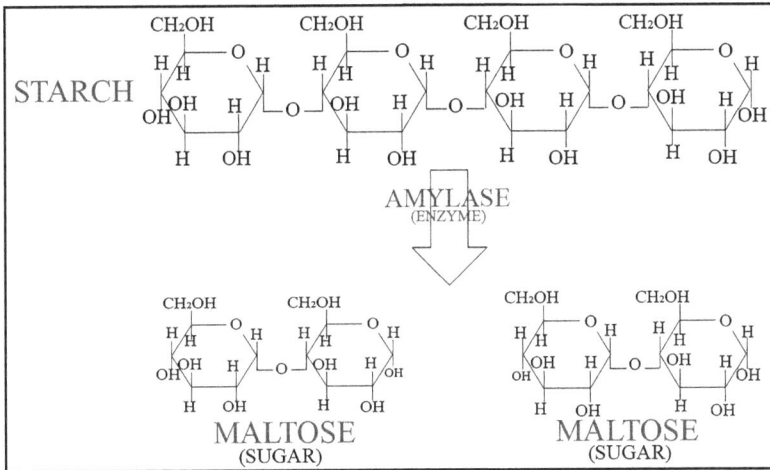

FIGURE 6.1 Reaction catalyzed by amylase enzyme.

6.2.2 DEHYDROGENASE

Dehydrogenase enzyme oxidizes soil organic matter in which it catalyzes transfer of protons and electrons from substrate to the acceptor. It is considered to be present in the living cell in its integral part and is not secreted out in soil (Das and Verma, 2011). The enzyme occurs in soil bacteria mainly in *Pseudomonas*, in which *Pseudomonas entomophilies* show the highest concentration. For dehydrogenase activity, bacterial host body is required, thus measuring the dehydrogenase level in soil directly measures the soil microbial activity (Walls-Thumma, 2000). By measuring the dehydrogenase concentration in soil, the soil quality could be determined and could be managed properly. Among all other soil enzyme indicators, dehydrogenase is considered as one of the most sensitive but important bioindicator related to soil quality. Its role is very significant in catalyzing biological oxidation of soil organic matter by transferring hydrogen ion to inorganic acceptor from the organic substrate (Zang et al., 2010). The enzyme is used to measure the microbial oxidative activities in soil, thus acting as an microbiological redox reaction indicator. It is also used to calculate any biological disruption in soil by pesticides, heavy metals, or other contaminants (Navnage et al., 2018; Karaca et al., 2011; Wolinska and Stepniewska, 2012). Various biotic and abiotic external factors like incubation time, temperature, soil aeration status, and moisture content influence the dehydrogenase enzyme activity. Enzyme activity decreases with the increase in the level of pollution. Many researchers found that in the reclaimed soil, enzyme activity increases. Polluted soil decreases the enzyme activity even after rewetting (Kumar. 2013).

6.2.3   UREASE

The main function of the urease enzyme is to degrade urea into simplest nitrogen compound, ammonia, which could be easily absorbed by plants. The stability of soil urease enzyme depends upon the soil type. The soil type determines the urease activity by protecting the urease enzyme from microbial decomposition and other processes, which could lead to degradation or inactivation of the enzyme (Dharmakeerti, 2013). Various reports concluded that the correlation between soil physiological properties with urease activity is very limited. The sandy and calcareous soils have low urease activity, whereas soil with high organic matter content tends to have high activity of urease enzyme (Baishya et al., 2015). Some work has also reported that increase in soil pH decreases the enzyme activity, whereas organic carbon content positively correlated with the urease enzyme activity. As the enzyme activity is related to soil pH and organic carbon content directly, thereafter, it is indirectly related to the type of vegetation at different scale (Gawde, 2017). The main sources of urease are microorganisms, plants, and animals. Urea hydrolysis bacteria are common in soil, which includes aerobes, microaerohilic, and anaerobes. Ureolytic bacterial community does not increase with the increase in substrate level. Plants are the rich sources of urease but no direct evidences are there to provide plant roots as a source of urease enzyme (Rahman et al., 2016). Urease is also present in animal intestine and excreta. Urease is the enzyme which is present in the soil in two different states. It may exist intracellularly or extracellularly.. The intracellular urease in soil occur as soil biomass form, whereas the extracellular enzyme is released into the soil through the metabolism or death of microbes and plant tissues. Urease exists in cell within the cytoplasm or attached with the cell membrane (Rahman, 2012). The main substrate for urease enzyme is urea (Fig. 6.2). Additionally, it could have the efficiency for hydrolyzing hydroxyurea, dihydroxyurea, and semicarbazide.

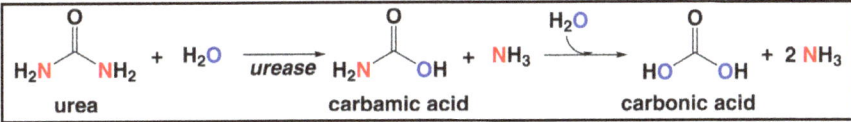

FIGURE 6.2   Reaction catalyzed by urease enzyme (Simon et al., 2020).

Source: Reprinted from Svane et al., 2020. © Springer. https://creativecommons.org/licenses/by/4.0/

6.2.4 PHENOL OXIDASE

Phenol oxidases are the group of enzymes having copper within. The enzyme is responsible for catalyzing the o-hydroxylation of monophenols to form o-diphenols. Additionally, they also catalyze the oxidation of o-diphenols to quinones when oxygen is present. These groups of enzymes are distributed broadly among microorganisms (in bacteria and fungi) and are also available in plants and animals (Araji et al., 2014). Phenol oxidase degrades the plant metabolite called phenolic compounds. These compounds decompose very slowly than that of the soil organic matter. Therefore, these phenolic compounds are considered as the slowest carbon pool in soil dynamic model (Min et al., 2015). Thus, they play a major role in carbon sequestration. Due to extensive substrate scope and biocatalytic applications, phenol oxidase is widely studied. These enzymes work on phenolic compounds and are derivatives of humic substances or other aromatic components of the soil (Kordon, 2010). These phenolic compounds are mainly degraded by fungi, for example, *Basidiomycetes* and *Ascomycetes* and bacteria, for example, *Pseudomonas* sp. which secrete extracellular phenolic oxidase enzyme (Sinsabaugh, 2010). Degradation of the phenolic compounds are affected by different environmental factors, including temperature, soil pH, oxygen and substrate level, they directly or indirectly affect the phenol oxidase activity. Sinsabaugh (2010) reported that phenol oxidase activity shows positive correlation with soil pH. Temperature relation with phenol oxidase is not clear as the oxygen content in different temperature directly affect the enzyme activity (He et al., 2009). Phenol oxidase activity is significantly correlated with seasonal trends and soil moisture, but they are not correlated significantly in short-term change in soil environment called as microclimate (Bell, 2011).

Laccase is considered as the largest class of phenol oxidase. It has the capacity to oxidize many small molecules that include humic compounds into stable radical in stepwise processes. These stable radicals are called as redox mediators, having redox potential greater than laccase themselves (Camerero et al., 2005). Laccase can also oxidize chelated manganese peroxide by interacting with Mn peroxidase. White rot fungi mainly Basidomycetes lacks peroxidase enzyme, thus use laccase enzyme to depolymerize lignin with a redox mediator.

6.2.5 PHOSPHATASE

Phosphorus is an essential element of life. It is the element of genetic information, cellular structure as well as of energy transport. The main source of phosphorus in soil is from weathering of rocks. The living organisms could assimilate only dissolved phosphate, an enzyme named phosphatase plays the most important role in transforming phosphate from soil organic matter into soluble form (Caldwell, 2005). The main sources of phosphatase enzyme are bacteria, fungi, and plant roots. Phosphatase enzyme cleaves phosphate from its substrate, transforming complex form into assimilated form. Thus, phosphatase enzyme synthesis is regulated by the availability of phosphate in soil in combination with its demand by living organisms. The rhizosphere region of soil which consists of both plant roots and mycorrhiza as well as root associated or free-living microorganisms secrete phosphatase enzyme in the soil. Therefore, in this zone, the soil contains both intracellular as well as extracellular phosphatase enzyme with very high concentration. Furthermore, the phosphatase enzyme gets stabilized on clay and iron or aluminum oxides as surface-reactive particles. Continuously through natural weathering, inorganic phosphate gets transformed into organic phosphate. After complete transformation, about 90% of phosphate in the soil comprised of organic phosphate which includes microbial phosphate in majority (Turner, 2013). Thus, in these weathered soil, phosphatase activity plays the most crucial role by cycling phosphate from soil organic matter to plant availability.

Several works have been done to identify the natural driver of phosphatase enzyme activity. In these experiments, fertilizers, temperature change, water availability, etc. under controlled conditions were taken under consideration. Markelie and Houlton (2011) concluded that nitrogen fertilization increases phosphatase activity whereas the same decreases by phosphate fertilization for different biome. Water stress is also one of the factors responsible for stimulating phosphatase activity. At Mediterranean sites, reduction in water availability by 21% decreases phosphatase activity by about 30%–40% (Sardans et al., 2004). Further phosphatase activity also has correlation with the components of soil microbial community and concentration of soil organic matters (Margalef et al., 2017). Its activity also depends upon soil depth and succession stage of ecology. Soil disturbance and heterogeneity in time and space of enzyme distribution has restricted phosphatase enzyme study under controlled manipulated conditions (Spohn and Kuzyakov, 2013). Soil phosphatase enzyme categorized into acid and alkaline phosphatase or several workers have studied phosphomonoesterase as they mineralize organic phosphate to inorganic phosphorus. They mainly hydrolyze the monoester bonds of mononucleotides or sugar phosphate and inositol phosphates of lower order. But they are unable to hydrolyze phosphate of phytic acid (myoinositol hexaphosphates) (Nannipeiri et al., 2010).

6.2.6 PROTEASE

Protease enzyme is widely distributed in soils and shows a range of activities. It is involved in the hydrolysis of proteins into simple amino acids. Hydrolytic degradation of proteins by proteolytic enzymes is an important part of nitrogen cycle. Protease is present in all types of organisms ranging from microorganisms, plants, and animals. The steps of degradation of protein include conversion of protein into polypeptides to oligopeptides to finally amino acids with in nitrogen cycle (Mosier, 2001). The rate limiting step of nitrogen cycling is the protein hydrolysis and thus increasing the soil fertility. The production and activity of soil protease are controlled by various external factors which include climate, soil properties, and the substrate level from plants and microbial origin. Various researches are being conducted on the regulation of protease enzyme by using low molecular weight molecules or organic compounds like amino acids, phytohormones, siderophores, and flavonoids (Vranova et al., 2013). Based on the sites of cleavage in protein molecule which protease catalyzes, protease is categorized into two major groups. The first group called exopeptidase targets the terminal ends of protein and the second group called endopeptidase targets within the protein. Endopeptidase is classified into different forms according to their catalytic mechanisms, such as aspartic endopeptidase, cysteine endopeptidase, glutamic endopeptidase, metalloendopeptidase, serine endopeptidase, and threonine endopeptidase. Apart from these, protease also has an additional type, which is reserved for those which specifically act on peptides. These are termed as oligopeptidase (Britannica, 2020).

6.2.7 CHITINASE

Chitin is the most abundant polysaccharide after cellulose in nature and basically found in the exoskeleton region of different insects, fungi, yeast, algae, and in the internal structures of vertebrates. Chitinase are the enzymes that degrade chitin thus contributing in carbon and nitrogen cycle. These enzymes have great importance in controlling pathogens of agricultural lands (Hamid et al., 2013). Saprophytic soil bacteria *Actinobacteria* are the most important taxa among soil microbial community for synthesizing Chitinase enzyme (Bai et al., 2016). In terms of Chitinase activity, *streptomyces* is the most abundant genera of *actinobacteria* (Brzezinska et al., 2013). Apart from these, *Bacillus thuringenesis* is also an important source of chitinase enzymes which is used as bioinsectiside in agriculture (Veliz et al., 2017). Chitin catabolism takes place in two steps; initially, chitin polymers are cleaved by chitinase into chitin oligosaccharides and then further cleaved by chitobiases into N-acetylglucosamine and monosaccharides (Chen et al., 2010). Chitinase have been categorized into two main groups; endochitinase and exochitinase. The endochitinase randomly cleaves chitin at the internal sites, thus forming dicetylchitobiose dimer and soluber multimers of GlcNAc, such as chitotriose and chitotetrose (Sasaki et al., 2006). Exochitinase has been further divided into two more subcategories; chitobiosidase, involved in catalyzing release of diacetylchitibiose progressively from the nonreducing end of chitin microfibril. The second one is 1,4-β-glucosaminidase, cleaving the oligomeric product of endochitinase and chitobiosidase into monomers of GlcNAc (Hamid et al., 2013).

6.2.8 CELLULASE

Cellulose, the most abundant polysaccharide present in the cell walls of plants that represents a significant contribution to the soil. Cellulase is an enzyme which acts on cellulose polymer. Cellulose on earth occurs abundantly in the form of wood, chips, and municipal waste. This enzyme activity is extensively studied in plant litters. During litter decomposition, soil microbes discharge cellulose as extracellular enzyme. Several external factors like temperature, moisture, pH, and cellulose concentration influence the cellulase activity in litter decomposition (Raju et al., 2016). Cellulase is a core enzyme consists of endo, exo, and β-glucosidase.

Endo-β-1,4-glucanase hydrolyzes 1,4-β-glycosidic bonds in cellulose, lichenin, and β-D glucan in cereals. Exo-β-1,4-glucanase hydrolyzes 1,3-β-D glycosidic linkages in arabinogalactan. β-Dglucosidase hydrolyzes terminal nonreducing β-D glucose which releases terminal glucose step-by-step (Raju et al., 2016).

Sethi et al. (2013) identified *Pseudomonas flourescens, Bacillus subtilis, E.coli,* and *Serratia marcescens* as cellulase producing bacteria from soil. According to the study, the optimum condition for cellulase activity was 40°C, pH 10 for both carbon and nitrogen sources.

6.3   MEASUREMENT OF SOIL ENZYME ACTIVITY

The enzyme activity is being measured by calculating the degradation level of substrate and the amount of product formed. Several methods include spectrophotometry, fluorescence, and radiolabeling have been used to measure enzyme activity based on mode of detection, reaction substrate, and other factors, such as temperature, buffer (pH), and time for reaction. The measurement could be carried out by two ways: in the first method, substrate is added in the soil and then the product is quantified, whereas in the second method, the particular enzyme taken under consideration was extracted out from the soil sample and then assayed. The former is considered as direct method whereas the latter is considered the indirect method of measurement. The direct method is generally preferred, as the indirect method has limitation in which the significant amount of soil enzymes is bounded with the soil component itself. As an advanced technique, substrate-induced respiration method has been developed for the measurement of enzymatic activity. In this method, specific substrate for the enzyme under study is added to the soil sample, and then consumption of oxygen and production of carbon dioxide in the reaction is measured. Multiwell plate system has been developed to study the substrate-induced respiration for multiple enzymes simultaneously (Shrinivasrao, 2017). Molecular biology aspects, such as genomics, transcriptomic, and proteomic approaches have been developed for the estimation of specific enzyme producing capacity by a microbial community (Nannipieri, 2012). The enzyme coding genes or the transcribed products' sequences were used for assaying enzyme activity. mRNA sequences were better considered for the study as it is closely related to the enzyme. However, for the persistence and turnover study for an enzyme, gene expression is taken under consideration (Burns et al., 2013). Geo-chips, quantitative PCR are used to quantify the enzyme coding genes for a specific organism at different conditions. Digital image analysis methods were also adopted for the measurement of enzyme activity (Spohn et al., 2013).

6.4   SPATIAL VARIABILITY OF SOIL ENZYMES

The main factor that supports the enzyme activity is the soil organic content level, which is positively correlated with soil microbial mass and activity (Bonanomi et al., 2011). There are many factors which influence the soil microbial activity and thus also the enzymatic activity, such as cropping and management practices, forest cleaning, agriculture waste burning, etc. Intensive soil management reduces the microbial activity. According to Cardoso et al. (2013) report, drastic reduction in fungal mycelium and other microbial communities has been observed in high input managed soil. Reduction in organic carbon content is mainly responsible for the decrease in the level of microbial biomass. Besides, there are also different factors that influence the spatial variability of soil enzymes and are represented in Table 6.2.

6.5   FACTORS FOR SPATIAL DISTRIBUTION OF SOIL ENZYMES

In terms of scale, at the largest scale use of land and the vegetation in dominant are the most important factors influencing soil enzyme activity (Stursova and Baldrian, 2011). As compared with agricultural land and grassland, forest soil contains more fungal biomass. Additionally, ectomycorrhizal fungi in forest soil form mycorrhizal association with plant roots. The dominant tree taxa that form the association, their litter production, rooting depth, and photosynthesis production directly influenced the soil enzyme distribution (Baldrian and Stursova, 2011). Changes in the soil enzyme both qualitatively and quantitatively determine the nutrient level of the soil. In agricultural land, different agricultural practices, such as cropping system, nutrient management, etc. influence the enzyme activity (Srinivasarao et al., 2014). Several workers have reported that in cultivated soil, low enzyme activity has been observed when compared with the forest and grasslands (Monreal and Bregstrom, 2000; Saviozzi et al., 2001). In the forest, the microbial extracellular enzymes as well as the enzymes from the plant litters are affected by the

TABLE 6.2 Spatial Variability of Soil Enzyme Activity (Rao et al., 2017).

Ecosystem/practices	C-cycling enzymes	N-cycling enzymes	Phosphatase	Dehydrogenase
Forest	Highest (β-galactosidase)	Highest (urease)	Highest (alkaline phosphatase)	—
Conservational	High (β-galactosidase)	High (urease and protease)	High	High
Organic residues	High (invertase)	High (urease and protease)	High (alkaline phosphatase)	High
Organic natural	High (β-galactosidase)	High (urease)	High (acidic phosphatase)	High
Rehabilitated (Over 50 years)	High (α and β-galactosidase)	High (protease)	High	High
Degraded	Low (cellulase)	—		Low
Mined	—	—		Low
Polluted	—	—	Low (alkaline and acidic phosphatase)	—

dominant tree species. Whereas in mixed forest, different tree species through their difference in litter quality, soil moisture absorption and rhizospheral effect regulate soil enzyme distribution and activity (Snajdr et al., 2013). At smaller scale, the soil physiology such as organic matter concentration, inorganic nutrients, and soil pH influences the enzyme activity. The pH of the soil determines the microbial community (Lauber et al., 2009). According to several workers, using Meta-analysis, it was found that organic matter content and the soil pH directly affect the soil enzyme activity (Rousk et al., 2010; Sinsabaugh et al., 2008).

With increase in organic matter content, the activity of cellulase, phosphatase, N-acetylglucosamine also increases (Nsabimana et al., 2004). The Chitinase activity found to be frequently decreases with increase in N content (Allison et al., 2008). Thus, from different studies, it could be concluded that the soil macronutrient level and the presence of humic acid in soil directly influences the soil enzyme activity (Baldrian, 2014). The moisture content of the soil not only influences the spatial distribution but also causes variation with time. In the forest ecosystem, soil moisture content also affects several enzymes for, for example, peroxidase cellulase, phenol oxidase, etc. (Baldrian et al., 2010). The moisture content directly influences the fungal and bacterial biomass ratio and other microbial biomass qualitatively as well as quantitatively in the litter and soil (Gomoryova et al., 2006).

Microbial biomass content shows positive correlation with soil enzyme activity as they were the producers for those extracellular enzymes, but with an exception in turnover rate because most of the enzymes remain in soil actively even after the death or inactivation of microbial biomass (Baldrain and Snajdr, 2011). Table 6.3 summarizes the factors affecting different enzymes in soil.

It is a fact that with differential spatial scales, the factors responsible for the distribution of enzymes vary. Therefore, it could be said that the most important factor responsible for spatial variability of soil enzyme is scale sampling.

6.6 CONCLUSION AND FUTURE PROSPECTS

Soil enzymes are a vast area of research. The soil enzymes' importance and role could be seen on soil quality, productivity, and in soil biogeochemical reactions. Soil enzymes are directly or indirectly influenced by various

TABLE 6.3 Factors Affecting Enzyme Activity (Rao, 2017).

Enzyme	Factors affecting enzyme activities
Amylase	Soil type, management practices, vegetation
Dehydrogenase	Management practices, pollution, moisture content, temperature, trace elements
Phenol oxidase	Management practices, organic matter content, pH, temperature, precipitation mean (annually), N compound level
Urease	Management practices, organic matter content, pH, temperature, heavy metals, depth of soil
Phosphatase	Management practices, organic matter content, pH, vegetation, pollution
Protease	Concentration of humic acid, C and N availability
Chitinase	N availability, soil depth, CO_2 level in atmosphere
Cellulase	Organic matter content, pH, temperature, moisture content, O_2 content

factors belongs to physical, chemical, and biological characteristics of soil. The enzyme activities are directly correlated to management practices as well as environmental factors. The environmental factors not only distribute these soil enzymes spatially but the time of sampling and storage before assaying the enzyme activities is also an important factor. Temporal and spatial enzyme activity assay required the knowledge of factors controlling the enzymes activities. With advancement in molecular biology techniques, it is possible to work on enzyme coding genes, transcript products, and protein production. Soil enzyme profiling will help in management practices and therefore could be used to increase productivity. It could also be used to control soil pollution crested directly or indirectly by different anthropogenic activities. Biotic and abiotic stresses induced by climate change also affect the soil enzyme activity. Advance and strategic research is required to study the dynamics of enzyme activity. Further researches should be carried out to minimize or control the change in enzyme activities due to such stresses. By combining the molecular technique of quantitative PCR with measurement of activities of soil enzymes, soil chemistry, and analyzing microbial community using next generation sequencing methods, a most probable future research could be conducted to avail knowledge related to soil enzyme distribution and activity. The most important challenge for assaying enzyme activity is the time, which should be considered during sampling and data analyzing.

In addition to adequate protocols, it is required to collect data of accurate definition based on spatial before generating any conclusion on the spatial factors for enzymatic activity.

KEYWORDS

- **enzyme activity**
- **nutrient turnover**
- **scales**
- **environment**
- **climate**

REFERENCES

Araji, S.; Grammer, T. A.; Gertzen, R. Novel Roles for the Polyphenol Oxidase Enzyme in Secondary Metabolism and the Regulation of Cell Death in Walnut. *Plant Physiol.* **2014,** *164* (3), 1191–1203.

Bai, Y.; Eijsink, V. G. H.; Kielak, A. M.Genomic Comparison of Chitinolytic Enzyme Systems from Terrestrial and Aquatic Bacteria. *Environ Microbiol.* **2016,** *18,* 38–49.

Baishya, L. K.; Rathore, S. S.; Singh, D.; Sarkar, D.; Deka, B. C. Effect of Integrated Nutrient Management on Rice Productivity, Profitability and Soil Fertility. *Ann. Plant Soil Res.* **2015,** *17,* 86–90.

Baldrian, P.; Štursová, M. *Enzymes in Forest Soils.* In: *Soil Enzymology;* Shukla, G.; Varma, A., Eds.; Springer-Verlag: Berlin, 2011; pp 61–73.

Baldrian, P.; Snajdr, J. Lignocellulose-Degrading Enzymes in Soils. In *Soil Enzymology;* Shukla, G.; Varma, A., Eds.; Springer-Verlag: Berlin. 2011; pp 167–186.

Baldrian, P. Distribution of Extracellular Enzymes in Soils: Spatial Heterogeneity and Determining Factors at Various Scales, soil. *Sci. Soc. Am. J.* **2014,** *78,* 11–18.

Baldrian, P.; Merhautová V.; Cajthaml, T.; Petránková, M.; Šnajdr, J. Small-Scale Distribution of Extracellular Enzymes, Fungal, and Bacterial Biomass in Quercus Petraea Forest Topsoil. *Biol. Fertil. Soils.* **2010,** *46,* 717–726.

Bell, T. H.; Henry, H. A. L. Fine Scale Variability in Soil Extracellular Enzyme Activity Is Insensitive to Rain Events and Temperature in a Mesic System *Pedobiologia* **2011,** *54* (2), 141–146.

Bonanomi, G.; D'Ascoli, R.; Antignani, V.; Capodilupo, M.; Cozzolino, L.; Marzaioli, R.; Puopolo, G.; Rutigliano, F. A.; Scelza, R.; Scotti, R.; Rao, M. A.; Zoina, A. Assessing Soil Quality under Intensive Cultivation and Tree Orchards in Southern Italy. *Appl. Soil Ecol.* **2011,** *47,* 184–194.

Brzezinska, M. S.; Jankiewicz, U.; Walczak, M. Biodegradation of Chitinous Substances and Chitinase Production by the Soil Actinomycete Streptomyces Rimosus. *Int. Biodeterior. Biodegradation.* **2013**, *84*, 104–110.

Burns, R. G.; DeForest, J. L.; Marxsen, J.; Sinsabaugh, R. L.; Stromberger, M. E.; Wallenstein, M. D.; Weintraub, M. N.; Zoppini, A. Soil Enzymes in a Changing Environment: Current Knowledge and Future Directions. *Soil Biol. Biochem.* **2013**, *58*, 216–234.

Caldwell, B. A. Enzyme Activities as a Component of Soil Biodiversity: A Review. *Pedobiologia (Jena).* **2005**, *49*, 637–644.

Camarero, S.; Ibarra, D.; Martinez, M. J.; Martinez, A. T. Lignin-Derivedcompounds as Efficient Laccase Mediators for Decolorization of Different Typesof Recalcitrant Dyes. *Appl. Environ. Microbiol.* 2005, *71*, 1775e1784.

Cardoso, E. J. B. N.; Vasconcellos, R. L. F.; Bini, D.; Miyauchi, M. Y. H.; dos Santos, C. A.; Alves, P. R. L.; de Paula, A. M.; Nakatani, A. S.; de Moraes Pereira, J.; Nogueira, M. A. Soil Health: Looking for Suitable Indicators. What Should Be Considered to Assess the Effects of Use and Management on Soil Health? *Sci. Agric.* **2013**, *70*, 274–289.

Chen, J. K.; Shen, C. R.; Liu, C. L. N-acetylglucosamine: Production and Applications. *Mar Drugs.* **2010**,*8*, 2493–2516.

Das, S. K.; Varma, A., Eds. *Role of Enzymes in Maintaining Soil Health. Soil Enzymology, Soil Biology 22*; Springer-Verlag: Berlin Heidelberg, 2011.

Dharmakeerthi, R. S.; Thenababu, M. W. Urease Activity in Soils: A Review. *J. Natl. Sci. Foundation Sri Lanka* **2013**, *24* (3), 159–195.

Gawde, N. Long Term Effect Of Integrated Nutrient Management on Soil Nutrient Status under Rice-Wheat Cropping System in Inceptisols. *Int. J. Chem. Stud.* **2017**, *5*, 1050–1057.

Gömöryová, E. J.; Pichler, G. V.; Gömöry, D. Spatial Patterns of Soil Microbial Characteristics and Soil Moisture in a Natural Beech Forest. *Biologia* **2006**, *61*, 329–333.

Hamid, R.; Khan, M. A.; Javed, S. Chitinases: An Update. *J. Pharma Bioallied Sci.* 2013, *5* (1), 21–29.

He, L.; Xiang, W.; Sun, X. Effects of Temperature and Water Level Changes on Enzyme Activities in Two Typical Peatlands: Implications for the Responses of Carbon Cycling in Peatland to Global Climate Change. In *Proceedings of the International Conference on Environmental Science and Information Application Technology (ESIAT '09)*, **2009**; pp 18–22.

http://clfs690.alivetek.org/CLFS690/glycolglucojmol/glucosephosphatase.htm

https://commons.wikimedia.org/wiki/File:Pyruvate_dehydrogenase_complex_reaction.PNG

https://sciencemusicvideos.com/ap-biology/module-9-energy-and-enzymes

https://slideplayer.com/slide/7328482/

https://www.britannica.com/science/proteolytic-enzyme, 2020

Ibitisam, K.Transformation of Refuse Banana Wastes to Value-Added Products, Technical Report, Research Gate, 2015.

Karaca, A.; Cetin, S. C.; Turgay, O. C.; Kizilkaya, R. Soil Enzymes as Indication of Soil Quality. In *Soil Enzymology*; Shukla, G.; Varma, A., Eds.; Springer-Verlag: Berlin, Heidelberg, 2011; pp 119–148.

Kordon, K.; Mikolasch, A.; Schauer, F. Oxidative Dehalogenation of Chlorinated Hydroxy-biphenyls by Laccase of White-Rot Fungi. *Int Biodeter Biodegr.* **2010**, *64* (3), 203–209.

Kumar, S.; Chaudhuri, S.; Maiti, S. K. Soil Dehydrogenase Enzyme Activity in Natural and Mine Soil—A Review. *Middle-East J. Sci. Res.* **2013**, *13* (7), 898–906.

Lauber, C. L.; Sinsabaugh, R. L.;, Zak, D. R. Laccase Gene Composition and Relative Abundance in Oak Forest Soil Is Not Affected by Short-Term Nitrogen Fertilization. *Microb. Ecol.* **2009**, *57*, 50–57.

Margalef, O.; Sardans, J.; Fernandez-Martinez, M.; Molowny Horas, R.; Janssens, A.; Ciais, P.; Goll, D.; Richter, A.; Obersteiner, M.; Asensio, D.; Penuelas, J. Global Patterns of Phosphatase Activity in Natural Soils. *Sci. Rep.* **2017**,*7* (1337), 1–13.

Marklein, A. R.; Houlton, B. Z. Nitrogen Inputs Accelerate Phosphorus Cycling Rates across a Wide Variety of Terrestrial Ecosystems. *New Phytol.* **2011**, *193* (3), 696–704.

Min, K.; Freeman, C.; Kang, H.; Choi, S. U.The Regulation by Phenolic Compounds of Soil Organic Matter Dynamics under a Changing Environment. *BioMed Res. Int.* **2015**, *2015*, Article ID 825098, 1–11.

Monreal, C. M.; Bergstrom, D. W. Soil Enzymatic Factors Expressing the Influence of Land Use, Tillage System and Texture on Soil Biochemical Quality. *Can. J. Soil Sci.* **2000**, *80*, 419–428.

Mosier, A. R. Exchange of Gaseous Nitrogen Compounds between Agricultural Systems and Ahe atmosphere. *Plant Soil* **2001**, *228*, 17–27.

Najafi, M. F.; Deobagkar, D. Protein Expression and Purification. **2005**, *41*, 349–354. DOI: 10.1016/j.pep.2005.02.015.

Nannipeiri, P.; Giagnonj, L.; Landi, L.; Renella, G. Role of Phosphatase Enzymes in Soil. *Phos. Action* 2010, *26*, 215–243.

Nannipeiri, P.; Giagnoni, L.;, Renella, G.; Puglisi, E.; Ceccanti, B.; Masciandaro, G.; Fornasier, F.; Moscatelli, M. C.; Marinari, S. Soil Enzymology: Classical and Molecular Approaches. *Biol. Fert. Soils* **2012**, *48*, 743–762.

Navnage, N. P.; Patle, P. N.; Ramteke, P. R. Dehydrogenase Activity (DHA): Measure of Total Microbial Activity and as Indicator of Soil Quality. *Int. J. Chem. Stud.* **2018**, *6* (1), 456–458.

Nsabimana, D.; Haynes, R. J.; Wallis, F. M. Size, Activity and Catabolic Diversity of the Soil Microbial Biomass as a-Ected by Land Use. *Appl. Soil Ecol.* **2004**, *26*, 81–92.

Rahman, F.; Rahman, M. M.; Rahman, G. M.; Saleque, M. A.; Hossain, A. S.; Miah, M. G. Effect of Organic and Inorganic Fertilizers and Rice Straw on Carbon Sequestration and Soil Fertility under a Rice–Rice Cropping Pattern. *Carbon Manage.* **2016**,*7* (1–2), 41–53.

Rahman, M. M. Potential Supplies and Use Efficiencies of Nutrients from Different Organic Wastes under Tomato Cultivation. *Ann. Bangladesh Agric.* **2012**, *16*, 25–39.

Raju, M. N.; Golla, N.; Vengatampalli, R. Soil Cellulase. *Soil Enzymes* **2016**, *P*, 25–30.

Rao, S.C., Grover, M., Kundo, S., Desai, S. Soil Enzymes, Encyclopedia of Soil Science, 3rd ed. (pp.2100–2107), 2017. Taylor & Francis.

Rousk, J. E.; Bååth, P. C.; Brookes, C. L.; Lauber, C.; Lozupone, J. G.; Caporaso, E. T. Soil Bacterial and Fungal Communities across a pH Gradient in an Arable Soil. *ISME J.* **2010**, *4*, 1340–1351.

Sardans, J.; Peñuelas, J. Increasing Drought Decreases Phosphorus Availability in an Evergreen Mediterranean Forest. *Plant Soil.* **2004**, *267*, 367–377.

Sasaki, C.; Vårum, K. M.; Itoh, Y.; Tamoi, M.; Fukamizo, T. Rice Chitinases: Sugar Recognition Specificities of the Individual Subsites. *Glycobiology* **2006**, *16*, 1242–1250.

Saviozzi, A.; Levi-Minzi, R.; Cardelli, R.; Riffaldi, R. A Comparison of Soil Quality in Adjacent Cultivated, Forest and Native Grassland Soils. *Plant Soil* **2001**, *233*, 251–259.

Sethi, S.; Datta, A.; Gupta, B. L.; Gupta, S. Optimization of Cellulase Production from Bacteria Isolated from Soil. *Hindawi Publishing Corporation ISRN Biotechnol.* **2013**, *2013*, 1–7.

Sinsabaugh, R. L.; Lauber, C. L.; Weintraub, M. N.; Ahmed, B.; Allison, S. D.; Crenshaw, C. Stoichiometry of Soil Enzyme Activity at Global Scale. *Ecol. Lett.* **2008**, *11*, 1252–1264.

Sinsabaugh, R. L. Phenol Oxidase, Peroxidase and Organic Matter Dynamics of Soil. *Soil Biol. Biochem.* **2020**, *42*, 391–404.

Snajdr, J.; Dobiášová, P.; Urbanová, M.; Petránková, M.; Cajthaml, T.; Frouz, J.; Baldrian, P. Dominant Trees Affect Microbial Community Composition and Activity in Post-Mining Afforested Soils. *Soil Biol. Biochem.* **2013**, *56*, 105–115.

Spohn, M.; Kuzyakov, Y. Phosphorus Mineralization Can Be Driven by Microbial Need for Carbon. *Soil Biol. Biochem.* **2013**, *61*, 69–75.

Spohn, M.; Carminati, A.; Kuzyakov, Y. Soil Zymography: A Novel in Situ Method for Mapping Distribution of Enzyme Activity in Soil. *Soil Biol. Biochem.* **2013**, *58*, 275–280.

Srinivasarao, C.; Lal, R.; Kundu, S.; Prasad Babu, M. B. B.; Venkateswarlu, B.; Singh, A. K. Soil Carbon Sequestration in Rainfed Production Systems in the Semiarid Tropics of India. *Sci. Total Environ.* **2014**, *487*, 587–603.

Štursová, M.; Baldrian, P. Effects of Soil Properties and Management on the Activity of Soil Organic Matter Transforming Enzymes and the Quantification of Soil-Bound and Free Activity. *Plant Soil.* **2011**, *338*, 99–110.

Svane, S.; Sigurdarson, J. J.; Finkenwirth, F.; Eitinger, F.; Etinger, T.; Karring, H. Inhibition of Urease Activity by Different Compounds Provide Insight into the Modulation and Association of Bacterial Nickel Import and Ureolysis. *Nat. Res.* **2020**, *10*, 8503.

Taniguchi, H.; Honnda, Y. *Encyclopedia of Microbiology*, 3rd ed., 2009. https://www.sciencedirect.com/topics/medicine-and-dentistry/amylase

Turner, B. L. Soil Microbial Biomass and the Fate of Phosphorus during Long-Term Ecosystem Development. *Plant Soil.* **2013**, *367*, 225–234.

Veliz, E. A.; Hidalgo, P. M.; Hirsch, A. M. Chitinase-Producing Bacteria and Their Role in Biocontrol. *AIMS Microbiol.* **2017**, *3*(3), 689–705.

Vranova, V.; Rejsek, K.; Formanek, P. Proteolytic Activity in Soil: A Review. *Appl. Soil Ecol.* **2013**, *70*, 23–32.

Walls-Thumma, D. Dehydrogenase Activity in Soil Bacteria. 2000. http://www.gardenguides.com/130633- dehydrogenase-activity-soil-bacteria.html.

Wolinska, A.; Stepniewska, Z. Dehydrogenase Activity in the Soil Environment. In *Dehydrogenases*; Canuto R. A., Ed.; Intech: Rijeka. http://www.ebook3000.com/2011.

Yoshida, M.; Kiyohikolgarashi, K. R.; Aida, K.; Samejiima, M. Differential Transcription of β-glucosidase and Cellobiose Dehydrogenase Genes in Cellulose Degradation by the Basidiomycete *Phanerochaete chrysosporium FEMS Microbiol. Lett.* **2004**, *235* (1), 177–182.

Zhang, N.; He, X.; Gao, Y.; Li, Y.; Wang, H.; Ma, D.; Zhang, R.; Yang, S. Pedogenic Carbonate and Soil Dehydrogenase Activity in Response to Soil Organic Matter in Artemisia Ordosica Community. *Pedosphere.* **2010**, *20*, 229–235.

CHAPTER 7

Exploration of Bacterial Siderophores for Sustainable Future

ANITA V. HANDORE[1*], S. R. KHANDELWAL[2], RAJIB KARMAKAR[3], and D. V. HANDORE[4]

[1]P. G. Department of Microbiology, HPT Arts and RYK Science College, Nashik, India

[2]H. A. L. College of Science & Commerce, Nashik, India

[3]Department of Agricultural Chemicals, Bidhan Chandra Krishi Viswavidyalaya, Directorate of Research, Research Complex Building, Nadia, West Bengal, India

[4]Research and Development Division, Sigma Winery Pvt. Ltd. Sinnar, Nashik, India

*Corresponding author. E-mail: avhandore@gmail.com

ABSTRACT

Iron is the fundamental cofactor for conducting numerous metabolic processes of organisms. It plays a vital role by regulating biosynthesis of vitamins, antibiotics, toxins, cytochromes, nucleic acid, and microbial biofilm formation by regulating the surface motility of microorganism. Even though abundant iron is available in the earth's crust and environment, its ferric form is insoluble and inaccessible at physiological pH (7.35–7.40). In such condition, the concentrations of dissolved iron is approximately 10^{-10}–10^{-9} M in precipitate form, whereas the essential level for living organisms is around 10^{-7}–10^{-5} M. Therefore, in order to survive under such iron-stress condition, most of the microorganisms synthesize microbial iron chelates, that is, organic compounds with low molecular weight below 1000 Daltons.Such compounds are called siderophores. Siderophores show ability to obtain iron from the surroundings and make it available to microbial cells by forming its soluble complexes. In plants, iron plays a prominent role in the biosynthesis of chlorophyll, thylakoid, and chloroplast development. Accordingly, iron deficiency in plants not only affects their growth but also increases the susceptibility towards various diseases and disorders resulting in poor micronutrient content and adverse impact on plant health as well as on the health of animal and humans. Siderophores shows a wide variety of applications, as they can efficiently act as plant growth promoter, biocontrol agent, biosensor, and can be utilized in biomedicine. They play an important role in bioremediation and also exhibit the potential to biodegrade petroleum hydrocarbons. Although, large number of siderophores produced by different microorganisms have been documented, amazing structural variations can be seen in bacterial siderophores and they show promising potential to overcome problems related to iron deficiency. This chapter presents the overview of bacterial siderophores, their classification, transport mechanism, and their production method. This chapter also discussesabout the qualitative and quantitative detection methods for different types of siderophores, such as chemical assay, bioassay by traditional as well as modified methods. It also proposed the method of their crystallization, identification, and characterization. Characterization of siderophores is performed by FTIR, LC-MS, ESI /MALDI Mass spectra and NMR, etc. This chapter has illustrated the elaborative applications of bacterial siderophores for sustainable future.

7.1 SIDEROPHORES

Fe^{+2} is an essential cofactor for conducting various metabolic activities. It also plays a vital role by regulating the biosynthesis of porphyrins, vitamins, antibiotics, toxins, cytochromes, pigments, aromatic compounds, nucleic acid synthesis, and microbial biofilm formation by regulating the surface motility of microorganism (Messenger and Barclay, 1983; Glick et al., 2010; Cai et al., 2010). At physiological pH (7.35–7.40), ferrous form of iron is soluble, while the ferric form is insoluble (Bou-Abdallah, 2010). It is reported that at this condition, the concentrations of dissolved ferrous iron is found to be approximately 10^{-10}–10^{-9} M in precipitate form, whereas the required level of ferrous iron by living organisms is around

10^{-7}–10^{-5} M (Poole and McKay, 2003; Matsumoto et al., 2004). Therefore, in order to survive under such iron-depleted situation, most of the micro-organisms produce low molecular masses, that is, siderophores (Ahmed and Holmstrom, 2014).

7.2 CLASSIFICATION AND CHEMICAL STRUCTURES

Siderophores are classified on the basis of coordinating groups that chelate the Fe^{+3}. It is reported that catecholates, hydroxamates, and carboxylates are the most common coordinating groups of siderophores (Ali and Vidhale, 2013). Whereas some siderophores have chemically different Fe^{+3} ion-binding group, including salicylic acid, oxazoline, or thizoline nitrogen. Some siderophores like pyoverdines are classified as "mixed ligands" having coordinating groups that fall into chemically different classes.

7.2.1 CATECHOLATE SIDEROPHORE

All the siderophores having binding groups of phenolate or 2, 3-dihydroxy benzoate belong to the catecholate type of siderophores. Each catecholate group supplies two oxygen atoms for chelation with iron in order to form a hexadentate octahedral complex. Certain bacteria like *E. coli, S. typhimurium, E. Herbicola, and K. pneumoniae* produce enterochelin, also known as enterobactin (Sah and Singh R. 2015). It is reported that Enterobactin with molecular formula $C_{30}H_{27}N_3O_{15}$ is one of the strongest catecholate type of siderophore exhibiting significant potential to chelate iron even from the environment where the concentration of iron is low (Raymond et al., 2003) (Fig. 7.1). It shows ability to bind ferric ion very tightly and this strong binding between enterochelin and iron can be used to determine even very low concentration of iron in the environmental sample. In nature, there are few bacteria exhibiting ability to produce either catecholate siderophore alone or mixed siderophores. It is stated that some bacteria like *E. carotovora* can produce only catecholate siderophore, whereas some members of *Pseudomonas* show the ability to produce mixed type of siderophore consisting of both catecholates and hydroxamates (Leong and Neilands, 1982).

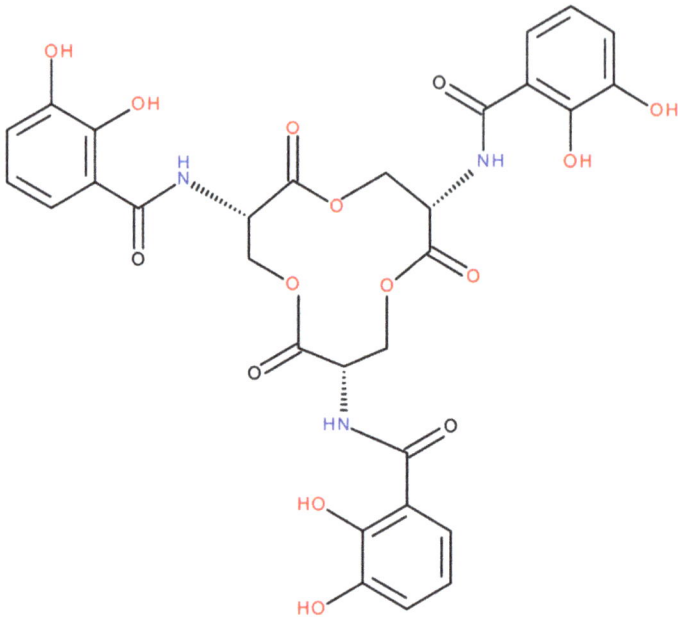

FIGURE 7.1 Structure of Enterobactin siderophore.

7.2.2 HYDROXIMATE SIDEROPHORE

It is observed that most of the hydroxamate groups consist of C (=O) N-(OH) R, where, "R" is either an amino acid or its derivative. In this, two oxygen molecules from each hydroxamate group forms a bidentate ligand with Fe^{+2}, as a result, each siderophore is able to form hexadentate octahedral complex with Fe^{3+}. The hydroxamates bind with ferric iron at binding constants in the range of 1022–1032 M^{-1} (Winkelmann, 2007). Hydroxamate siderophores formed hexadentate, tetradentates, and bidentate ligands. This siderophore family includes rhodotorulic acid, dimerium acid, alcaligens and putribactin. They are further divided into the subclasses like, desferrioxamine, ferrichrome and aerobactin.

7.2.2.1 DEFEROXAMINE SIDEROPHORE

Deferoxamine is the linear trihydroxamate siderophore with a molecular formula of $C_{25}H_{48}N_6O_8$ (Fig. 7.2) and is found to be produced by *S. pilosus*.

It is frequently used to treat hemochromatosis and to reduce mortality in persons with sickle-cell disease or β-thalassemia who are transfusion-dependent (Ballas et al., 2018). This agent is included in WHO's list of essential medicine in basic health system (World Health Organization, 2019).

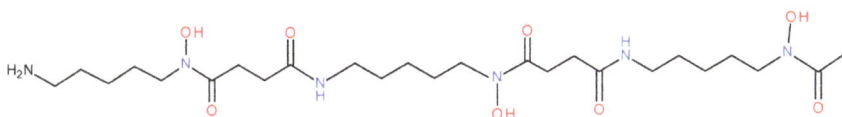

FIGURE 7.2   Structure of desferrioxamine siderophore.

7.2.2.2   *FERRICHROME SIDEROPHORE*

Ferrichromes are cyclic trihydroxamate siderophore comprising large family of hydroxamate siderophores with molecular formula $C_{27}H_{42}FeN_9O_{12}$. These are composed of three *N*-acyl-*N*-hydroxyl-L-ornithine, two variable amino acids linked by way of peptide bonds (Fig. 7.3). Certain ferrichrome derivatives (ferrioxamines) display antibiotic activity and have been designated as ferrimycines (Sah and Singh, 2015).

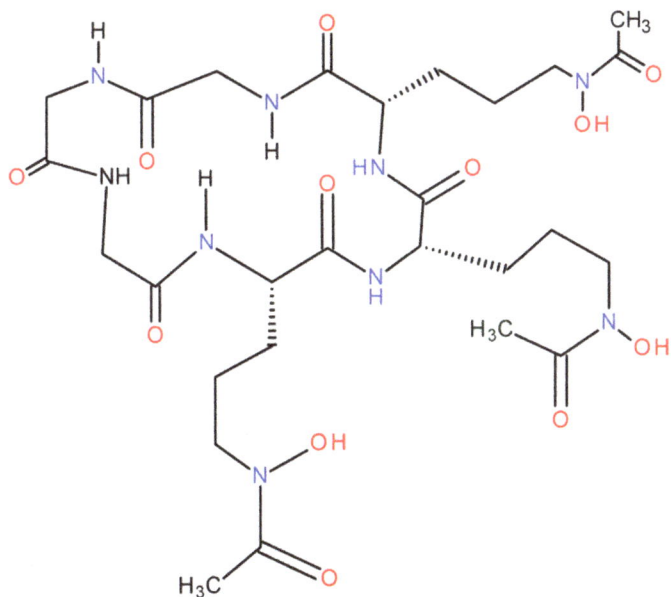

FIGURE 7.3   Structure of ferrichrome siderophore.

7.2.2.3   AEROBACTIN SIDEROPHORE

Aerobactin siderophore is an exogeneous siderophore of *Pseudomonas* (marine origin), *K. pneumonia, A. aerogenes, E. coli* and few other bacteria (Buyer et al., 1991; Sah and Singh, 2015). This siderophore has molecular formula of $C_{22}H_{36}N_4O_{13}$ (Neilands, 1995) (Fig. 7.4).

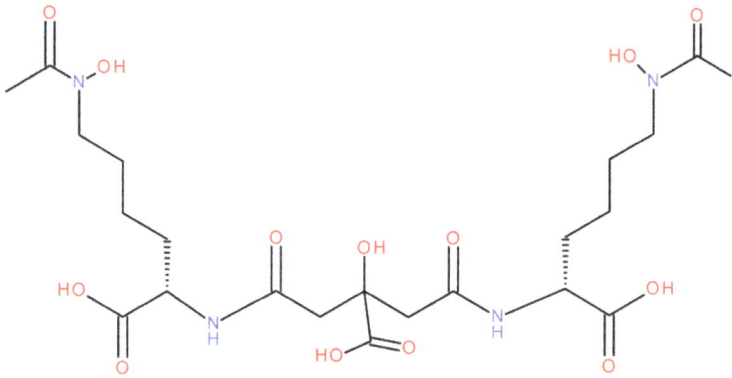

FIGURE 7.4   Structure of aerobactin siderophore.

7.2.3   CARBOXYLATE SIDEROPHORE

The novel class of siderophore whose members neither possess hydroxamate nor phenolate ligands rather iron binding is achieved by hydroxyl carboxylate and carboxylates (Schwyn and Neiland, 1987). It is found that rhizoferrin isthe polycarboxylate or complexone-type siderophore, originally isolated from *R. microsporus,* represents the common siderophore within the Zygomycetes (Thieken and Winkelmann, 1992**).** Rhizoferrin has the molecular formula, $C_{16}H_{24}N_2O_{12}$ and composed of diaminopropane symmetrically acylated with citric acid via amine bonds to terminal carboxylate of citric acid (Fig. 7.5**)** (Drechsel et al., 1992). Some bacteria can also synthesize Enantiorhizoferrin S, S-rhizoferrin, etc. (Munzinger et al., 1999). However, the best-characterized carboxylate type siderophore is rhizobactin produced by *R. meliloti* strain DM4 and is an amino poly (carboxylic acid) with ethylene diamine dicarboxyl and hydroxycarboxyl moieties as iron-chelating groups. Staphyloferrin is another member of this class produced by *S. hyicus* (Ali and Vidhale, 2013).

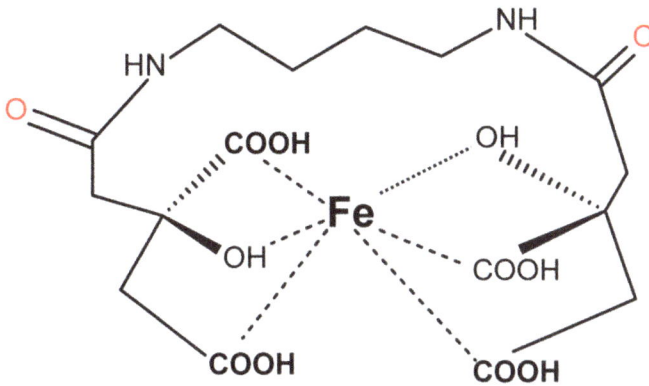

FIGURE 7.5 Structure of rhizoferrin siderophore.

7.2.4 SIDEROPHORE WITH MIXED LIGAND

There are certain siderophores having the mixed ligands of lysine, ornithine, and histamine derivatives (Sah and Singh, 2015).

7.2.4.1 LYSINE DERIVATIVE (MYOBACTIN)

These are the 2-hydroxy phenyl oxazoline having siderophore molecules for iron acquisition. *M. tuberculosis* has been reported to produce two types of siderophores with molecular formulae $C_{27}H_{37}N_5O_{10}$ (Sah and Singh, 2015). The only chemical difference between them is the nature of the N-acyl chain on hydroxylated lysine in the middle of the molecule. Myobactin siderophore isolated from Myobacteria showed the presence of two hydroxamate, a phenolate, and oxazoline nitrogen (Fig. 7.6). It is reported that *M. tuberculosis* produces only the mycobactin class of siderophore comprising salicylic acid-derived moiety. Whereas, *M. smegmatis* produces exochelin, that is, extracellular siderophore (De Voss et al., 1999; Varma and Podila, 2005; Sah and Singh, 2015)

7.2.4.2 ORNITHINE DERIVATIVE (PYOVERDINES/PSEUDOBACTINS)

Among all the bacterial siderophores, Pyoverdines are found to be unique due to the diversity between the *Pseudomonas* sp. and still the specific features are conserved among the hundreds of structures. They have molecular formula, $C_{56}H_{88}N_{18}O_{22}$ (Cezard et al., 2014). *Pseudomonas* sp. synthesizes various fluorescent chromo peptide siderophore-like pseudobactin and pyoverdines, etc. In *Pseudomonas* sp., two important siderophore-mediated iron uptake systems like one involving pyoverdin and other is pyochelin are found (Abddalah, 1991; Meyer, 2000; Meneely and Lamb, 2007). The structure of peptide differs among pseudomonads and more than 40 structures have been described, while the chromophore, (1S)-5-amino-2, 3-dihydro-8, 9-dihydroxy-1H-pyrimido [1, 2-a]quinoline-1 carboxylic acid is the same with the exception of azobactin from *A. vinelandii* having extra urea (Kloepper et al., 1980). Pyoverdines are fluorescent siderophore having dihydroxyquinoline derivative. The chromophore derived from 2, 3-diamino-6, 7-dihydroxyquinoline are linked to a peptide chain exhibiting two hydroxamate groups or one hydroxamate and one hydroxycarboxylate group (Berthold, 2011).

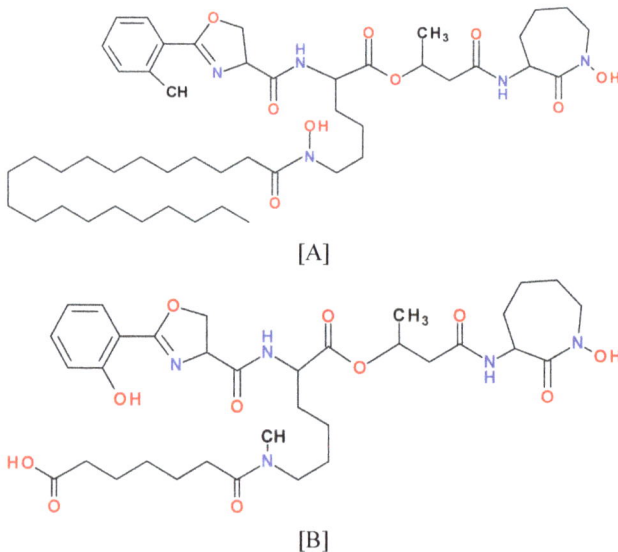

[A]

[B]

FIGURE 7.6   (A, B) Structure of myobactin siderophore.

7.2.4.3   *HISTAMINE DERIVATIVE: ANGUIBACTIN*

These siderophores are isolated from the marine pathogen, that is, *V. anguillarum* and plays a vital role in iron uptake from vertebrate tissues and can obtain iron from other siderophores and ferric hydroxide as well removes ferric ion from aqueous solutions, including cell culture media (Fig. 7.7) (Sah and Singh, 2015).

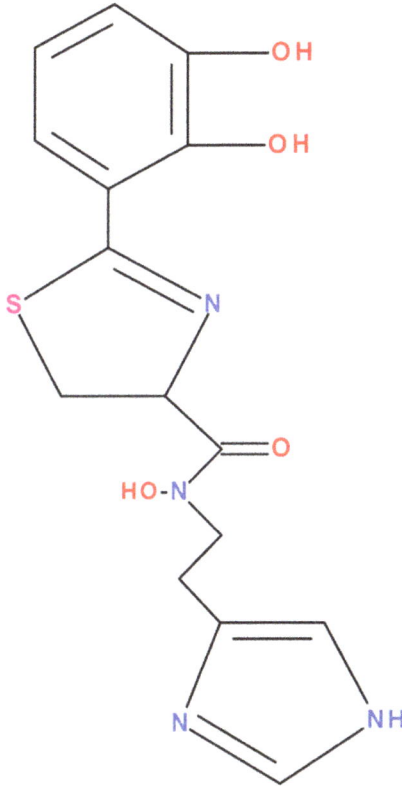

FIGURE 7.7 Structure of anguibactin siderophore.

7.3 SIDEROPHORE TRANSPORT MECHANISM

It is found that siderophores chelate the traces of iron in stable complex form that are internalized into the cell by specific cell receptors (Sah and Singh, 2015; Winkelmann, 1991; Neilands, 1982). It is observed that bacteria capture iron-loaded siderophores at cell surface and transport them into the cytosol to transport iron in the cell cytoplasm. The binding constants of siderophores for Fe^{+3} are too high, indicating that these compounds can effectively chelate Fe^{+3} from a variety of complexes found in natural environment (Sah and Singh, 2015; Stintzi et al., 2000; Bernd and Rehm, 2008). It is found that siderophores are part of a multicomponent system for transporting Fe^{+3} into a cell. Other components include a specific outer membrane receptor protein Fec A, Fep A, and TonB-ExbB-ExbD protein complex in the inner membrane, a periplasmic-binding protein, and inner membrane ATP-dependent Fec CDE-Fep CDE protein (Ali and Vidhale, 2013). Gram-positive bacteria have thick cell membrane and Gram-negative bacteria possess two layers, that is, outer and inner membrane. The porines present on the outer membrane permits siderophore (below 500 Da) to enter this barrier (Nikaido, 2003). Although, siderophores with higher molecular weight have to be absorbed via specific receptor. This absorption system consists of outer membrane receptor, a periplasmic-binding protein and cytoplasmic membrane protein belonging to ATP-binding cassette transporter (ABC-transporter) system. As soon as the siderophores reach the cytoplasm, Fe^{+3} carried by the siderophores get reduced, and iron-free siderophore gets degraded or recycled by releasing through an efflux pump system. In the case of Gram-negative bacteria, the transport system requires a combination of protein pattern, that is, outer membrane receptor, which specifically binds the ferric–siderophore complex and transfer it into the periplasm (Sah and Singh, 2015). The protein complex containing TonB transduces energy from the proton motive force into transport-proficient receptor. The binding protein located in the periplasm transfers the siderophore bound iron into cytoplasmic-membrane-associated transporter (Sah and Singh, 2015). It is reported that once the siderophore is released into the periplasm, it is rapidly bounded by specific periplasmic-binding protein FhvA (hydroxamate siderophores) (Coulton et al., 1986), FepB (Enterobactin), and FecB (ferric dicitrate) (Pressler et al., 1988).

7.4 SIDEROPHORE PRODUCTION

7.4.1 DEFERRATED MEDIA

Preparation of Mayer and Abdullah (MA) media is to be carried out by using monopotassium phosphate (6 gm), dipotassium phosphate (3 gm), ammonium sulfate (1 gm), magnesium sulfate (0.2 gm), and succinic acid (4 gm) in 1000 mL. D/W. Media is to be deferrated by the addition of 8– hydroxyquinoline (2%) dissolved in chloroform (5 mL/L). Phase separation is to be carried out by shaking the media vigorously in the separating funnel. After the phase separation, chloroform layer should be removed followed by repeatedly washing the media with chloroform to ensure the complete removal of iron complexes and any residues of 8-hydroxyquinoline, which could inhibit the growth (Khandelwal et al., 2002; Manwar et al., 2004; Chincholkar et al., 2000; Handore et al., 2019).

7.4.4.1 PREPARATION OF FERMENTATION MEDIUM FOR SIDEROPHORE PRODUCTION

Bacterial culture is to be grown in the nutrient broth and inoculated in autoclaved iron-deficient succinate medium followed by incubation at 28°C on a rotary shaker (110 rpm).

7.4.4.2 PREPARATION OF CAS (CHROME AZUROL SULPHONATE) REAGENT

- **Preparation of CAS Stock:** 2 mM CAS stock is to be prepared by the addition of 0.121 gm CAS in 100 mL DW.
- **Preparation of Ferrous Stock Solution:** (1 mM ferrous stock is to be prepared by the addition of $FeCl_3.6H_2O$ (0.027 g) into 100 mL of HCl (10 mM HCl).
- **Preparation of Piperazine Buffer:** Piperazine (4.307 g) is to be dissolved in 30 mL D/W and pH is to be adjusted to 5.6.
- **Preparation of HDTMA Solution:** HDTMA (hexadecyltrimethyl-ammonium) (0.0219 g) is to be dissolved in 50 mL D/W.
- **CAS Reagent Preparation:** In the CAS solution (7.5 mL), freshly prepared ferrous solution (1.5 ml) is to be slowly added and properly mixed. Then the mixture is slowly added to HDTMA (30 mL) in the mixing cylinder (gentle stirring should be carried out so that there should be no foaming). Then, piperazine solution is to be slowly added to make volume up to 100 mL with double D/W. The prepared reagent should be stored in fridge by covering with black paper (Khandelwal et al., 2002; Manwar et al., 2004; Chincholkar et al., 2000; Handore et al., 2019).

7.5   SIDEROPHORE DETECTION

Traditionally, CAS assay uses test tubes to carry out qualitative and quantitative detection of siderophores. Microtiter plate with 96 wells is to be used for modified Microplate method.

7.5.1   *QUALITATIVE DETECTION OF SIDEROPHORES*

7.5.1.1   *CHEMICAL ASSAYS*

Fermented broth of bacteria is to be centrifuged (3500 rpm, 15 min) and after 3–4 days, the cell-free supernatant is subjected to CAS assay for siderophore. Detection of siderophores can be carried w.r.t. Blue colored CAS reagent. Development of orange or pink coloration indicates the production of siderophore (Khandelwal et al., 2002; Manwar et al., 2004; Chincholkar et al., 2000; Handore et al., 2019)

7.5.1.2   *BIOASSAY BY CAS AGAR PLATE METHOD*

This assay can be performed by modified method of Hu and Xu (2011). CAS agar plates should be prepared by mixing 100 mL CAS reagent in 900 mL sterilized LB/NB agar medium. Inoculation of selected bacterial strains is to be carried out on CAS agar plate, the uninoculated plate is considered as control. After inoculation, plates are incubated for 1 week at 28°C. Formation of pink/orange zone around the bacterial colonies indicates siderophore production by the test organisms (Louden et al., 2011; Sanaz Rajabi et al., 2020).

7.5.2   *QUANTITATIVE ESTIMATION IF SIDEROPHORES (SCHWYN AND NEILANDS, 1987)*

Quantitative estimation of siderophore by test bacterial strain can be carried out by both traditional method and modified microplate method. In traditional method, quantitative estimation of siderophore is to be carried out by taking cell-free supernatant of bacterial cultures in test tubes. The percent siderophore unit is to be determined by the addition of cell-free supernatant to CAS solution (1:1) followed by 1 h incubation of mixture at 28°C and optical density of each sample is to be recorded at 630 nm (Arora and Verma, 2017). In the case of modified method, estimation of siderophore production is done using microtiter plate. Maximum 96 samples of cell-free culture can be added in separate wells of microplate followed by the addition of CAS solution (1:1). Mixture is allowed to incubate at 28°C for 1 h. Using the microplate reader, optical density of each sample in the well of microplate is recorded at 630 nm (Arora and Verma, 2017). Percentage of siderophore units can be calculated using the following formula,

$$\% \text{ Siderophore Unit} = [A_r - A_s/A_r] \times 100,$$

where A_r is Absorbance of reference (Uninoculated media with CAS reagent) (Handore et al., 2019).

In the modified method, siderophore production by several samples (Max. 96 nos) can be estimated both qualitatively and quantitatively at one go, saving time, chemicals, making it less tedious, and cheaper in comparison to traditional method. As a result, modified microplate (96 well) method is found to be comparatively rapid and is an efficient method.

7.6 EXTRACTION AND PURIFICATION OF SIDEROPHORE

Cell-free supernatant is to be acidified to pH 3.0 with 12 M HCl and extracted first with 1/5 volumes of ethyl acetate (three times) followed by phenol: chloroform (1:1) extraction. Ethyl acetate layer should be collected, combined and dried using rotary evaporator (Ahire et al., 2011). The extract is to be re-suspended in deionized water (1 mg/mL). Similarly, aqueous layer should be concentrated (five times) and evaporated to dryness. The deferrated extract can be further fractionated through Sep-Pak C_{18} cartridge and the supernatant can be loaded, followed by washing with small quantity of D/W. Two fractions should be collected. (First fraction is recoverable from filtrate and the second fraction is eluted with acetonitrile. Then, the CAS-reactive acetonitrile fraction can be evaporated to dryness and checked. Siderophore purification can be carried out using ion exchange resins. For this purpose, XAD-2 column is to be used. The supernatant should be slowly passed through the column. The resulting two bands are to be eluted with (1:1, v/v) methanol: water that can yield two fractions of pure desferric siderophores that can be used for further identification and structural elucidation (Sayyed, 2006; Budzikiewicz, 1993).

7.7 CHARACTERIZATION AND IDENTIFICATION OF SIDEROPHORE

Identification of the functional groups of crystalline siderophores can be carried out w.r.t. the standard siderophores by FTIR analysis. Crystal formation is accelerated using saturated $FeSO_4$ solution to cell-free culture and pH is adjusted to 3.0 using H_2SO_4. Deproteinization is carried out by using ammonium sulfate solution (50%) to concentrate aqueous phase under vacuum and then set aside in the cold for crystallization. Later, the filtrate is neutralized, reduced to dryness and extracted in dry hot methanol. The crystals can then be separated using Whatman filter paper (Tank et al. 2012). It is proposed to use HPLC−electrospray ionization mass spectrometry for the separation and detection of siderophores (Cormack et al., 2003). Structural characterization of siderophores can also be carried out by LC-MS/MS (Baars and Perlman, 2016). The ferric and non-ferric forms of siderophores can be observed under the electrospray ionization (ESI) or matrix-assisted laser desorption/ionization (MALDI) mass spectra. The iron signature can be inferred even from low resolution mass spectrum due to stable isotope profile (Tomas Pluhacek et al., 2014). Furthermore, NMR spectrometer is to be proposed to get the molecular architecture of siderophore by observing and measuring the interaction of nuclear spins when placed under magnetic field. The ^1H NMR spectra of purified siderophore can be collected on NMR spectrometer equipped with triplet resonance probe and triple axis gradients. XAD-purified siderophore sample is to be dissolved in 0.75 mL deuterated DMSO and injected through 5-mm one-dimensional NMR tube to record the nuclear magnetic resonance, with solvent signal as internal reference (Patel AK et al., 2009; Murugappan R et al., 2011).

7.8 APPLICATIONS OF BACTERIAL SIDEROPHORES

7.8.1 PLANT GROWTH PROMOTER

Bacterial siderophore plays an important role in agrobiotech. They can promote the plant growth by direct or indirect mechanism. Several Gram-positive and Gram-negative bacteria show significant ability to produce a variety of siderophores (Tomas Pluhacek et al., 2016; Ahemad and Khan, 2010a, 2010b). *Pseudomonas* species can boost the plant growth by producing pyoverdine siderophores (Kloepper et al., 1980; Gamalero and Glick, 2011). It has been proved that *E. coli* from endorhizosphere of sugarcane (*Saccharum* sp.) and rye grass (*Lolium perenne*) exhibited plant growth because of maximum siderophore production (Gangwar and Kaur, 2009). Siderophores of endophytic *Streptomyces* sp. isolated from the roots of Thai jasmine rice have been reported to enhance plant growth w.r.t. biomass and increased lengths of roots and shoots (Rungin et al., 2012).

7.8.2 BIOCONTROL AGENT

It is proved that siderophores can act as biocontrol agent. It is reported that siderophores can tightly bind to iron and reduce the bioavailable iron for phytopathogens (Badrul Haider et al., 2018; Beneduzi et al., 2012; Ahmed and Holmstrom, 2014). Such siderophores can inhibit various phytopathogenic fungi, viz. *P. parasitica* (Seuk et al., 1988), *P. ultimum* (Hamdan et al., 1991), *F. oxysporum veri dianthi* (Buysens et al., 1996) and *S. sclerotiorum* (McLoughlin et al., 1992). It is found that various siderophore-producing bacteria isolated from the intestinal tracts of fishes inhibit fish pathogens, such as *A. logei, V. ichthyoenteri, V. anguillarum, V. Splendidus,* and *A. salmonicida* (Sugita et al., 2012). Different strains of *P. fluorescens* are known to produce remarkable production of sidero-phore (Kumar V. et al., 2017; Kloepper et al., 1980). It is documented that *P. fluorescens* has been used as probiotics in fish farming due to its significant potential of siderophore production with noteworthy ability to inhibit the growth of fish pathogens, such as *V. anguillarum, V. ordalii, A. salmonicida, L. garvieae, S. iniae, F. Psychrophilum,* and *C. ruckeri*) (Gram et al., 2001; Brunt et al., 2007; Dimitroglou et al., 2011). Plant pathogens of potato, peanuts, and maize can be strongly controlled by pyoverdine siderophores produced by *Pseudomonads* (Schippers et al., 1987; Pal et al., 2001). It is reported that siderophores produced by *A. indica* chelates Fe^{+3} from soil and suppresses the growth of several fungal pathogens (Verma et al., 2011).

7.8.3 BIOREMEDIATION OF ENVIRONMENTAL POLLUTANTS

Heavy metal contamination to water and soil causes negative impacts on ecosystem. However, siderophores show ability to bioremediate metals by solubilizing and increasing the mobility of a wide range of metals, such as Cd, Cu, Ni, Pb, Zn, and the actinides Th (IV), U(IV), and Pu(IV). Their capability depends on the functionality of ligand w.r.t. selectivity or affinity for specific metals other than "Fe" w.r.t. Stability constant of metal–sidero-phore complex (Katarzyna et al., 2018; Schalk et al., 2011; Hernlem et al., 1999). Pyochelin siderophore produced by *P. aeruginosa* can chelate a wide range of metals like, Ag^+, Al^{3+}, Cd^{2+}, Co^{2+}, Cr^{2+}, Cu^{2+}, Eu^{3+}, Ga^{3+}, Hg^{2+}, Mn^{2+}, Ni^{2+}, Pb^{2+}, Sn^{2+}, Tb^{3+}, Ti^+ and Zn^{2+} Etc. The uptake process found to assimilate Fe^{3+} as compared to other metals (Braud et al., 2009a). There are a number of metals like Fe, Ni, and Co etc. which have been mobilized from waste material, that is, acid-leached ore of Uranium mine produced by the siderophores of *P. fluorescens* (Edberg et al., 2010) It is reported that pyoverdines mobilized U (VI), Np(V) and other metals from Uranium mine waste (Behrends et al., 2012; Ahmed and Holmstrom, 2014). Moreover, *A. radiobacter* exhibited noticeable ability to remove approximately 54% of Arsenic from metal contaminated soil (Wang et al., 2011).

7.8.4   BIODEGRADATION OF PETROLEUM HYDROCARBONS

Oil spill is the major environmental issue exhibiting negative impact on biodiversity closest to spill area. It is reported that siderophores application is one of the effective remedy to clean up the oil spill. Microbial siderophores play vital role in the biodegradation of petroleum hydrocarbons through indirect mechanism, by facilitating the "Fe" acquisition for degraded microorganisms under iron-stress conditions. Generally, petrobactin sulfonate, isolated from oil-degrading marine bacterium, *M. Hydrocarbonclasticus* are the first structurally characterized siderophore (Barbeau et al., 2002; Hickford S et al., 2004).

7.8.5   BIOGEOCHEMICAL CYCLING OF IRON IN WATER ECOSYSTEM

Iron is one of the essential trace micronutrient for marine organisms. Even with its low concentration in the ocean, it controls the productivity and community structure of phytoplankton (Gledhill and Buck, 2012). It is found that there is competition between the marine bacteria and phytoplankton for "Fe" by producing different types of siderophores which plays a significant role in biogeochemical cycling of "Fe" in the ocean (Boyd et al., 2007;

Cordero et al., 2012). In the surface water, siderophores contributed to photochemical cycling of 'Fe' by forming Fe^{+3} siderophore complexes that increase the availability of 'Fe' for the phytoplanktons (Barbeau et al., 2001; Hunter and Boyd, 2007; Ahmed and Holmström, 2014).

7.8.6 BIOSENSOR

Biosensors are the integrated devices generating specific quantitative or semiquantitative analytical information by means of biological recognition element attached to a transducer. The siderophore-based biosensor are cheap, user-friendly, ecofriendly, and comparatively simple device (Thevenot et al., 1999; Gupta et al., 2008). The biosensors are selective for Fe^{+3} and exhibited the ability to detect and analyze samples ranging from 3 to 10 ng/mL either in solution or in the immobilized form, respectively. Due to its noteworthy stability, it can be used for at least 3 months or over 1000 determinations. Besides, it can be utilized for the determination of iron in various water samples (Barrero et al., 1993). Normally, Pyoverdines are considered as a promising agent for the construction of optical biosensors due to their ability to form a strong complex with Fe^{+3} and weak or negligible affinity towards Fe^{+2} (Pesce and Kaplan, 1990; Barrero et al., 1993; Kurtz and Crouch, 1991).

7.8.7 TROJAN HORSE ANTIBIOTICS

Most of the MDR (multidrug-resistant) are found to be resistant to numerous antibiotics, exerting the challenge to modern medicine w.r.t. discovery of new antibiotics or strategies for combating the MDR, especially Gram-negative bacteria. Application of bactericide compounds by means of siderophores can be a promising strategy to enhance the drug efficacy (Isabelle et al., 2018). It is proved that siderophore shows ability to mediate selective delivery of antibiotics to antibiotic-resistant bacteria by means of Trojan horse strategy. This technique is based on the utilization of iron transport potential of siderophores for transporting the drugs into cells by forming the conjugates with antimicrobial agents (Huang et al., 2013). Such drug conjugate normally consists of a linker for attaching drug to siderophore that allows controlled, chemical or enzymatic release of drugs in the target bacterial cell. After reaching into the cell, the siderophore–drug conjugate destroys the cell by releasing the drug or by acting as an intact antibacterial agent or by blocking the further iron acquisition (Nagoba and Vedpathak, 2011). Bacterial species can produce and use specific siderophores like aerobactin, yersiniabactin, enterobactin, and salmochelin for *K. pneumonia*; enterobactin, salmochelin, aerobactin, and yersiniabactin for *E. coli*; and acinetobactin, fimsbactin, and baumannoferrins for *A. Baumanii*, etc. (Isabelle et al., 2018). In this context, there is possibility of selectively targeting bacterial species using appropriately selected siderophores for vectorizing the antibiotic. Such narrow selectivity reduces the risk of antibiotic resistance. Although, there are number of siderophore–antibiotic conjugates, catechol or hydroxamate siderophore analogues are found to be effective (Ghosh et al., 2017; Isabelle et al., 2018).

7.8.8 AGAINST INFECTION

Iron is metabolically selected by components of human body leading unavailability of free iron for pathogenic bacteria. Most aerobic, faculta-tive anaerobic, and saprophytic microorganism exhibited the ability to synthesize siderophores and contributed to bacterial virulence (Litwin and Calderwood, 1993; Payne, 1993; Wooldridge and Williams, 1993; Ali and Vidhale, 2013). Some siderophores possess antimalarial activity against *P. falciparum* (Tsafack et al., 1996), for example, siderophore produced by *K. pneumoniae* (Gysin et al., 1991) and siderophore desferrioxamine B, produced by *S. pilosus,* have antimalarial activity against *P. falciparum* (Nagoba and Vedpathak, 2011). It is reported that desferrioxamine B enters inside the parasite and resulted into iron sequestration. It is found to conjugate with methyl anthranilic acid and exhibited 10-fold greater in vitro activity against *P. falciparum*, which can be further enhanced by using nalidixic acid as a conjugate (Gysin et al., 1991; Loyevsky et al., 1993, 1999).

7.8.9 CANCER THERAPY

Iron is a vital nutrient for cells under the iron-deficient conditions, cells are unable to transit from G1-phase to S-phase in cell cycle. Therefore, iron depletion by iron-chelating agents may deprive the rapidly dividing malignant cells of DNA precursors required for replication and resulted into inhibition of their proliferation (Richardson, 2002; Cheng et al., 2012). Desferrioxamines have been reported to reduce the growth of aggressive tumors significantly (Neuroblastoma or Leukemia; Buss et al., 2003; Lovejoy and Richardson, 2003). Besides, desferioxamine E produced by *Actinobacterium* was reported to reduce the viability of malignant melanoma cells significantly (Blatt et al., 1987; Blatt and Stitely, 1988; Nakouti et al., 2013). Normally, enterobactin, that is, catecholate-type siderophore is found to exhibit anticancer potential as it could hinder the generation of reactive oxygen species in the mitochondria and disrupts its function, promoting the cancer cell apoptosis (Piu Saha et al., 2019).

7.9 CONCLUSION

In earth's crust and environment, there is remarkable abundance of iron. However, its ferric form is insoluble and inaccessible at physiological pH (7.35–7.40). In order to survive under such iron stress environment, most of the microorganisms synthesize siderophores, which show the ability to chelate iron and convert it into its soluble complexes and making it available to microbial cells. Among the siderophore-producing microorganisms, amazing structural variations have been observed in bacterial siderophores. Siderophores have a wide variety of applications in various fields. They can efficiently act as plant growth promoter, biocontrol agent, biosensor, and biomedicine. They play an important role in bioremediation as well as exhibit capability to biodegrade petroleum hydrocarbons. They can be used for selective drug delivery by Trojan horse strategy in biomedicine. Moreover, these compounds can be exploited in the treatment of hemochromatosis, sickle-cell anemia, thalassemia, dialysis encephalopathy, cancer, malaria, removal of transuranic elements, such as aluminum and vanadium, etc. The chapter presents the overview of bacterial siderophores, classification, transport mechanism, and their production method. It provides qualitative and quantitative detection methods for different types of siderophores, such as chemical assay, bioassay by traditional as well as modified methods. It also proposed method of crystallization of siderophores, their identification and characterization by FTIR, LC-MS, ESI /MALDI Mass spectra and NMR, etc. This chapter has illustrated elaborative applications of bacterial siderophores for sustainable future.

ACKNOWLEDGMENT

The authors are grateful to Prin. V. N. Suryavanshi and P. G. Department of Microbiology, H. P. T. Arts and R. Y. K. Science College, Nashik, India and Sigma Wineries Pvt. Ltd. Nashik for valuable assistance. Mr. V. C. Handore and Mrs. Hira V. Handore for consistent support.

KEYWORDS

- **siderophores**
- **iron deficiency**
- **hydroxamate**
- **catecholate**
- **carboxylate**
- **cost effective**

REFERENCES

Abdallah, M. A. Pyoverdines and Pseudobactins. In *CRC Handbook of Microbial Iron Chelates*; Winkelmann, G., Ed.; Boca Raton, FL: CRC Press, 1991; pp 139–152.

Ahemad, M.; Khan, M. S. Comparative Toxicity of Selected Insecticides to Pea Plants and Growth Promotion in Response to Insecticide-Tolerant and Plant Growth Promoting Rhizobium Leguminosarum. *Crop Prot.* **2010a**, *29*, 325–329.

Ahemad, M.; Khan, M. S. Phosphatesolubilizing and Plant-Growth Promoting Pseudomonas Aeruginosa PS1 Improves Green Gram Performance in Quizalafop-p-Ethyl and Clodinafop Amended Soil. *Arch. Environ. Contam. Toxicol.* **2010b**, *58*, 361–372.

Ahire, J. J.; Patil, K. P.; Chaudhari, B. L.; Chincholkar, S. B. Bacillus spp. of Human Origin: A Potential Siderophoregenic Probiotic Bacteria. *Appl. Biochem. Biotechnol.*, **2011**, *164*, 386–400. https://doi.org/10.1007/s12010-010-9142-6.

Ahmed, E.; Holmström, S. J. Siderophores in Environmental Research: Roles and Applications. *Appl. Microbiol., Microbial Biotechnol.* **2014**, *7*, 196–208.

Ali, S. S.; Vidhale, N. N. Bacterial Siderophore and Their Application: A Review. *Int. J. Curr. Microbiol. Appl. Sci.* **2013**, *2* (12), 303–331.

Andrea, T. G.; Winkelmann, G. Rhizoferrin: A Complexone Type Siderophore of the Mocorales and Entomophthorales (Zygomycetes). *FEMS Microbiol. Lett.* **1992**, *94* (1), 37–41.

Arnow, L. E. Colorimetric Determination of the Components of 3, 4-dihydroxyphenylalanine-tyrosine Mixtures. *J. Biol. Chem.* **1937**, *118*, 531–537.

Arora, N. K. Verma, M. Modified Microplate Method for Rapid and Efficient Estimation of Siderophore Produced by Bacteria. *3 Biotech.* **2017,** *7,* 380–381.

Badrul, H.; Ferdous, M.; Babry, F.; Ahlan, S.; Mohammad, R. I.; Khan, H. Population Diversity of Bacterial Endophytes from Jute (Corchorus Olitorius) and Evaluation of Their Potential Role as Bioinoculants. *Microbiol. Res.* **2018,** *208,* 43–53. doi: 10.1016/j. micres. 2018.01.008.

Ballas, S. K.; Zeidan, A. M.; Duong, V. H.; DeVeaux, M.; Heeney, M. M. The Effect of Iron Chelation Therapy on Overall Survival in Sickle Cell Disease and β-thalassemia: A Systematic Review. *Am. J. Hematol.* **2018,** *93* (7), 943–952.

Barbeau, K.; Rue, E. L.; Bruland, K. W.; Butler, A. Photochemical Cycling of Iron in the Surface Ocean Mediated by Microbial Iron (II)-Binding Ligands. *Nature* **2001,** *413,* 409–413.

Barbeau, K.; Zhang, G. P.; Live, D. H.; Butler, A. Petrobactin, a Photoreactive Siderophore Produced by the Oil-Degrading Marine Bacterium Marinobacter Hydrocarbonoclasticus. *J. Am. Chem. Soc.* **2002,** *124,* 378–379.

Barrero, J. M.; Moreno Bondi, M. C.; Pérez Conde, M. C.; Cámara, C. A Biosensor for Ferric Ion. *Talanta* **1993,** *40,* 1619–1623.

Behrends, T.; Krawczyk-Bärsch, E.; Arnold, T. Implementation of Microbial Processes in the Performance Assessment of Spent Nuclear Fuel Repositories. *Appl. Geochem.* **2012,** *27,* 453–462.

Beneduzi, A.; Ambrosini, A.; Passaglia, L. M. Plant Growth-Promoting Rhizobacteria (PGPR): Their Potential as Antagonists and Biocontrol Agents. *Genet. Mol. Biol.* **2012,** *35,* 1044–1051.

Bernd, H.; Rehm, A. Biotechnological Relevance of Pseudomonads. In *Pseudomonas. Model Organism, Pathogen, Cell Factory*; Bernd, H.; Rehm, A., Eds.; Wiley-Vch Verlag Gmbh And Co. Kgaa: Weinheim, Germany, 2008; p 377.

Berthold, F. M. Iron Transport: Siderophores. *Encyclopedia Inorg. Chem.,* **2011.** https:// doi. org/10.1002/9781119951438. eibc0110.

Blatt, J.; Stitely, S. Antineuroblastoma Activity of Desferoxamine in Human Cell Lines. *Cancer Res.* **1987,** *47,* 1749–1750.

Blatt, J.; Taylor, S. R.; Stitely, S. Mechanism of Antineuroblastoma Activity of Deferoxamine in Vitro. *J. Lab. Clin. Med.* **1988,** *112,* 433–436.

Bou-Abdallah, F. The Iron Redox and Hydrolysis Chemistry of the Ferritins. *Biochim. Biophys. Acta Gen. Subj.* **2010,** *1800* (8), 719–731.

Boyd, P. W.; Jickells, T.; Law, C. S.; Blain, S.; Boyle, E. A.; Buesseler, K. OMesoscale Iron Enrichment Experiments 1993–2005: Synthesis and Future Directions. *Science* **2007,** *315,* 612–617.

Braud, A.; Jézéquel, K.; Bazot, S.; Lebeau, T. Enhanced Phytoextraction of an Agricultural Cr-and Pb-Contaminated Soil by Bioaugmentation with Siderophore-Producing Bacteria. *Chemosphere.* **2009a,** *74,* 280–286.

Brunt, J.; Newaj-Fyzul, A.; Austin, B. The Development of Probiotics for the Control of Multiple Bacterial Diseases of Rainbow Trout, Oncorhynchus Mykiss (Walbaum). *J Fish Dis.* **2007,** *30,* 573–579.

Budzikiewicz, H. Secondary Metabolites from Fluorescent Pseudomonads. *FEMS Microbiol. Rev.* **1993,** *204,* 209-228.

Buss, J. L.; Torti, F. M.; Torti, S. V. The Role of Iron Chelation in Cancer Therapy. _Curr. Med. Chem._ **2003**, _10_, 1021–1034.

Buyer, J. S.; Lorenzo, V. D.; Neilands, J. B. Production of the siderophore aerobactin by a halophilic Pseudomonad. In Applied and Enviromental Microbiology, **1991**, 57(8), 2246–2250.

Buysens, S.; Heungens, K.; Poppe, J.; Hofte, M. Involvement of Pyochelin and pioverdin in suppression of Pseudomonas aeruginosa 7NSK2. Applied and Environmental Microbiology. **1996**, 62(3), 865- 871.

Cai, Y.; Wang, R.; An, M. M.; Bei-Bei, L. Iron-depletion prevents biofilm formation in Pseudomonas aeruginosa through twitching motility and quorum sensing. Braz J Microbiol. **2010**, 41(1), 37–41.

Cézard, C.; Farvacques, N.; Sonnet, P. Chemistry and Biology of Pyoverdines, Pseudomonas Primary Siderophores. _Curr. Med. Chem._ **2014**, _21_ (1), 1–22.

Chincholkar, S. B.; Chaudhari, B. L.; Talegaonkar, S. K.; Kothari, R. M. Microbial Iron Chelators: A Sustainable Tool for the Biocontrol of Plant Diseases. In _Biocontrol Potential and Its Exploitation in Sustainable Agriculture_; Upadhyay, R. K.; Mukerji, K. G.; Chamola, P. C., Eds.; Kluwer Academic/Plenum Publishers: USA **2000**, _1_, 49–70.

Cordero, O. X.; Ventouras, L. A.; DeLong, E. F.; Polz, M. F. Public Good Dynamics Drive Evolution of Iron Acquisition Strategies in Natural Bacterioplankton Populations. _Proc. Natl. Acad. Sci. USA_ **2012**, _109_, 20059–20064.

Cotton, J. L.; Tao, J.; Carl, J.; Balibar, C. J. Identification and Characterization of the Gene Cluster Coding for Staphyloferrin A. _Biochemistry._ **2009**, _48_(5), 1025–1035.

Coulton, J. W.; Mason, P.; Cameron, D. R.; Carmel, G.; Jean, R.; Rode, H. N. Protein fusions of Beta-Galactosidase to the Ferrichrome-Iron Receptor of Escherichia Coli K-12. _J. Bacteriol._ **1986**, _165_ (1), 181–192.

De Voss, J. J.; Rutter, K.; Schroeder, B. G.; Barry, Iii. Iron Acquisition and Metabolism by Mycobacteria. _J. Bacteriol._ **1999**, _181_, 4443–4451.

Dimitroglou, A.; Merrifield, D. L.; Carnevali, O.; Picchietti, S.; Avella, M.; Daniels, C. Microbial Manipulations to Improve Fish Health and Production–A Mediterranean Perspective. _Fish Shellfish Immunol._ **2011**, _30_, 1–16.

Drechsel, H., Jung, G. Winkelmann. G. Stereochemical Characterization of Rhizoferrin and Identification of Its Dehydration Products. _Bio-Metals._ **1992**, _5_, 141–148.

Edberg, F.; Kalinowski, B. E.; Holmström, S. J.; Holm, K. Mobilization of Metals from Uranium Mine Waste: The Role of Pyoverdines Produced by Pseudomonas Fluorescens. _Geobiology_ **2010**, _8_, 278–292.

Gamalero, E.; Glick, B. R. Mechanisms Used by Plant Growth-Promoting Bacteria. In _Bacteria in Agrobiology_: _Plant Nutrient Management_; Maheshwari, D. K., Ed.; Springer Verlag; Berlin, Heidelberg, Germany, 2011; pp 17–46.

Gangwar, M.; Kaur, G. Isolation and Characterization of Endophytic Bacteria from Endorhizosphere of Sugarcane and Ryegrass. _Internet J. Microbiol._ **2009**, _7_, 139–144.

Ghosh, M.; Miller, P. A.; Mollmann, U.; Claypool, W. D.; Schroeder, V. A.; Wolter, W. R.; Suckow, M.; Yu, H.; Li, S.; Huang, W.; Zajicek, J.; Miller, M. J. Targeted Antibiotic Delivery: Selective Siderophore Conjugation with Daptomycin Confers Potent Activity against Multidrug Resistant Acinetobacter baumannii Both in Vitro and in Vivo. _J. Med. Chem._ **2017**, _60_ (11), 4577–4583.

Gledhill, M.; Buck, K. N. The Organic Complexation of Iron in the Marine Environment: A Review. *Front Microbiol.* **2012**, *3*, 1–17.

Glick, R.; Gilmour, C.; Tremblay, J.; Satanower, S.; Avidan, O.; Deziel, E.; Greenberg, E. P.; Poole, K, Banin, E. Increase in Rhamnolipid Synthesis under Iron-Limiting Conditions Influences Surface Motility and Biofilm Formation in Pseudomonas Aeruginosa. *J. Bacteriol.* **2010**, *192* (12), 2973–2980.

Gram, L.; Lovold, T.; Nielsen, J.; Melchiorsen, J.; Spanggaard, B. In Vitro Antagonism of the Probiont Pseudomonas Fluorescens Strain AH2 against Aeromonas Salmonicida Does Not Confer Protection of Salmon against Furunculosis. *Aquaculture* **2001**, *199*, 1–11.

Gupta, V.; Saharan, K.; Kumar, L.; Gupta, R.; Sahai, V.; Mittal, A. Spectrophotometric Ferric Ion Biosensor from Pseudomonas Fluorescens Culture. *Biotechnol Bioeng.* **2008**, *100* (2), 284–296.

Gysin, J.; Crenn, Y.; Pereira, Da.; Silva, L.; Breton, C. Siderophores as Anti Parasitic Agents. US Patent (US 5192807 A) 5, 192–807, March 9, 1993.

Hamdan, H.; Weller, D.; Thomashow, L. Relative Importance of Fluorescenssiderophores and Other Factors in Biological Control of Gaeumannomycesgraminis var. Tritici by Pseudomonas Fluorescens 2-79 and M4-80R. *Appl. Environ. Microbiol.* **1991**, *57* (11), 3270–3277.

Handore, A. V.; Khandelwal, S. R.; Bholay, A. D.; Ex-situ Conservation of Siderophore Producing Endophytes: The Way Towards Sustainable Agriculture. *Int. J. Res. Analyt. Rev.* **2019**, 152–158.

Haselwandter, K. *Identification and Characterization of Siderophores of Mycorrhizal Fungi, in Mycorrhza Manual*; Varma, A., Ed.;Springer Verlag: USA, 1998; p 243.

Hernlem, B. J.; Vane, L. M.; Sayles, G. D. The Application of Siderophores for Metal Recovery and Waste Remediation: Examination of Correlations for Prediction of Metal Affinities. *Water Res.* **1999**, *33*, 951–960.

Hickford, S. J.; Küpper, F. C.; Zhang, G.; Carrano, C. J.; Blunt, J. W.; Butler, A. Petrobactin Sulfonate, a New Siderophore Produced by the Marine Bacterium Marinobacter Hydrocarbonoclasticus. *J. Nat. Prod.* **2004**, *67*, 1897–1899.

Hu, Q. P.; Xu J. G. A Simple Double-Layered Chrome Azurol S Agar (SDCASA) Plate Assay to Optimize the Production of Siderophores by a Potential Biocontrol Agent Bacillus. *Afr. J. Microbiol. Res.* **2011**, *5*, 4321–4327.

Huang, Y.; Jiang, Y.; Wang, H.; Wang, J.; Shin, M. C.; Byun, Y.; He, H.; Liang, Y.; Yang, V. C. Curb Challenges of the "Trojan Horse" Approach: Smart Strategies in Achieving Effective Yet Safe Cell-Penetrating Peptide-Based Drug Delivery. *Adv. Drug Deliv. Rev.* **2013**, *65* (10), 1299–1315.

Hunter, K. A.; Boyd, P. W. Iron-Binding Ligands and Their Role in the Ocean Biogeo-chemistry of Iron. *Environ. Chem.* **2007**, *4*, 221–232.

Isabelle, J. S.; Gaëtan, L.; Mislin, A. Bacterial Iron Uptake Pathways: Gates for the Import of Bactericide Compounds. *J. Med. Chem.* **2017**, *60*, 4573–4576.

Katarzyna, H.; Michal, Z.; Tomasz, K.; Christel, B.; Boguslaw, B. Efficiency of Microbially Assisted Phytoremediation of Heavy-Metal-Contaminated Soils. *Environ. Rev.* **2018**, *26*, 316–332. https://doi. org/10.1139/er-2018-0023.

Khandelwal, S. R.; Manwar, A. V.; Chaudhari, B. L.; Chincholkar, S. B. Dynamic Activity of Siderophoregenic Bradyrhizobium Japonicum NCIM 2746 in Growth Enhancement and Yield Improvement of Soybean. *Front. Microbial Biotechnol. Plant Pathol.*; Manoharachary, C.; Purohit, D. K.; Ram Reddy, S.; Singara Charya, M. A.; Girishan, S., Eds.; Scientific Publishers: Jodhpur, 2002; pp 215–220.

Kloepper, J. W.; Leong, J.; Teintze, M.; Schiroth, M. N. Enhanced Plant Growth by Siderophores Produced by Plant Growth Promoting Rhizobacteria. *Nature* **1980**, *286*, 885–886.

Kumar, V., et al., Eds. *Probiotics and Plant Health, Springer Science and Business Media LLC.* 2017. DOI 10.1007/978-981-10-3473-2_9.

Kurtz, K. S.; Crouch, S. R. Design and Optimization of a Flow-Injection System for Enzymatic Determination of Galactose. *Anal. Chim. Acta* **1991**, *254*, 201–208.

Leong, S. A.; Neilands, J. B. Siderophore Production by Phytopathogenic Microbial Species. *Arch. Biochem. Biophys.* **1982**, *281*, 351–359.

Litwin, C. M.; Calderwood, S. B. Role of Iron in Regulation of Virulence Genes. *Clin. Microbiol. Rev.* **1993**, *6*, 137–149.

Louden, B. C.; Haarmann, D.; Lynne, A. M. Use of Blue Agar CAS Assay for Siderophore Detection. *J. Microbiol Biol. Educ.* **2011**, *12* (1), 51–53.

Lovejoy, D. B.; Richardson, D. R. Iron Chelators as Anti-Neoplastic Agents: Current Developments and Promise of the PIH Class of Chelators. *Curr Med Chem.* **2003**, *10*, 1035–1049.

Loyevsky, M.; John, C.; Dickens, B.; Hu, V.; Miller, J. H.; Gordeuk, V. R. Chelation of Iron within the Erythrocytic Plasmodium Falciparum Parasite by Iron Chelators. *Mol. Biochem. Parasitol.* **1999**, *101*, 43–59.

Loyevsky, M.; Lytton, S. D.; Mester, B.; Libman, J.; Shanzer, A.; Cabantchik, Z. I. The Antimalarial Action of Desferal Involves a Direct Access Route to Erythrocytic (Plasmodium Falciparum) Parasites. *J. Clin. Investig.* **1993**, *91*, 218–224.

Manawa, A. V.; Khandelwal, S. R.; Chaudhari, B. L.; Meyer, J. M.; Chincholkar, S. B. Siderophore Production by a Marine Pseudomonas Aeruginosa and Its Antagonistic Action against Phytopathogenic Fungi. *Appl. Biochem. Biotechnol.* **2004**, *118*, 243–251.

Matsumoto, K.; Ozawa, T.; Jitsukawa, K.; Masuda, H. Synthesis, Solution Behavior, Thermal Stability, and Biological Activity of an Fe (III) Complex of an Artificial Siderophore with Intramolecular Hydrogen Bonding Networks. *Inorg Chem.* **2004**, *43*, 8538–8546.

McLoughlin, T.; Quinn, J.; Bettermann, A.; Bookland, R. Pseudomonas Cepacia Suppression of Sunflower. Pseudomonas Cepacia. Wilt Fungus and Role of Antifungal Compounds in Controlling the Disease. *Appl. Environ. Microbiol.* **1992**, *58* (3), 1760–1763.

Meneely, K. M.; Lamb, A. L. Biochemical Characterization of an FAD-Dependent Monooxygenase, the Ornithine Hydroxylase from Pseudomonas Aeruginosa, Suggests a Novel Reaction Mechanism. *Biochemistry.* **2007**, *46*, 11930–11937.

Messenger, A. J.; Barclay, R. Bacteria, Iron and Pathogenicity. *Biochem. Educ.* **1983**, *11* (2), 54–63.

Meyer, J. M. Pyoverdines: Pigments, Siderophores and Potential Taxonomic Markers of Fluorescent Pseudomonas Species. *Arch. Microbiol.* **2000**, *174* (3), 135–142.

Munzinger, M.; Taraz, K.; Budzikiewicz, H. Ss-Rhizoferrin (Enantio-Rhizoferrin) A Siderophore of Ralstonia (Pseudomonas) Pickettii Dsm 6297–The Optical Antipode of R, R-Rhizoferrin Isolated from Fungi. *Biometals.* **1999**, *12*, 189–193.

Murugappan, R. M.; Aravinth, A.; Karthikeyan, M.; Chemical and Structural Characterization of Hydroxamate Siderophore Produced by Marine Vibrio Harveyi. *J. Indus. Microbiol. Biotechnol.* **2011**, *38* (2), 265–273. doi:10.1007/s10295-010-0769-7.

Nagoba, B. and Vedpathak, D. Medical Applications of Siderophores. *Eur. J. Gen Med.* **2011**, *8* (3), 229–235.

Nakouti, I.; Sihanonth, P.; Palaga, T.; Hobbs, G. Effect of a Siderophore Producer on Animal Cell Apoptosis: A Possible Role as Anti-Cancer Agent. *Int. J. Pharm. Med. Bio. Sci.* **2013**, *2* (2), 1–5.

Neilands, J. B. Iron Absorption and Transport in Microorganisms. *Annu. Rev. Nutr.* **1981**, *1*, 27–46.

Neilands, J. B. Microbial Iron Transport Comounds. *Annu. Rev. Microbiol.* **1982**, *36*, 285–309.

Neilands, J. B. Siderophores: Structure and Function of Microbial Iron Transport Compounds. *J. Biol. Chem.* **1995**, *270* (45), 26723–26726.

Nico, K.; Etienne, W.; Ahmed, Gaballa.; Anja, G.; Monica, H.; Pierre, C. Detection and Differentiation of Microbial Siderophores by Isoelectric Focusing and Chrome Azurol S Overlay. *Biometals.* **1994**, *7*, 287–291.

Nikaido, H. Molecular Basis of Bacterial Outer Membrane Permeability Revisited. *Microbiol. Mol. Biol. Rev.* **2003**, *67* (4), 593–656.

Oliver, B.; David, H. P. Small Molecule LC-MS/MS Fragmentation Data Analysis and Application to Siderophore Identification. Applications from Engineering with MATLAB Concepts, Jan Valdman, IntechOpen. https://www. intechopen. com/books/applications-from-engineering-with-matlab-concepts/small-molecule-lc-ms-ms-fragmentation-data-analysis-and-application-to-siderophore-identification (accessed July 7, 2016).

Pal, K. K.; Tilak, K. V.; Saxena, A. K.; Dey, R.; Singh, C. S. Suppression of Maize Root Diseases Caused by Macrophomina Phaseolina, Fusarium Moniliforme and Fusarium Graminearum by Plant Growth Promoting Rhizobacteria. *Microbiol Res.* **2001**, *156*, 209–223.

Patel, A. K.; Deshattiwar, M. K.; Chaudhari, B. L.; Chincholkar, S. B. Production, Purification and Chemical Characterization of the Catecholate Siderophore from Potent Probiotic Strains of Bacillus spp. *Biores Technol.* **2009**, *100*, 368–373.

Paul, Mc Cormack; Paul, J. W.; Martha, G. Separation and Detection of Siderophores Produced by Marine Bacterioplankton Using High-Performance Liquid Chromatography with Electrospray Ionization Mass Spectrometry. *Analyt. Chem.* **2003**, *75* (11), 2647–2652.

Pérez-Miranda, S.; Cabirol, N.; George-Téllez, R.; Zamudio-Rivera, L. S.; Fernández, F. J. O-CAS, a Fast and Universal Method for Siderophore Detection. *J. Microbiol. Methods* **2007**, *70* (1), 127–131.

Pesce, A. J.; Kaplan, L. A. *MPtodosQubnicuClinica.* Medica Panamericana Ed: Buenos Aires, 1990.

Piu, S.; Yeoh, B. S.; Xiao, X.; Golonka, R. M.; Kumarasamy, S.; Vijay, K. M. Enterobactin, an Iron Chelating Bacterial Siderophore, Arrests Cancer Cell Proliferation. *Biochem Pharmacol.* **2019**, *168*, 71–81.

Poole, K.; McKay, G. A. Iron Acquisition and Its Control in Pseudomonas Aeruginosa: Many Roads Lead to Rome. *Front Biosci.* **2003**, *8*, 661–686.

Pressler, U.; Staudenmaier, H.; Zimmermann, L.; Braun, V. Genetics of the Iron Dicitrate Transport System of Escherichia Coli. *J. Bacteriol.* **1988**, *170*, 2716–2724.

Raymond, K. N.; Emily, A. D.; Sanggoo, S. K. Enterobactin: An Archetype for Microbial Iron Transport. *Proc. Natl. Acad. Sci. USA* **2003**, *100* (7), 3584–3588.

Richardson, D. R. Iron Chelators as Therapeutic Agents for the Treatment of Cancer. *Crit. Rev. Oncol. Hematol.* **2002**, *42* (3), 267–281.

Rungin, S.; Indananda, C.; Suttiviriya, P.; Kruasuwan, W.; Jaemsaeng, R.; Thamchaipenet, A. Plant Growth Enhancing Effects by a Siderophore-Producing Endophytic Streptomycete Isolated from a Thai Jasmine Rice Plant (Oryza sativa L. cv. KDML105). *Antonie Van Leeuwenhoek.* **2012**, *102* (3), 463–472.

Sah, S.; Singh, R. S. Structural and Functional Characterisation: A Comprehensive Review. *Agriculture (Pol'nohospodárstvo).* **2015**, *61* (3), 97–114.

Sanaz, R. K.; Abdolrazagh, D. S.; Mohammad, R. H.; Stress Tolerance in Flax Plants Inoculated with Bacillus and Azotobacter Species under Deficit Irrigation *Physiologia Plantarum.* **2020**, 13154.

Sayyed, R. Z.; Purification of Siderophores of Alcaligenes Faecalis on Amberlite XAD. *Bioresour. Technol.* **2006**, *97* (8), 1026–1029.

Schalk, I. J.; Hannauer, M.; Braud, A. Minireview New Roles for Bacterial Siderophores in Metal Transport and Tolerance. *Environ Microbiol.* **2011**, *13*, 2844–2854.

Schippers, B.; Bakker, A. W.; Bakker, P. A. Interactions of Deleterious and Beneficial Rhizosphere Microorganisms and the Effect of Cropping Practices. *Ann Rev Phytopathol.* **1987**, *25*, 339–358.

Schwyn, B.; Neilands, J. B. Universal Chemical Assay for the Detection and Determination of Siderophores. *Analyt. Biochem.* **1987**, *160* (1) 47–56.

Seuk, C.; Paulita, T.; Baker, R. Attributes Associate with Increased Biocontrol Activity of Fluorescent Pseudomonads. *J. Plant Pathol.* **1988**, *4* (3), 218–225.

Snow, G. A. Mycobactin. A Growth Factor for Mycobacterium Jhnei. II. Degradation and Identification of Fragments. *J. Chem. Soc.* **1954**, *10*, 2588–2596.

Stintzi, A.; Barnes, C.; Xu, J.; Kenneth, N. Raymond Microbial Iron Transport via a Sidero-phore Shuttle: A Membrane Ion Transport Paradigm. *Proc. Natl. Acad. Sci.* **2000**, *97* (20), 10691–10696.

Sugita, H.; Mizuki, H.; Itoi, S. Diversity of Siderophore-Producing Bacteria Isolated from the Intestinal Tracts of Fish along the Japanese Coast. *Aquac Res.* **2012**, *43*, 481–488.

Tank, N.; Narayanan, R.; Patel, B.; Saraf, M. Evaluation and Biochemical Characterization of a Distinctive Pyoverdin from a Pseudomonas Isolated from Chickpea Rhizosphere. *Brazilian J. Microbiol.*. **2012**, 639–648.

Thevenot, D. R.; Toth, K.; Durst, R. A.; Wilson, G. S. Electrochemical Biosensors: Recom-mended Definitions and Classification. *Pure Appl Chem.* **1999**, *71* (12), 2333–2348.

Tomas, P.; Karel, L.; Ghosh, D.; David, Milde, Novak, J.; Havlıcek, V. Characterization of Microbial Siderophores by Mass Spectrometry. *Mass Spectrometry Rev.*, **2014**, 1–12.

Tsafack, A.; Libman, J.; Shanzer, A.; Cabantchik, Z. I. Chemical Determinants of Antimalarial Activity of Reversed Siderophores. *Antimicrob Agents Chemother.* **1996**, *40*, 2160–2166.

Varma A., Podila G.K.,Chincholkar S.B., Rane M.R., Chaudhari B.L. Siderophores: Their Biotechnological Applications. In *Biotechnological Application of Microbes.* I.K. International: New Delhi, India; 2005. pp. 177–198.

Verma, V. C.; Singh, S. K.; Prakash, S. Bio-Control and Plant Growth Promotion Potential of Siderophore Producing Endophytic Streptomyces from Azadirachta indica A. *Juss. J Basic Microbiol.* **2011**, *51*, 550–556.

Virginie, V.; Nolwenn, W.; Muriel, J.; Martine, S.; Magali, A.; Laurent, D.; Anne, M. D. Siderophores in Cloud Waters and Potential Impact on Atmospheric Chemistry: Production by Microorganisms Isolated at the Puy de Dôme Station. *Environ. Sci. Technol.* **2016**, *50* (17), 9315–9323. DOI: 10.1021/acs. est. 6b02335.

Vogel, A. L. *Class Reactions* (Reactions for Functional Groups); CBS Publishers: New Delhi, India. 1992; **p** 190.

Wang, Q.; Xiong, D.; Zhao, P.; Yu, X.; Tu, B.; Wang, G. Effect of Applying an Arsenic-Resistant and Plant Growth-Promoting Rhizobacterium to Enhance Soil Arsenic Phytoremediation by Populus Deltoides LH05–17. *J. Appl. Microbiol.* **2011,** *111*, 1065–1074.

Winkelmann, G. Ecology of Siderophores with Special Reference to Fungi. *Biometals.* **2007,** *20*:379–392.

Winkelmann, G. Specificity of Iron Transport in Bacteria and Fungi. In *Handbook of Microbial Iron Chelates*; Winkelmann, G., Ed.; Boca Raton, FL: CRC Press, 1991; pp 65–106.

World Health Organization Model List of Essential Medicines: 21st List 2019. World Health Organization: Geneva, 2019. hdl:10665/325771. WHO/MVP/EMP/IAU/2019.06. License: CC BY-NC-SA 3.0 IGO.

CHAPTER 8

Forest Cuttings and the Impact on Edaphic and Regeneration Performance of Tree Species in the Temperate Forests of Kashmir Himalaya

MUSHEERUL HASSAN[1], SHIEKH MARIFATUL HAQ[2], AMRINA SHAFI[3], and HUMA HABIB[4*]

[1]Clybay Private Limited, Bangalore, India

[2]University of Kashmir, Srinagar, India

[3]Education Department, Government of Jammu & Kashmir, Srinagar, India

[4]Islamia College of Science and Commerce, Srinagar, India

*Corresponding author. E-mail: huma99@gmail.com

ABSTRACT

The regeneration potential of the composing tree species dramatically influences the structure of a forest ecosystem. The current study analyzes the regeneration status of tree species concerning edaphic and disturbance gradient in the temperate forests of Kashmir Himalaya. Based on the degree of disturbances, two comparable sites, that is, low-disturbed (LD) and high-disturbed (HD) sites, were selected for the sampling of vegetation. Preliminary results showed that the seedling/tree ratio of LD sites (15.33) was higher than that of the HD sites (6.26). A sapling to tree ratio followed the same trend, that is, higher (10.98) at LD sites and lower (4.16) at HD sites. The regeneration performance of the tree species was not very much satisfied because most of the species had low seedling: tree and sapling: tree ratios. In the case of HD site, 17%, 33%, 33%, and 17% of the tree species had moderate, absent, establishing, and hampered regeneration performance respectively while at LD site moderate, sufficient and high regeneration performance was observed as 50%, 25% and 25% of tree species. The higher nutrient content and disturbances at the HD sites make light and nutrients available, thereby facilitating resources for partitioning among colonizing plant species. Anthropogenic disturbances usually limit natural regeneration recovery of forest ecosystems. The researchers, land managers, and the common public need to highlight the issue, and the local government needs to reduce the intensity of anthropogenic disturbances so that their regeneration performance is improved.

8.1 INTRODUCTION

Forests are renewable, because of their ability to regenerate (Hjeljord et al., 2014; Malik, 2014; Tripathi and Khan, 2007). It is the process of silvigenesis by which trees and forests survive over time (Malik, 2014). The regeneration potential of trees often predicts the possible changes in forest vegetation in the future (Henle et al., 2004). Regeneration and geographic distribution of plant species are being restricted to a specific set of environmental gradients (Gonçalves et al., 2018; Grubb, 1977). The forests are not static but are dynamic entities. The seeds of different tree species in a forest germinate to form seedlings that grow and compete with adult trees for the limited available resources. Some of them survive for number of years, while others perish during the struggle to flourish because of many reasons (Malik, 2014). Growth of any plant community requires a sustained regeneration plus growth of whole species in the presence of older plant species (Francos et al., 2018; Taylor and Zisheng, 1988). The effective regeneration of a tree species depends upon the surviving and flourishing potential of its sapling and seedling (Good and Good, 1972). The regeneration is one of the precarious events in the management of forest ecosystems as it maintains the species composition and strengthening after various disturbances (Khumbongmayum et al., 2005). Detail functions and subsequent values delivered by forest habitats are capricious that depend on biotic and abiotic factors (Mohandass et al., 2018; Schoenholtz et al., 2000). The indices of forest soil eminence integrate soil physical and chemical properties. They are the utmost readily adopted and are sensitive. As a result, regeneration changes will be easily measured and relevant across sites adaptable for specific ecosystems. Nevertheless, there is little direct evidence for altered nutrient availability in disturbances within forests (Vitousek and Denslow, 1986).

In forest ecosystems, trees produce and maintain the overall physical structure of habitats (Jones et al., 1994). In a forest ecosystem, the tree is a fundamental component as it influences the resources and habitats for almost all other forest organisms (Rawat et al., 2018). The quantification of woody plant species is a significant characteristic while studying the impact of disturbances on forest ecosystems. Indeed, woody species represent a dominant life form that provides resources and habitat for many animal species and is easy to count (Sagar and Singh, 2004). The regeneration of tree species has an important consequence in terms of natural forest management (Dekker and De Graaf, 2003).

In general, it has been observed that mountain forest ecosystems have poor regeneration capacity (Kräuchi et al., 2000), and the same is true for Himalayan forests (Singh et al., 2016). In the Himalayan region, the main reasons for poor regeneration are different anthropogenic disturbances, such as the construction of roads, overgrazing, fire, lopping of tree species for fodder and fuel, and removal of litter from the forest floor (Malik et al., 2016). Numerous researchers presented quantitative tree regeneration status from a different region of the Himalayas like Western Himalaya (Gairola et al., 2012; Malik et al., 2016; Singh et al., 2016), Central Himalaya (Pant and Samant, 2012), and Eastern Himalaya (Rawat et al., 2018). But there is a paucity of the quantitative information of tree regeneration status in the Kashmir Himalayan region. Keeping in mind the facts mentioned above, the present study was carried out to study the regeneration analysis in relation to edaphic factors along the disturbance gradient in the forests of Kashmir Himalaya.

8.2 MATERIALS AND METHODS

8.2.1 STUDY AREA

The representative for Himalayan Dry Temperate forest from the Keran valley was selected as the study area. The region falls in the remote Tehsil of District Kupwara of Union Territory of Jammu and Kashmir, located between 34° 34' 0" N - 34° 42' 30" N and 73° 55' 15" E - 74° 17' 05" E (Fig. 8.1), and altitude ranging from 1500 to 3450 (masl). The Keran valley is mostly inhabited by Pahari and Gujjar tribes, with a total population of 12,026 persons. The study area comprises of three Panchayats (village administrative units): Keran, Mundiyan, and Pathran. The vegetation is mainly composed of mixed forests composed of broad-leaved and conifers tree species.

8.2.2 VEGETATION AND DRIVERS OF ANTHROPOGENIC DISTURBANCES

The Keran valley is a natural site of conifers and mixed forests. These forests have great importance to the local inhabitants to fulfill their basic needs of timber, fodder, fuelwood, and other non-timber forest product (i.e., litter, humus, and ethnomedicinal plants). Due to climatic harshness, the absence of alternative energy resources and pitiable socioeconomic conditions, people are forced to depend upon local forest resources and hence create high anthropogenic pressure on these forests (Table 8.1).

TABLE 8.1   Vegetation and Resource Use Pattern in the Keran Valley.

Forest site	Main vegetation	Anthropogenic disturbances (Resource use pattern)[a]
LD site	*Cedrus deodara, Pinus wallichiana, Aesculus indica, Prunuscornuta, Parrotiopsis jacquemontiana, Vibrurnum grandiflorum, Rosa webbiana, Dyropteris stewartii, Stipa sibirica, Fragaria nubicola, Sambucus wightiana, Pteris creta, Polygonum amplexicaule and Adiantum capilusveneris.*	LTL, LG, LSC LCNTFP, LFE, LAI, NRC
HD site	*Cedrus deodara, Pinus wallichiana, Quercus incana, Quercus baloot, Aesculus indica, Morus nigra, Parrotiopsis jacquemontiana, Vibrurnum grandiflorum, Sorbaria tomentosa, Stipa sibirica, Fragaria nubicola, Digitalis purpurea, Leucanthemum vulgare, Polygonum amplexicaule* and *Asplenium ofeliae.*	HTL, HG, HST, HCNTFP, LFE, HAI, RC

[a]*HAI, High alien invasion; HCNTFP, heavy collection of non-timber forest products; HFE, High fire effect; HG, Heavy grazing; HST, heavy stem cutting; HTL, heavy tree lopping; LAI, Low alien invasion; LCNTFP, Low collection of non-timber forest products; LFE, Low fire effect; LG, low grazing; LSC, low stem cutting; LTL, Low tree lopping; NRC, No road connectivity; RC, Road connectivity.*

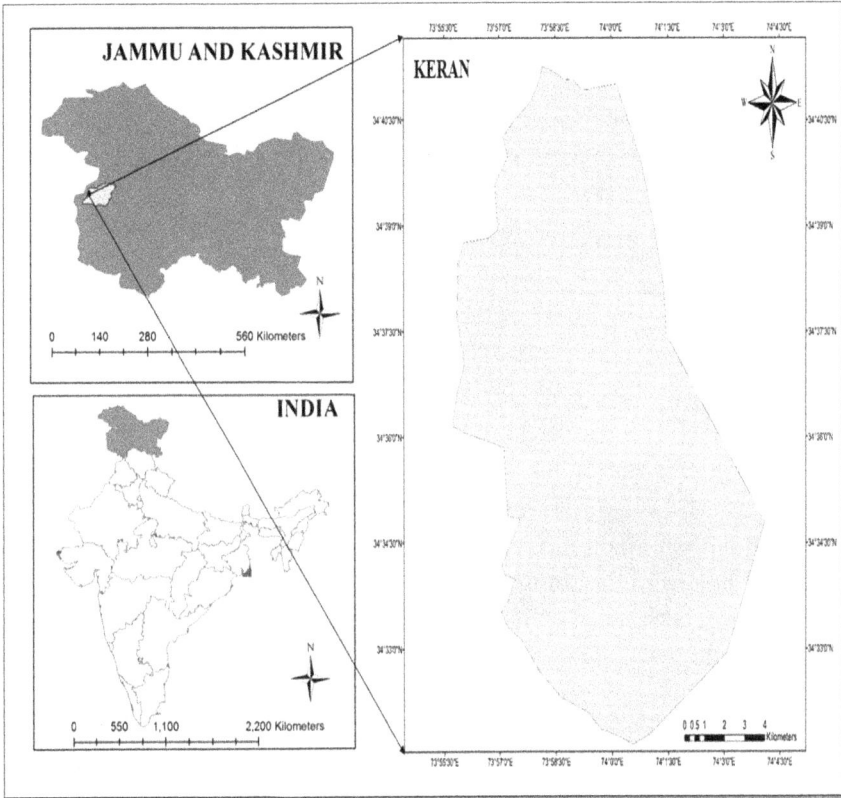

FIGURE 8.1 GIS Map showing the location of the study area (Kashmir Himalaya).

8.3 METHODOLOGY

Field reconnaissance surveys were carried out in 2018 to get a general idea about the features of the terrain, general vegetation, species composition, accessibility, and distribution of different forest types. The working plan of the Kehmil Forest Division was consulted for authenticating geographical location, administrative jurisdiction, and forest vegetation types. Based on the intensity of disturbances, two differently disturbed sites, High-Disturbed (HD) and Low-Disturbed (LD) sites were selected for sampling (Table 8.2). In these selected sites, the comparative disturbances were characterized according to Seipel et al. (2012). The disturbance drivers, such as stem cutting (CT), tree lopping (LP), livestock grazing (GZ),

degradation (DG), road connectivity (RC), distance from human settlements (DS), and forest fire (FR) were taken into consideration for charac-
terizing disturbance level of the sampling sites. The three-point scale (0 =
none or low, 1 = moderate, 2 = high) was used to record disturbance levels
based upon a visual assessment in the vicinity of each sampling site.

TABLE 8.2   Details of the Study Area.

Parameter	LD site	HD site
Geo-coordinates	34° 39' 06.2" 73° 57' 18.6"	34° 38' 52" 73° 57' 59"
Altitudinal range	1550–1650 masl	1530–1630 masl
Slope (°)	34 ± 15.24	38 ± 8.04
Crown density (%)	53 ± 8.38	48 ± 5.06
Stem/stump ratio	4.9	1.7

The quadrant method of vegetation sampling was used to record forest
structure data from the study area. In each of the selected forest types (LD
and HD), we outlined four sample plots, each of 31.6×31.6 m ($\cong 0.1$ ha)
size in all the four directions, that is, northeast, three northwests, three
southeast, and three southwest directions, respectively were placed in each
super-plot. At each of the selected forests, four permanent quadrats of 5×5
m size were established with each 0.1 ha plot. Thus, a total of 96 quadrats
(4 quadrats $\times$ 12 (0.1 ha) = 48 quadrats $\times 2$ (forest types) =96 quadrats)
plots were laid at a site. The tree species composition and diameter at
breast height (DBH) depicting growth potential were recorded within each
of the four sample plots. All the plants were identified and density of seedling (<20 cm height), sapling (<30 cm collar circumference at the base
and >20 cm in height), and tree species (>30 cm DBH) were determined.
The species regeneration status was determined based on sapling plus
population size of seedling (Khan et al., 1987; Shankar, 2001). Regeneration performance was determined by the ratio of seedling and saplings to
trees (Gairola et al., 2012). Soil samples were collected from each plot
and sieved via a 2 mm mesh screen. Soil pH and electrical conductivity
were measured using digital pH and electrical conductivity meter, organic
carbon via wet digestion, available phosphorus through ascorbic acid
method and nitrogen via the Kjeldahl nitrogen method. Whereas, other
physicochemical parameters were determined by standard procedure of
soil, plant, water, and fertilizer analysis (Gupta, 2000).

8.4   DATA ANALYSIS

The results determined from the analysis of soil properties were compared between different forest sites soil samples. The data were standardized to zero, mean, and unit variance, and the analysis was done on the correlation matrix. Principal component analysis (PCA) for different soil parameters among the sampled forest sites was done using R statistical software (Team, 2014) (Borůvka et al., 2007).

8.5   RESULTS

8.5.1   SOIL CHARACTERISTICS

The variations in soil physicochemical features at HD and LD regions are presented in Table 8.3; Figure 8.2. The mean pH value was lower at the LD site (6.26) as compared with that at HD site (6.47). The conductivity value observed at the HD site (198μs) was higher than that of the LD site (153 μs). The mean value of organic carbon (OC) at the HD region (4.23%) was higher as compared with that at the LD site (3.14%). Mean values of available phosphorus (P in μg/g) reported at HD and LD sites were 17 and 15, respectively. The mean total nitrogen (%) and available nitrogen (kg/ha) reported at HD and LD sites are 0.82 and 0.50; 155 and 130, respectively (Table 8.1; Figure 8.2). Whereas mean exchangeable concentrations of potassium (K), Calcium (Ca), sodium (Na) and magnesium (Mg) at the HD site were 6.5, 22.4, 134, and 187 mg/kg and that of LD site were 14 and 4.7, and 123 and 112 4.1 mg/kg, respectively. The mean soil moisture was maximum at the LD site as compared to the HD site. The average water holding capacity at HD and LD sites was 61.7% and 47.4% respectively. The coefficient of determination between the two sites LD and HD sites, $R^2=0.93$. This fact is also supported by the PCA which showed distinct segregation based on variation in soil characteristics (Fig. 8.3).

8.5.2   REGENERATION PERFORMANCE

The ratio of seedling and saplings to trees indicated the regeneration performance of the tree species in two forest sites (Table 8.4). The

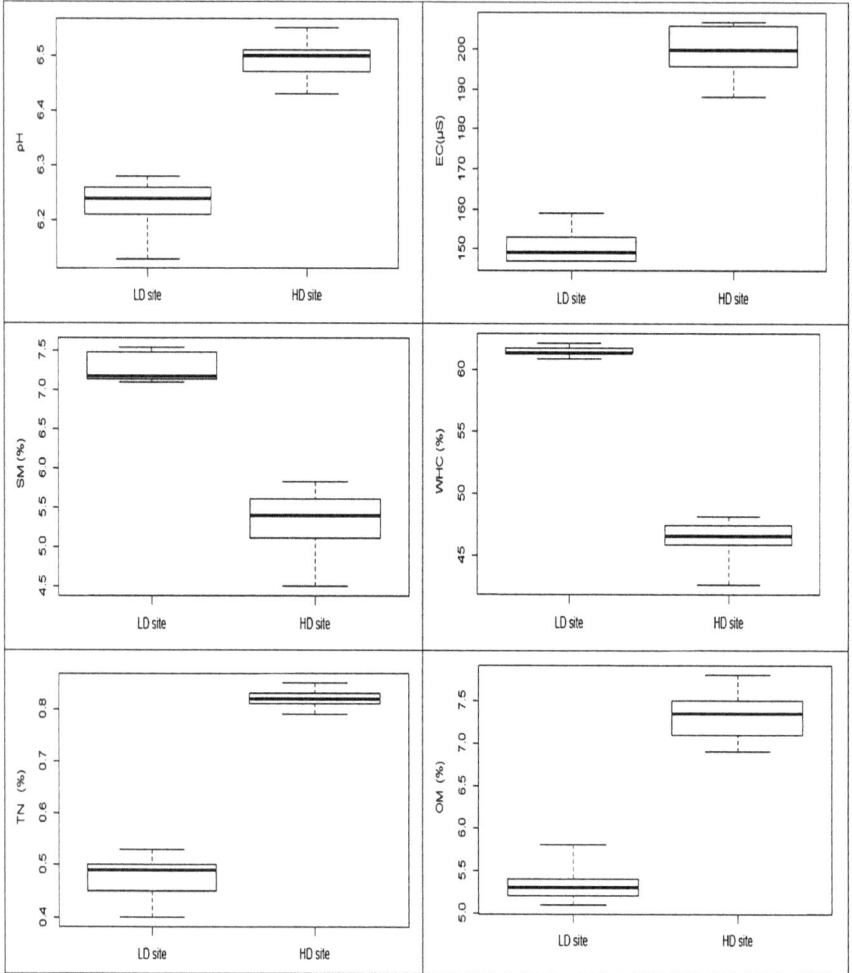

(Continued)

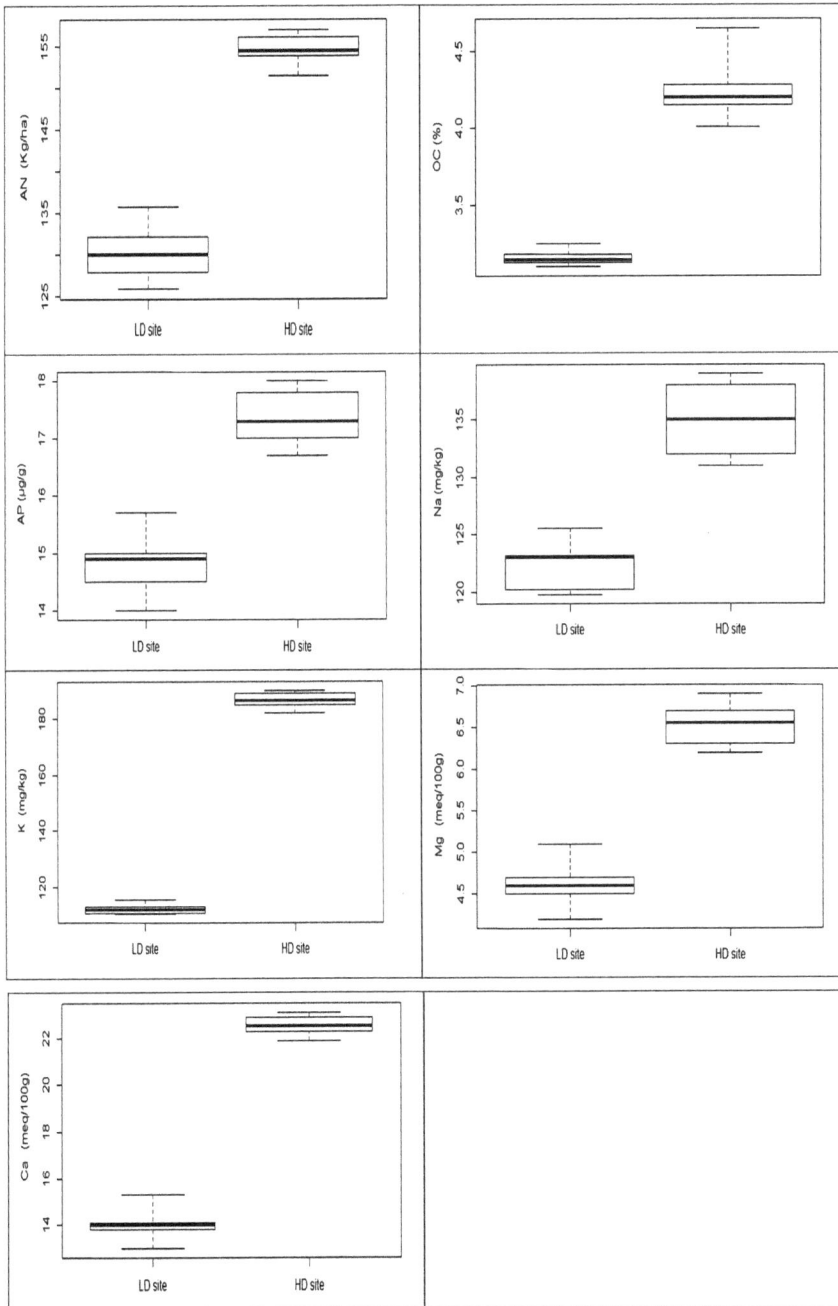

FIGURE 8.2 Boxplots of soil parameters in the LD site and HD site. (a) ph. (b) organic carbon (OC) (C) Cond (EC).

seedling: tree ratio of the LD site (15.33) was higher than that of the HD site (6.26). A sapling:tree ratio also followed the same trend, that is, higher (10.98) at (LD site) and lower (4.16) at HD site. The number of trees represented by seedling and seedling stages differed in the two sites (Table 8.4). The seedling density at the LD site (3235 seedlings/ha) was higher than that of the HD site (1200 seedling/ha). Sapling density also followed the same trend, that is, higher (2317 ha^{-1}) at the LD site and lower (800 saplings/ha) at the HD site. In the case of HD site, 17%, 33%, 33%, and 17% had moderate, absent, establishing, and hampered regeneration performance, respectively (Fig. 8.4). *Cedrus deodara* had hampered regeneration performance with a low value of Se : T ratio and the absence of sapling stage (Se:T=0.84; Sa:T=0/118). *Pinus wallichiana* showed moderate regeneration performance (Table 8.4). *Aesculus indica* and *Morus nigra* (both with Se/T=0/3; Sa/T=0/3) did not show any regeneration performance. At HD site, the "establishing" regeneration performance was reported for the two species of *Quercus viz., Q. baloot* and *Q. incana* (Table 8.4).

TABLE 8.3   Mean Soil Characteristics (0–10 cm depth) at LD and HD Forest Sites.

Parameters	LD site	HD site
PH	6.26±0.09	6.47±0.16
Cond (EC)	153±27	198±54
SM (%)	7.54±1.58	5.62±1.76
WHC (%)	61.7±1.76	47.4±1.76
OC (%)	3.14±0.78	4.23±0.93
OM (%)	5.4±1.23	7.3±1.36
Avail P(μg/g)	15±6.05	17±4.6
Avail N(kg/ha)	130±13.6	155±15.9
TN (%)	0.5±0.03	0.82±0.13
Ca (meq/100 g)	14±5.9	22.4±2.1
Mg (meq/100 g)	4.7±0.6	6.5±1.1
Na (mg/kg)	123±17	134±6.5
K (mg/kg)	112±14.6	187±5.9

Cond (EC), Electrical conductivity; Ca, calcium; K, potassium; Mg, magnesium; N, nitrogen; Na, sodium; OC, organic carbon; OM,organic matter; P, phosphorus; SM, soil moisture; TN, total nitrogen; WHC, water holding capacity.

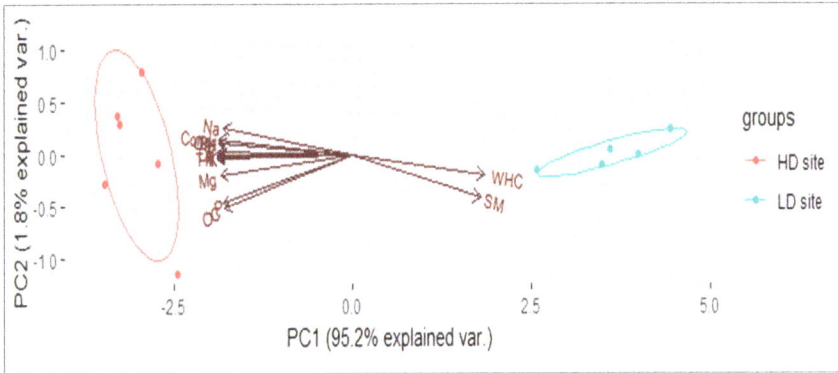

FIGURE 8.3 Principal component analysis (PCA).

At the LD site, *Cedrus deodara* showed "sufficient" regeneration performance with seedlings to tree ratio (Se/T=8.18) and saplings to tree ratio (Sa/T=5.45). *Pinus wallichiana* had a "high" regeneration performance due to high values of Se/T (24.73) and Sa/T (18.27). The moderate regeneration performance was shown by 50% of the tree species. These species include *Aesculus indica* and *Prunus cornuta* (Table 8.4 and Figure 8.4).

8.6 DISCUSSION

Anthropogenic disturbances usually limit natural regeneration recovery of forest ecosystems, even if the rates of biomass removal are within the carrying capacity of the forest (Gimmi et al., 2008). Regardless of the magnitude of disturbance, the forest ecosystems are affected either by destroying vegetation or by exposing bare soil (Clark, 1990; Laurance et al., 1998). The wealth of forest ecosystems depends upon the promising regeneration performance of tree species composing the forest stand (Jones et al., 1994). Therefore, periodic assessment and monitoring of forest regeneration potential are crucial for long-term forest management and policy-making (Chakraborty et al., 2018; Malik, 2014).

The current study revealed that the seedling density ranged from 1200 (HD) to 3235 (LD) –Ind/ha, whereas sapling density ranged from 800 (HD) to 2317 (LD) Ind/ha. Malik and Bhatt (2016) while studying the regeneration status and seedling survival in the Western Himalaya reported the seedling and sapling densities ranging between 1670–7485

TABLE 8.4 Regeneration Performances Indicated as the Ratio of Seedling and Sapling to Tree Densities in the Studied Forests.

Forest sites Tree species	LD site				HD site			
	Densities/ha Se, Sa and T	Ratios (Se, Sa/T)	LD site	Regeneration performance	Densities/ha Se, Sa and T	Ratios (Se, Sa/T)	HD site	Regeneration performance
Cedrus deodara	900 (Se), 600 (Sa)	Se/T	8.18	Sufficient	100 (Se) 0(sa)	Se/T	0.84	Hampered
	110 (T)	Sa/T	5.45		118 (T)	Sa/T	0/118	
Pinus wallichiana	2300 (Se), 1700 (Sa)	Se/T	24.73	High	200 (SE), 100 (Sa)	Se/T	4.41	Moderate
	93 (T)	Sa/T	18.27		68 (T)	Sa/T	2.94	
Aesculus indica	20 (Se), 9 (Sa)	Se/T	4	Moderate		Se/T	0/3	Absent
	5 (T)	Sa/T	1.8		3 (T)	Sa/T	0/3	
Prunus cornuta	15 (Se), 8(Sa)	Se/T	5	Moderate		Se/T	-	
	3 (T)	Sa/T	2.6		–	Sa/T	-	
Quercus baloot	–	Se/T	–	–	600 (Se), 500 (Sa)	Se/T	600/0	Establishing
Morus nigra	–	Sa/T	–			Sa/T	500/0	
		Se/T	–	–		Se/T	0/3	Absent
	–	Sa/T	–		3 (T)	Sa/T	0/3	
Quercus incana	–	Se/T	–	–		Se/T	200/0	Establishing
	–	Sa/T	–			Sa/T	100/0	
Total	**3235(Se; 2317(Sa)**	**Se/T**	**15.33**		**1200(Se); 800 (Sa)**	**Se/T**	**6.26**	
	211 (T)	**Sa/T**	**10.98**		**192 (T)**	**Sa/T**	**4.16**	

Ind/ha and 850–5600 Ind/ha respectively. In forests of the Phakot and Pathri watersheds, Himalaya, Pokhriyal et al. (2010) recorded comparable sapling density of 370–701 Ind/ha, and seedling density ranging from 2321to 3496 Ind/ha. Similar values have also been reported by Samant et al. (2007) from Mornaula Reserve Forests of central Himalaya where they reported seedling and sapling densities ranging from 266–1571 Ind/ha to 340–2277 Ind/ha respectively Singh et al. (2016) reported the seedling and sapling densities to be 1550–9600 and 167–1296 individuals/ha respectively from different oak-dominated forests of Garhwal. Singh et al. (2016) while describing the forest composition, structure in the montane zone of the western Ramganga valley, Uttarakhand Himalaya found the average seedling and sapling densities to be 628 and 1283 Ind/ha respectively. Recently, Rawat et al. (2018) reported a very high mean seedling density (32278±3717) from Eastern Himalaya.

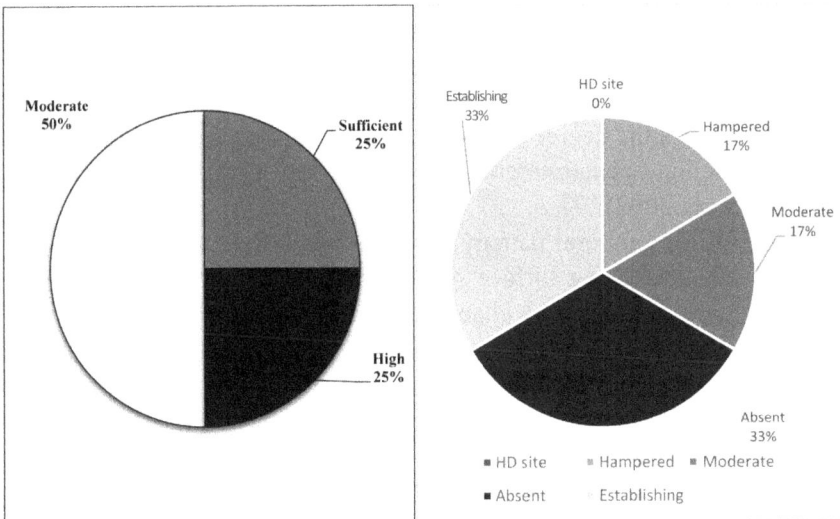

FIGURE 8.4 Regeneration performance of tree species in the LD site (A) and HD site (B).

The reason for lower densities of seedlings and saplings in the present study as compared with those reported by Malik and Bhatt (2016), Singh et al. (2016), and Rawat et al. (2018) may be the different anthropogenic disturbances and poor management which are evident by the presence of stem cuttings and charcoal making area. The development of human settlements with road and footpath including road near to forest edge, logged, and grazing also might be the reason to count a low number of seedlings and saplings in the HD forest stand. The reason for high seedling and sapling densities in the prior studies is the "protected" nature of the study area. The various anthropogenic disturbances like fodder and fuel collection, grazing, and lopping are not allowed in these areas.

In the present study, forest disturbance indicators show a significant effect on forest regeneration. The overall picture of the regeneration performance of trees in studied forests is shown in Figure 8.3 and Table 8.4. In the HD site, 17%, 33%, 33%, and 17% had moderate, absent, establishing, and hampered regeneration performance, respectively (Fig. 8.3). The frequent lopping of mature tree species for fuelwood and leaf fodder results in the reduction of canopy cover. Hence, it has a direct effect on seed production and regeneration (Saxena and Singh, 1984; Vetaas, 2000). The disturbance regime and the level of variation in the soil properties are system-level characteristics of a forest. The ultimate effects of disturbance on regeneration are expressed at the level of individual tree species. The agents and their potential effects change throughout the life of an individual plant. Seedlings can be affected by essentially all types of disturbances. As the degree of disturbances increases, canopy unevenness will increase light penetrating laterally in the forest (Canham et al., 1990; Franklin et al., 2002).

This shift of temporal pattern and quantity of light depletes the water resources by decreasing surface soil moisture and may increase daytime soil temperature, which may change the microclimate circumstances, and light accessibility that affect the growth of the plants (Xu et al., 2008). Lower moisture content, higher phosphorus (Denslow, 1980), nitrogen concentrations (Vitousek and Denslow, 1986), and high availability of exchangeable cations (Hwang and Son, 2006) obstruct the regeneration of seedlings growth (Uhl et al., 1988), Disturbances significantly affect the carbon (C) cycle of forest ecosystems. They lead to an increase in decomposition, primarily due to warmer soil conditions. However, an established regeneration retards decomposition due to its modulating effect on soil temperature (Mayer et al., 2017). Organic carbon (OC) and organic matter (OM) content were higher in HD site than in the LD site (Reth et al., 2005), higher grazing at HD site has paid to increase organic carbon level due to animal excreta, however, higher level of exchangeable cations at HD site may be due to absence of the canopy cover (Boerner and Sutherland, 1997). Altered nutrient availability, at least of nitrogen, is plausible disturbances such as deforestation generally lead to increased nitrogen mineralization and nitrification. The occurrence of only seedlings and sapling of *Quercus baloot* and *Quercus incana* indicated that these plants have come newly to forest and likely to become canopy species in the near future. It is suggested by Denslow (1980) that the disturbances increase light and nutrients availability in the soil, and these resources are partitioned among colonizing plant species.

The anthropogenic disturbances are the main drivers altering forest structure, creating landscape mosaics, and setting the initial conditions for successional dynamics and structural development (Swanson et al., 2011). Disturbances like deforestation, development of human settlements, road accessibility, cattle grazing, trampling, litter collection, fire, and seasonal drought are the most important factors affecting forest regeneration (Gerhardt, 1993; Lieberman and Li, 1992; Saxena and Singh, 1984). Under immense anthropogenic pressure, heavy plant extraction from the forest, with the limited area available for grazing livestock that results in threatening the seedling of tree species (Alam and Ali, 2010; Oza, 1980, 1982, 2003). The mountain roads provide an important anthropogenic impact in the forest ecosystem via changing plant species composition, hydrology, nutrient availability, and funneling anthropogenic influence into the most pristine ecosystem (Forman et al., 2003; Müllerová et al., 2011). Road access is a key factor by bringing population and investment and in increasing the profitability of deforestation, whether it is for agricultural production or land speculation (Laurance, 2015). There is also pressure from loggers to extract timber in the remaining forest areas, which affects forest diversity. The regeneration of forest will benefit forest diversity in the form of providing habitat for species adaptation to primary successional stages and via increasing connectivity and decrease fragmentation of forest fragments (Jantz et al., 2014).

8.7 CONCLUSIONS

Assessment of the regeneration performance of tree species is very significant for their sustainable utilization, conservation, and management and it was not very much satisfactory in the present study because most of the species had low Se/T and Sa/T ratios. Various types of anthropogenic disturbances like overgrazing, lopping, and forest fires adversely affect the regeneration performance of Himalayan forests. The tree species that showed either "absent" or "hampered" regeneration performance would be in trouble in the future. The researchers, policymakers, land managers, common public, and the local government need to take care of these forests so that their regeneration performance is improved. The results of the current study highlighted that there is an urgent need for objective measures to be taken primarily by the local government to improve the regeneration performance of these forests so that they can be conserved for sustainable utilization. The future ecological implications and impacts on the flow of ecosystem services due to poor regeneration status of the resident tree species as against newly colonizing *Quercus* species need to be studied in detail to develop systematic management plans for the mountainous forest landscapes of Kashmir Himalaya.

ACKNOWLEDGMENT

This work was supported by the Department of the Botany, University of Kashmir Srinagar, Jammu and Kashmir.

KEYWORDS

- disturbance
- Kashmir Himalaya
- regeneration potential
- seedling/tree ratio
- soil properties

REFERENCES

Alam, J.; Ali, S. I. Contribution to the Red List of the Plants of Pakistan. *Pak. J. Bot.* **2010,** *42* (5), 2967–2971.

Boerner, R. E.; Sutherland, E. K. The Chemical Characteristics of Soil in Control and Experimentally Thinned Plots in Mesic Oak Forests Along a Historical Deposition Gradient. *Appl. Soil Ecol.* **1997,** *7* (1), 59–71.

Borůvka, L.; Mládková, L.; Penížek, V.; Drábek, O.; Vašát, R. Forest Soil Acidification Assessment Using Principal Component Analysis and Geostatistics. *Geoderma* **2007,** *140* (4), 374–382.

Canham, C. D.; Denslow, J. S.; Platt, W. J.; Runkle, J. R.; Spies, T. A.; White, P. S. Light Regimes Beneath Closed Canopies and Tree-Fall Gaps in Temperate and Tropical Forests. *Can. J. Forest Res.* **1990,** *20* (5), 620–631.

Chakraborty, A.; Shukla, R.; Sachdeva, K.; Roy, P. S.; Joshi, P. K. The Climate Change Conundrum and the Himalayan Forests: The Way Forward into the Future. *Proc. Natl. Acad. Sci., India B: Biol. Sci.* **2018,** *88* (3), 837–847.

Clark, D. B. The Role of Disturbance in the Regeneration of Neotropical Moist Forests. *Reprod. Ecol. Trop. Forest Plants* **1990,** *7,* 291–315.

Dekker, M.; De Graaf, N. R. Pioneer and Climax Tree Regeneration Following Selective Logging with Silviculture in Suriname. *Forest Ecol. Manage.* **2003,** *172* (2–3), 183–190.

Denslow, J. S. Patterns of Plant Species Diversity during Succession under Different Disturbance Regimes. *Oecologia* **1980,** *46* (1), 18–21.

Forman, R. T.; Sperling, D.; Bissonette, J. A.; Clevenger, A. P.; Cutshall, C. D.; Dale, V. H.; … Jones, J. *Road Ecology: Science and Solutions*; Island Press, 2003.

Francos, M.; Ubeda, X.; Pereira, P. Long-Term Forest Management after Wildfire (Catalonia, NE Iberian Peninsula). *J. Forest. Res.* **2020,** *31* (1), 269–278.

Franklin, J. F.; Spies, T. A.; Van Pelt, R.; Carey, A. B.; Thornburgh, D. A.; Berg, D. R.; … Bible, K. Disturbances and Structural Development of Natural Forest Ecosystems with Silvicultural Implications, Using Douglas-Fir Forests as an Example. *Forest Ecol. Manage.* **2002,** *155* (1–3), 399–423.

Gairola, S.; Sharma, C. M.; Ghildiyal, S. K.; Suyal, S. Regeneration Dynamics of Dominant Tree Species Along an Altitudinal Gradient in Moist Temperate Valley Slopes of the Garhwal Himalaya. *J. Forest. Res.* **2012,** *23* (1), 53–63.

Gerhardt, K. Tree Seedling Development in Tropical Dry Abandoned Pasture and Secondary Forest in Costa Rica. *J. Veg. Sci.* **1993,** *4* (1), 95–102.

Gimmi, U.; Bürgi, M.; Stuber, M. Reconstructing Anthropogenic Disturbance Regimes in Forest Ecosystems: A Case Study from the Swiss Rhone Valley. *Ecosyst.* **2008,** *11* (1), 113–124.

Gonçalves, F. M.; Revermann, R.; Cachissapa, M. J.; Gomes, A. L.; Aidar, M. P. Species Diversity, Population Structure and Regeneration of Woody Species in Fallows and Mature Stands of Tropical Woodlands of Southeast Angola. *J. Forest. Res.* **2018,** *29* (6), 1569–1579.

Good, N. F.; Good, R. E. Population Dynamics of Tree Seedlings and Saplings in a Mature Eastern Hardwood Forest. *Bull. Torrey Bot. Club* **1972,** 172–178.

Grubb, P. J. The Maintenance of Species-Richness in Plant Communities: The Importance of the Regeneration Niche. *Biol. Rev.* **1977,** *52* (1), 107–145.

Gupta, P. K. *Soil Plant Water and Fertilizer Analysis*; Agrobios Pub: Bikaner. India, 2000.

Henle, K.; Sarre, S.; Wiegand, K. The Role of Density Regulation in Extinction Processes and Population Viability Analysis. *Biodivers. Conserv.* **2004,** *13* (1), 9–52.

Hjeljord, O.; Histøl, T.; Wam, H. K. Forest Pasturing of Livestock in Norway: Effects on Spruce Regeneration. *J. Forest. Res.* **2014,** *25* (4), 941–945.

Hwang, J.; Son, Y. Short-Term Effects of Thinning and Liming on Forest Soils of Pitch Pine and Japanese Larch Plantations in Central Korea. *Ecol. Res.* **2006,** *21* (5), 671–680.

Jantz, P.; Goetz, S.; Laporte, N. Carbon Stock Corridors to Mitigate Climate Change and Promote Biodiversity in the Tropics. *Nat. Clim. Change* **2014,** *4* (2), 138–142.

Jones, R. H.; Sharitz, R. R.; Dixon, P. M.; Segal, D. S.; Schneider, R. L. Woody Plant Regeneration in Four Floodplain Forests. *Ecol. Monogr.* **1994,** *64* (3), 345–367.

Khan, M. L.; Rai, J. P. N.; Tripathi, R. S. Population Structure of Some Tree Species in Disturbed and Protected Subtropical Forests of North-East India. *ACTA OECOL. (OECOL. APPL.)* **1987,** *8* (3), 247–255.

Khumbongmayum, A. D.; Khan, M. L.; Tripathi, R. S. Survival and Growth of Seedlings of a Few Tree Species in the Four Sacred Groves of Manipur, Northeast India. *Curr. Sci.* 2005, 1781–1788.

Kräuchi, N.; Brang, P.; Schönenberger, W. Forests of Mountainous Regions: Gaps in Knowledge and Research Needs. *Forest Ecol. Manage.* **2000,** *132* (1), 73–82.

Laurance, W. F. Emerging Threats to Tropical Forests. *Ann. Missouri Bot. Garden* **2015,** 100, 159–169.

Laurance, W. F.; Ferreira, L. V.; Rankin-De Merona, J. M.; Laurance, S. G.; Hutchings, R. W.; Lovejoy, T. E. Effects of Forest Fragmentation on Recruitment Patterns in Amazonian Tree Communities. *Conserv. Biol.* **1998,** *12* (2), 460–464.

Lieberman, D.; Li, M. Seedling Recruitment Patterns in a Tropical Dry Forest in Ghana. *J. Veg. Sci.* **1992,** *3* (3), 375–382.

Malik, Z. A. Phytosociological Behaviour, Anthropogenic Disturbances and Regeneration Status along an Altitudinal Gradient in Kedarnath Wildlife Sanctuary (KWLS) and Its Adjoining Areas. *Garhwal, Uttarakhand: PhD Thesis HNB Garhwal University Srinagar.*

Malik, Z. A.; Pandey, R.; Bhatt, A. B. Anthropogenic Disturbances and Their Impact on Vegetation in Western Himalaya, India. *J. Mountain Sci.* **2016,** *13* (1), 69–82.

Mayer, M.; Matthews, B.; Rosinger, C.; Sandén, H.; Godbold, D. L.; Katzensteiner, K. Tree Regeneration Retards Decomposition in a Temperate Mountain Soil after Forest Gap Disturbance. *Soil Biol. Biochem.* **2017,** *115*, 490–498.

Mohandass, D.; Campbell, M. J.; Davidar, P. Impact of Patch Size on Woody Tree Species Richness and Abundance in a Tropical Montane Evergreen Forest Patches of South India. *J. Forest. Res.* **2018,** *29* (6), 1675–1687.

Müllerová, J.; Vítková, M.; Vítek, O. The Impacts of Road and Walking Trails Upon Adjacent Vegetation: Effects of Road Building Materials on Species Composition in a Nutrient Poor Environment. *Sci. Total Environ.* **2011,** *409* (19), 3839–3849.

Oza, G. M. Threat to Chir and Fir Forests of Kashmir. *Environ. Conserv.* **1980,** *7* (1), 31–32.

Oza, G. M. Save Trees, Save Our Biosphere! *Environ. Conserv.* **1982,** *9* (3), 255–256.

Oza, G. M. Destruction of Forests and Wildlife in the Kashmir Wilderness. *Environmentalist* **2003,** *23* (2), 189–192.

Pant, S.; Samant, S. S. Diversity and Regeneration Status of Tree Species in Khokhan Wildlife Sanctuary, North-Western Himalaya. *Tropical Ecol.* **2012,** *53* (3), 317–331.

Pokhriyal, P.; Uniyal, P.; Chauhan, D. S.; Todaria, N. P. Regeneration Status of Tree Species in Forest of Phakot and Pathri Rao Watersheds in Garhwal Himalaya. *Curr. Sci.* **2010,** 171–175.

Rawat, D. S.; Dash, S. S.; Sinha, B. K.; Kumar, V.; Banerjee, A.; Singh, P. Community Structure and Regeneration Status of Tree Species in Eastern Himalaya: A Case Study from Neora Valley National Park, West Bengal, India. *Taiwania* **2018,** *63* (1), 16–24.

Reth, S.; Reichstein, M.; Falge, E. The Effect of Soil Water Content, Soil Temperature, Soil pH-Value and the Root Mass on Soil CO_2 Efflux–a Modified Model. *Plant Soil* **2005,** *268* (1), 21–33.

Sagar, R.; Singh, J. S. Local Plant Species Depletion in a Tropical Dry Deciduous Forest of Northern India. *Environ. Conserv.* **2004,** 55–62.

Samant, S. P. S. Assessment of Plant Diversity and Prioritization of Communities for Conservation in Mornaula Reserve Forest. *Appl. Ecol. Environ. Res.* **2007,** *5* (2), 123–138.

Saxena, A. K.; Singh, J. S. Tree Population Structure of Certain Himalayan Forest Associations and Implications Concerning Their Future Composition. *Vegetatio* **1984,** *58* (2), 61–69.

Schoenholtz, S. H.; Van Miegroet, H.; Burger, J. A. A Review of Chemical and Physical Properties as Indicators of Forest Soil Quality: Challenges and Opportunities. *Forest Ecol. Manage.* **2000,** *138* (1–3), 335–356.

Seipel, T.; Kueffer, C.; Rew, L. J.; Daehler, C. C.; Pauchard, A.; Naylor, B. J.; … Cavieres, L. A. Processes at Multiple Scales Affect Richness and Similarity of Non-Native Plant Species in Mountains Around the World. *Global Ecol. Biogeogr.* **2012,** *21* (2), 236–246.

Shankar, U. A Case of High Tree Diversity in a Sal (Shorearobusta)-Dominated Lowland Forest of Eastern Himalaya: Floristic Composition, Regeneration and Conservation. *Curr. Sci.* **2001,** 776–786.

Singh, S.; Malik, Z. A.; Sharma, C. M. Tree Species Richness, Diversity, and Regeneration Status in Different Aak (Quercus spp.) Dominated Forests of Garhwal Himalaya, India. *J. Asia-Pacific Biodiversity* **2016,** *9* (3), 293–300.

Swanson, M. E.; Franklin, J. F.; Beschta, R. L.; Crisafulli, C. M.; DellaSala, D. A.; Hutto, R. L.; … Swanson, F. J. The Forgotten Stage of Forest Succession: Early-Successional Ecosystems on Forest Sites. *Front. Ecol. Environ.* **2011,** *9* (2), 117–125.

Taylor, A. H.; Zisheng, Q. Regeneration Patterns in Old-Growth Abies-Betula Forests in the Wolong Natural Reserve, Sichuan, China. *J. Ecol.* **1988,** 1204–1218.

Team RC. *R: A Language and Environment for Statistical Computing*; R Foundation for Statistical Computing: Vienna, Austria, 2014.

Tripathi, R. S.; Khan, M. L. Regeneration Dynamics of Natural Forests. *Proc. -Indian Natl. Sci. Acad.* **2007,** *73* (3), 167.

Uhl, C.; Clark, K.; Dezzeo, N.; Maquirino, P. Vegetation Dynamics in Amazonian Treefall Gaps. *Ecology* **1988,** *69* (3), 751–763.

Vetaas, O. R. The Effect of Environmental Factors on the Regeneration of Quercussemecarpifolia Sm. in Central Himalaya, Nepal. *Plant Ecol.* **2000,** *146* (2), 137–144.

Vitousek, P. M.; Denslow, J. S. Nitrogen and Phosphorus Availability in Treefall Gaps of a Lowland Tropical Rainforest. *J. Ecol.* **1986,** 1167–1178.

Xu, Q.; Jiang, P.; Xu, Z. Soil Microbial Functional Diversity under Intensively Managed Bamboo Plantations in Southern China. *J. Soils Sediments* **2009,** *8* (3), 177.

CHAPTER 9

Assessment of Standing Biomass, Forest Structure, and Edaphic Properties in the Temperate Forest of Kashmir Himalaya

SHIEKH MARIFATUL HAQ[1], MUSHEERUL HASSAN[2*], HUMA HABIB[3], and GULZAR AHMED RATHER[4]

[1]*University of Kashmir, Srinagar, India*

[2]*Clybay Private Limited, Bangalore, India*

[3]*Islamia College of Science and Commerce, Srinagar, India*

[4]*Sathyabama Institute of Science and Technology, Chennai, India*

Corresponding author. E-mail: musheer123ni@gmail.com

ABSTRACT

Holding one of the world's most abundant carbon (C) pools, the forests represent opportunities for economic climate change mitigation, still on average, millions of hectares of forests disappear annually, often ruining the welfare of local communities. Here, we investigated the standing biomass, forest structure, and microbial activities in the temperate forests of the Kashmir Himalaya. Total tree biomass was documented to be 292.03 ± 139.25 mg/ha, with above-ground biomass as 217.82 ± 103.97 mg/ha and belowground biomass as 74.21 ± 35.28 mg/ha. The highest (220.53 mg/ha) and lowest (0.04 mg/ha) biomass values were recorded for *Cedrus deodara* and *Morus nigra,* and thereby contributing 75.51% and 0.01% to the total tree biomass, respectively. The average tree density was recorded as 190 trees/ha. The most dominant species was *Cedrus deodara* with an average density of 117.5 trees/ha, followed by *Pinus wallichiana* (67.5 trees/ha), *Morus nigra* (2.5 trees/ha), and *Aesculus indica* (2.5 trees/ha). All the individuals of the tree species had the average diameter at breast height (DBH) as 78.65 cm. Meanwhile, the highest and lowest average DBH values of 175.43 and 5.5 cm were recorded in *Cedrus deodara* and *Morus nigra*, respectively. The enzyme activity of dehydrogenase, alkaline phosphatase, and acid phosphatase was observed as 247.45 ± 4.3, 57.02 ± 2.7, and 45.92 ± 1.5, respectively. The results showed that biomass of the forest is a composite variable that integrates diverse structural and functional attributes, thereby linking "growing stock volume density", basal area, height, and wood density, and hence plays a notable role in carbon (C) accumulation. The outcome of the present study provides the baseline data for effective management and policymaking to lessen regional warming and mitigate the elevated level of atmospheric carbon (C) by planting tree species having high carbon storage potential.

9.1   INTRODUCTION

Among the terrestrial ecosystems, the forests act as the carbon sink. They exert a vital influence in regulating the global carbon cycle (Coulston et al., 2015; Espírito-Santo et al., 2014; Pan et al., 2011; Beer et al., 2010). Forest ecosystems play a critical role in various carbon sequestration pools and accumulate carbon (C) quickly (Pan et al., 2011), thereby proving to be a pioneering step for preventing global warming. Thus, it makes forest conservation critically important for mitigating climate change (Agrawal et al., 2011). The Himalayan forests are viewed to be one among the world's most threatened ecosystems (Duke, 1994; Schickhoff, 1995; Shaheen et al., 2011). It has been ascribed to enormous pressure from the rising human population, which is associated with land-use variations, socioeconomic transformation, and unsustainable utilization of the forest resources (Upadhyay et al., 2005). In the current context of climate change and its mitigation, forests have attracted particular attention because of their significant role in sequestering carbon from the atmosphere (Bellassen et al., 2014). When the forests are disturbed by human or natural causes, they become a source of carbon (C) (Luyssaert et al., 2008). The anthropogenic activities like deforestation, land degradation (Bryan et al., 2010), agricultural intensification (Dale et al., 1997), logging (Nepstad et al., 1999; Putz et al., 2001), fire (Barbosa and Fearnside, 1999), land-use changes (Rashid et al., 2017), human settlements (Yanai et al., 2017) have led to an increase in atmospheric carbon dioxide (CO_2) resulting in global warming and climate change (Buizer et al., 2014).

Several attributes, such as geographical features, ecological differences, climatic conditions, tree species composition, stem density, sampling strategies, and seasonal variations in forest structure (Melkania, 2009), and also basal area, height, and wood density values, age structure, disturbance frequencies and use of generalized allometric models for biomass estimation (Franklin et al., 2002; Rosenfield and Souza, 2013), and forest type (Wei et al., 2013; Zhang et al., 2013) influence the forest biomass and carbon (C). Forest biomass is a complex property. Although it is partly limited by tree height, it also integrates various ecological and functional processes as well as numerous structural attributes, thereby linking growing stock density, basal area and height, and wood density (Pan et al., 2013). The maximum capacity for aboveground biomass and carbon (C) repository in forests is generally found within the tree biomass (Son et al., 2001; Peichl and Arain, 2006). Tree species composition is essential for carbon storage in regions of the same climate range (Chen et al., 2011). Timber harvesting, land use, climate change, and several other naturals, as well as anthropogenic-generated interferences notably, affect the forest biomass (Canadell et al., 2007). At the national and international levels, intensified apprehension for climate change has led to enhanced attention on sustainable carbon management in forestry (FAO, 2010). Therefore, the scientific studies on temporal changes in forest biomass assume priority to predict the future carbon (C) accumulation in these forest ecosystems.

Biomass and carbon inventories of forests are significantly important for regions in which scientific studies are scarce, especially in "Jammu and Kashmir" (northwestern part of the Himalayan region in India). With the intense focus on the increasing levels of atmospheric carbon dioxide (CO_2) and the potential for global climate change, there is a grave need to evaluate the possibility of managing forest ecosystems to sequester and accumulate carbon (C) (Johnson and Kern, 2002). Estimation of carbon (C) pools in various forest systems could help to initiate proper carbon (C) management within forest ecosystems.

Soil reflects and arbitrates humankind's effects on the health and ramification of the ecosystem. Continuous efforts are seen throughout the globe to develop indicators that could give the information of soil quality, effects of pollutions on soils, the activity of forests, and agriculture based on the potential of soil, biogeochemical ecosystem processes in soil. Soil enzyme activity is a potential indicator that is closely related to various aspects like organic matter, microbial activity, and it provides information about soil health change. Enzyme activity can be done any time in a laboratory, so it is less complicated than the physical and biological measurements that require complex procedures. The concept of enzyme activity is based on the observation of biotic (micro-organisms) and abiotic components. Abiotic components like humid soil inhabit the biotic components, expression of enzymes have been related to the degree of recovery of corresponding microbial life (Dick, 2005). The most notable enzyme known for the microbial indication is dehydrogenase (Dick, 2005). These enzymes (dehydrogenases) are intracellular and mostly play a key role in the initial phases of oxidation of soil organic matter (Januszek et al., 2015). The soil microorganisms produce proteases (enzyme/extracellular), which are widely distributed in the soil (Ladd et al., 1972). Proteases catalyze the breakdown of protein components into simple amino acids. Phosphatases are a group of catalytic enzymes, known for the catalysis of anhydrides and esters of phosphoric acid (H_3PO_4). Soil phosphates (acid phosphate and alkaline phosphate) are responsible for the organic phosphorous (P) changes in soil (Eichler et al., 2004).

It is in this context that the current study was carried out to study the assessment of standing biomass, forest structure, and edaphic properties in the temperate forest of Kashmir Himalaya.

9.2   MATERIALS AND METHODS

9.2.1   STUDY AREA

District Kupwara is one of the 10 districts in the Kashmir Division of Jammu & Kashmir, India (Fig. 9.1) with the geographical area $2379km^2$ and mean altitude of 5300 feet. It represents the backward frontier of Kashmir Himalayas with the line of control (LOC) bordering its north-western part, and from the south is situated District Baramulla. Outer areas of the District are traversed by the river Kishen-Ganga, which originates from the Himalayas and maintains its flow from east to west. The dominant vegetation of the area is represented by semi-evergreen to evergreen coniferous forests. Shrub forest vegetation intersperses the coniferous evergreen forests at higher altitudes. Patches of grassland meadows are also quite common here.

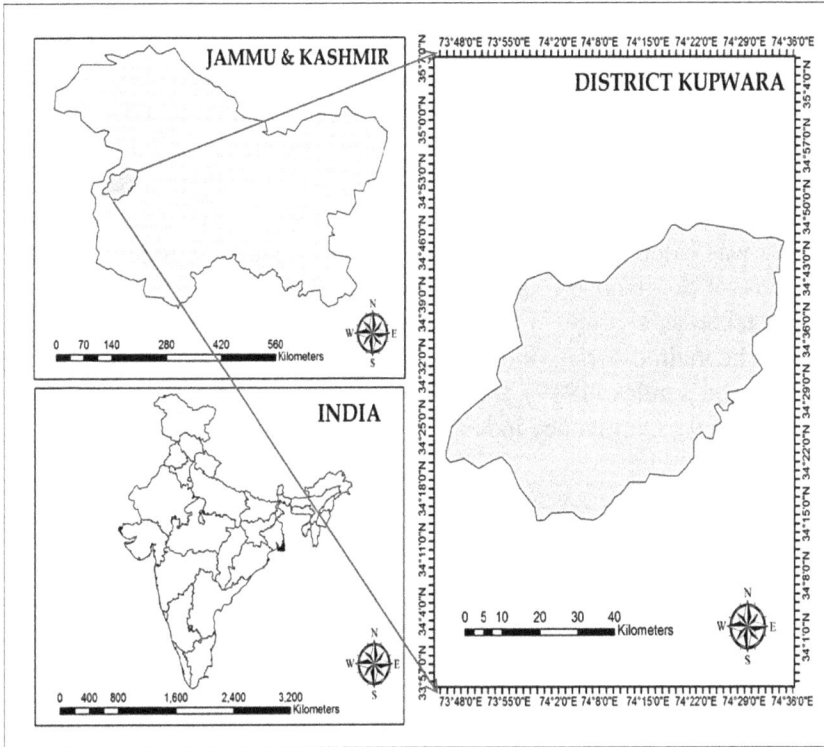

FIGURE 9.1 Map depicting the GPS location of the present study.

9.2.2 SAMPLING DESIGN AND MEASUREMENTS

To get an understanding of the nature of the terrain, species composition, accessibility, and distribution of different forest types in the study area, field reconnaissance surveys were conducted. For authenticating geographical location, administrative jurisdiction, and forest vegetation type, the working plan of the Kehmil Forest Division was consulted. The quadrant method of vegetation sampling was used to record floristic diversity data of the study area. In each of the selected forest types, four sample plots, each of the plots measuring 31.6×31.6m ($\cong$ 0.1ha) size in all the four directions, that is, northeast (NE), northwest (NW), southwest (SW), and southeast (SE), respectively were laid for trees sampling. The density of live stem plus dead (loss by stump cuttings) was recorded in each sample plot. Accommodation of the differences in vegetation growth caused by variation in slope and other aspects was ensured to minimize any sampling error as far as possible. GPS device named Garmin, GPS map76cs was used for the assessment of altitude as well as the measurement of forest sites.

The importance value index (IVI) was used to determine the dominance of the plant species. Based on the relative frequency, relative density, relative abundance, the IVI of the shrub and herb layer was estimated using the formula given by Curtis and McIntosh (1951). The IVI of tree species was calculated as the sum of relative frequency, relative density, and relative dominance (Naidu and Kumar, 2016). Species richness was simply taken as a count of the total number of species in that particular forest. The main diversity indices like the Shannon–Wiener index (1948), the Simpson's index (1949), the Margalef richness index, Evenness Index (Pielou, 1969), Dominance index were calculated.

9.2.3   SOIL COLLECTION

The vegetation sampling was followed by soil sampling on the same locations (plots of 31.6×31.6m ($\cong$ 0.1ha). Soil samples were transported to the laboratory (Department of Botany, University of Kashmir, India) following the standard procedure. The soil samples were fractionated into two parts: one part was air-dried and selected for chemical analysis while another fraction was stored for biological analysis. The soil samples were stored at a temperature of 4°C. Colorimetric method (Maurya et al., 2011) was adopted for estimating dehydrogenase and phosphate activity, and 2-3-5 triphenyl tetrazolium chloride (TPC) was selected as the substrate for dehydrogenase activity while p-nitrophenyl phosphate was selected as the substrate for phosphatase activity. The protease activity was evaluated by determining the total amount of tyrosine released from 1g oven-dry soil sample when incubated with 5mL, 50mM tris buffer (pH 8.1) and 5mL 2% of Na-caseinate at 50°C for 2h (Ladd et al., 1972). Soil samples were taken from each plot and were sieved via a 2mm mesh screen. Electrical conductivity and soil pH were measured using electrical conductivity meter and digital pH meter respectively, organic carbon was measured via wet digestion, available phosphorous (P) was determined using the ascorbic acid method, nitrogen was determined by the Kjeldahl nitrogen method. Whereas other physicochemical parameters were determined by following the methodology adopted by Gupta (2008).

9.2.4 ESTIMATION OF BIOMASS/CARBON

At each site, the measurements of "diameter at breast height" (DBH) and tree heights of all the individual trees with $\geq$10 cm DBH at 1.37 m height aboveground were sampled. Clinometer and Ravi multimeter were used respectively to calculate the slope angle and the tree height. The growing stock volume density (GSVDm^3/ha) was evaluated using volume equations developed for the individual tree species (FSI, 1996). For tree species for which the volume equations were unavailable (e.g., *Aesculus indica*), the GSVD was calculated with the help of volume equations of congeneric species with comparable, form, canopy, growth rate, and height. The Aboveground Biomass Density (AGBD) of the tree parts (stem, branches, twigs, and leaves) was calculated by multiplying GSVD of the sample plot with appropriate biomass expansion factor (BEF) (Brown et al., 1999).

Below given are the equations used to calculate BEF's for hardwood, pine species (Sharma et al., 2010). The equation used for the hardwood species (*Morusnigra* and *Aesculus indica*) is given below:

$$\text{Hardwood: for GSVD} \leq 200\text{m}^3/\text{ha, BEF} = \exp$$
$$\{1.91{-}0.34 \times \ln(\text{GSVD})\}$$
$$\text{for GSVD} > 200\text{m}^3/\text{ha, BEF} = 1.0$$
$$\text{Spruce–Fir: for GSVD} \leq 160\text{m}^3/\text{ha, BEF} = \exp \{1.77{-}0.34 \times \ln(\text{GSVD})\}$$
$$\text{for GSVD} > 160\text{m}^3/\text{ha, BEF} = 1.0$$
$$\text{Pine: for GSVD} < 10\text{m}^3/\text{ha, BEF} = 1.68$$
$$\text{for GSVD} = 10{-}100\text{m}^3.\text{ha, BEF} = 0.95$$
$$\text{for GSVD} > 100\text{m}^3/\text{ha, BEF} = 0.81.$$

The equation that was used for spruce-fir was applied to other conifer tree species (e.g., *Cedrus deodara*). The belowground density (BGBD); (fine and coarse roots) was estimated for each tree species by using the regression equation (Cairns et al., 1997): BGBD = exp $\{-1.059 + 0.884 \times \ln(\text{AGBD}) + 0.284\}$. Summation of AGBD to BGBD was used to estimate the Total Biomass Density (TBD); and the Total Carbon Density (TCD) was calculated by using the formula: TCD (C mg/ha) = TBD (mg/ha) $\times$ Carbon %. Carbon Stock in a forest stand is generally considered as 50% of the biomass (Brown et al., 1995). Therefore, to estimate the carbon stock (C) of the sampled site, total biomass values were multiplied with 0.5.

9.3   RESULTS

9.3.1   FOREST COMMUNITY DESCRIPTION

9.3.1.1   CEDRUS DEODARA–PARROTIOPSIS JACQUEMONTIANA–STIPA SIBIRICA COMMUNITY

A total of 28 species were recorded at the Site-I of the HDTF. The tree layer comprised of four species with *Cedrus deodara* as the dominant tree having the highest IVI value of 178.44. The codominant species in terms of decreasing values of IVI were represented by *Pinus wallichiana* (98.84), *Aesculus indica* (11.36), and *Morus nigra (*11.34*)*. In the case of the shrub layer, three species were recorded. *Parrotiopsis jacquemontiana* had the highest IVI value of 143.38, followed by *Vibrurnum grandiflorum* (95.79), and *Sorbariato mentosa* (60.82). The herbaceous layer comprised of 21 species with *Stipa sibirica* as the dominant species with the highest IVI of 36.73. Some of the codominant species were *Fragaria nubicola* (34.99) and *Digitalis purpurea (*32.25*)*; while other herbaceous species were quite rare such as *Pteracanthus alatus (*5.96*)*, *Chenopodium album (*3.62*)* (Table 9.1).

9.3.2   SPECIES DISTRIBUTION PATTERN

The abundance of frequency ratio for the tree layer of *Cedrus deodara* and *Pinus wallichiana* indicated that they exhibited a contagious distribution pattern, whereas *Aesculus indica* and *Morus nigra* showed a random pattern. In the case of the shrub layer, *Parrotiopsis jacquemontiana* and *Sorbaria tomentosa* exhibited contagious distribution pattern; *Vibrurnum grandiflorum* which showed a random pattern. In herb layers, all the species showed a contagious pattern expect *Digitalis purpurea*, *Polygonum amplexicaule,* and *Viola odorata* which showed a random pattern (Table 9.1).

9.3.3   SPECIES RICHNESS, DOMINANCE, DIVERSITY, AND EVENNESS

Species richness based on the number of species present was 28. The value of dominance based on the Simpson index was 0.06. The value for species diversity based on the Shannon–Wiener index was 2.99. The value of species evenness was 0.71.

TABLE 9.1 Quantitative Phytosociological Attributes of Individual Species.

Name of the species	AB	FR	AB/FR ratio	DN	DM	RDM	RDN	RFR	IVI
				Tree layer					
Cedrus deodara	11.75	100	0.11	11.75	27.39	76.61	61.84	40	178.44
Pinus wallichiana	6.75	100	0.06	6.75	8.33	23.31	35.52	40	98.84
Aesculus indica	1	25	0.04	0.25	0.01	0.04	1.31	10	11.36
Morus nigra	1	25	0.04	0.25	0.01	0.02	1.31	10	11.34
				Shrub layer					
Parrotiopsis jacquemontiana	6.5	100	0.06	6.5	58.75	63.68	43.33	36.36	143.38
Viburnum grandiflorum	5.5	100	0.05	5.5	21	22.76	36.66	36.36	95.79
Sorbaria tomentosa	6	75	0.08	3	12.5	13.55	20	27.27	60.82
				Herb layer					
Chenopodium album	3	20	0.15	0.6	2	1.03	0.58	2	3.62
Digitalis purpurea	4.2	100	0.04	4.2	35	18.13	4.11	10	32.25
Dyropteris stewartii	10.5	40	0.26	4.2	15	7.77	4.11	4	15.88
Fragaria rubicola	31.67	60	0.52	19	20	10.36	18.62	6	34.99
Geranium napalense	8	60	0.13	4.8	3	1.55	4.71	6	12.26
Hedera helix	3	20	0.15	0.6	4	2.07	0.58	2	4.66
Hedera nepalensis	7.5	40	0.18	3	3	1.55	2.94	4	8.49
Oxalis acetosella	22.67	60	0.37	13.6	5	2.59	13.33	6	21.92
Phytolacca acinosa	3	20	0.15	0.6	12	6.21	0.58	2	8.81
Polygonum amplexicaule	3	80	0.03	2.4	15	7.77	2.35	8	18.12
Pteracanthus alatus	7	20	0.35	1.4	5	2.59	1.37	2	5.96

TABLE 9.1 *(Continued)*

Name of the species	AB	FR	AB/FR ratio	DN	DM	RDM	RDN	RFR	IVI
Rumex nepalensis	2.5	40	0.06	1	8	4.14	0.98	4	9.12
Stellaria media	8.5	40	0.21	3.4	3	1.55	3.34	4	8.88
Stipa sibirica	24.75	80	0.30	19.8	18	9.32	19.41	8	36.73
Trifolium pretense	6	20	0.3	1.2	3	1.55	1.17	2	4.73
Viola odorata	4.25	80	0.05	3.4	2	1.03	3.33	8	12.36
Asplenium ofeliae	9.25	80	0.11	7.4	15	7.78	7.25	8	23.02
Sambucus wightiana	2	20	0.1	0.4	15	7.77	0.39	2	10.16
Adiantum capilus veneris	15	60	0.25	9	4	2.07	8.82	6	16.89
Clinopodium vulgare	4.5	40	0.11	1.8	2	1.03	1.76	4	6.81
Cynoglossum lanceolatum	3	20	0.15	0.2	4	2.07	0.19	2	4.26

AB, Abundance; DM, dominance; DN, density; FR, frequency; IVI, Importance value index; RDM, relative dominance; RDN, relative density; RFR, relative frequency.

9.3.4 TREE CHARACTERISTICS

Average tree density was recorded as 190 trees/ha. The average density of *Cedrus deodara* was found to be 117.5 trees/ha (most dominant species), followed by *Pinus wallichiana* 67.5 trees/ha, *Morus nigra* and *Aesculus indica* 2.5 trees/ha each. The average DBH value of all the individuals of tree species was 78.65 cm, and the highest and lowest average tree DBH values of 175.43 and 5.5cm were recorded in *Cedrus deodara* and *Morus nigra*, respectively. The average height value of all the individuals of tree species was 16.58 m, and the tallest and smallest tree height values of 40.12 and 1.75 m were recorded in *Cedrus deodara* and *Morus nigra,* respectively (Table 9.2).

9.3.5 FOREST BIOMASS

The total tree biomass was recorded to be 292.03 ±139.25 mg/ha, with aboveground biomass as 217.82 ±103.97 mg/ha and belowground biomass as 74.21 ±35.28 mg/ha. The highest and lowest biomass values of 220.53 and 0.04 mg/ha were recorded for *Cedrus deodara* and *Morus nigra,* and thereby contributing 75.51% and 0.01% to the total tree biomass, respectively (Table 9.2). The low density of mature tree individuals was recorded in lower and middle girth classes (18%), which contributes only 2% of biomass and maximum density in the higher girth class was 82%, which contributes 98% of biomass (Fig. 9.2).

9.3.6 FOREST CARBON STOCK

With the contribution of AGCD and BGCD as 100.21 ±47.82 and 34.13 ±16.22 mg/ha, respectively, the average total carbon stock (TCD) was estimated to be 134.33 ± 64.05 mg/ha (range: 71.17–195.19). The average carbon dioxide (CO_2) assimilation value was estimated to be 492.56 ±234.86 mg CO_2/ha (range: 260.98–715.71) (Table 9.3). DBH was lowest (111.12cm) on the NW aspect and highest (163.57 cm) on the SE aspect. AGBD varied between a minimum of 115.58 mg/ha on the SW aspect and a maximum of 316.91 mg/ha on the NE aspect. Values of BGBD ranged from 39.15 mg/ha on the SW aspect to 107.42 mg/ha on the NE aspect.

TABLE 9.2 Density, Diameter at Breast Height (DBH), Height and Species Wise Contribution in Aboveground Biomass Density (AGBD), Belowground Biomass Density (BGBD), Total Biomass Density (TBD), Total Carbon Density (TCD), and CO_2 assimilation.

Tree species	Density/ ha	DBH (cm)	Height (m)	AGBD	BGBD	TBD	TCD	CO_2 assimilation	Percentage contribution in biomass
Cedrus deodara	117.5	175.43	40.12	164.84	55.68	220.53	101.44	371.96	75.51
Pinus wallichiana	67.5	121.45	21.34	52.06	18.12	70.19	32.28	118.39	24.03
Morus nigra	2.5	5.5	1.75	0.01	0.03	0.04	0.02	0.07	0.01
Aesculus indica	2.5	12.25	3.12	0.91	0.36	1.26	0.58	2.14	0.43
Average	–	**78.65**	**16.58**	–	–	–	–	–	–
Total	**190**	–	–	**217.82**	**74.21**	**292.03**	**134.33**	**492.56**	**100**

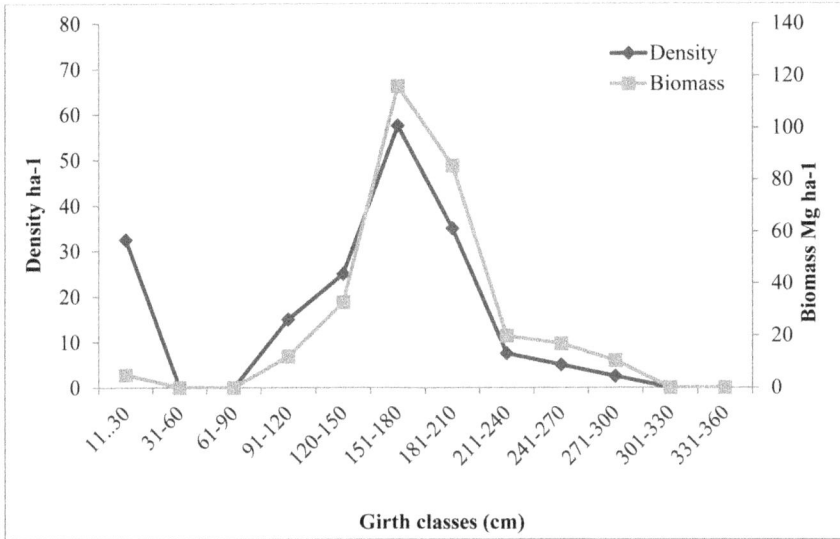

FIGURE 9.2 Comparative girth class wise contribution of density and biomass in the study area.

The total biomass density (TBD) was minimum on the SW aspect (154.73Mg/ha) and maximum on the NE aspect (424.33 mg/ha). The total tree carbon density (TCD) was found to be minimum on the SE aspect (71.17CMg/ha) and found to be maximum on the NE aspect (195.19CMg/ha). The carbon dioxide (CO_2) assimilation was recorded maximum on the NE aspect as 715.71MgCO_2/ha and a minimum of 260.98 on the SW aspect. The decreasing trend was NE>NW >SE>SW (Table 9.3).

9.3.7 ENZYMATIC ACTIVITY

Enzymatic activities in the forests of Keran valley (34°34' 0" –34°42' 30" N & 73°55' 15"–74°17' 05" E) varied to some extent across selected plots (Table. 9.4). Maximum enzymatic activity in terms of dehydrogenase was observed in Plot.1 followed by Plot. 2, Plot. 3, and Plot. 4, respectively. Compared with dehydrogenase activity, acid phosphatase activity was found to be very low. Maximum activity (acid phosphate) was observed in Plot. 2, meanwhile Plot. 3, Plot. 4, and Plot. 1 followed respectively. Compared with dehydrogenase activity, alkaline phosphatase activity was low while it was seen higher than acid phosphatase activity. Enzymatic

TABLE 9.3 Diameter at Breast Height (DBH), Height, and Plot-Wise Contributions in Aboveground Biomass Density (AGBD), Belowground Biomass Density (BGBD), Aboveground Carbon Density (AGCD), Belowground Carbon Density (BGCD), Total Biomass Density (TBD), Total Carbon Density (TCD), and CO_2 Assimilation.

Forest Stand	DBH (cm)	Height (m)	AGBD	BGBD	AGCD	BGCD	TBD	TCD	CO_2 assimilation
Plot 1 (NE)	124.99	26.08	316.91	107.42	145.78	49.41	424.33	195.19	715.71
Plot 2 (NW)	111.12	24.62	297.36	101.59	136.78	46.73	398.95	183.51	672.91
Plot 3 (SE)	163.57	36.32	141.45	48.66	65.06	22.38	190.12	87.45	320.67
Plot 4 (SW)	114.79	22.82	115.58	39.15	53.16	18.01	154.73	71.17	260.98
Average	128.62	27.46	217.82	74.21	100.21	34.13	292.03	134.33	492.56
	±24.02	±6.05	±103.97	±35.28	±47.82	±16.22	±139.25	±64.05	±234.86

activity (alkaline phosphatase) was found higher in Plot. 3. Plot. 4 was followed by Plot. 1 and Plot. 2, respectively. In the case of proteases, Plot. 1 showed the highest enzymatic activity followed by the other Plots (2, 3, and 4) respectively. On average, from all four selected plots, dehydrogenase activity was 247.5±8.53, followed by protease activity (67.37±5.57), alkaline phosphatase activity (57.025±5.24), and acid phosphatase activity (45.925±1.56).

TABLE 9.4 Enzyme Activities of Microbes in the Different Plots of the Study Site.

Enzyme activity	Plot 1	Plot 2	Plot 3	Plot 4	Average/STDEV
Dehydrogenase	258.7	248.7	243.7	238.7	247.45 ±8.53
Acid phosphatase	44.1	47.9	46.1	45.6	45.925±1.56
Alkaline phosphatase	53.2	52.5	63.7	58.7	57.025±5.24
Protease	72.3	69.5	68.3	59.4	67.375±5.57

9.4 DISCUSSION

The disturbance is the main force that changes the forest structure, creates landscape mosaics, and sets initial conditions for the successional phase and structural development (Swanson et al., 2011; Pan et al., 2013). The traits that describe a forest structure are individual structural characteristics of trees, spatial patterns (vertical or horizontal) components, and tree-size distribution (Franklin et al., 2002). The development of forest ecosystem functioning and contribution to the structural complexity are formed by the structural characteristics of trees (living or dead), together with other biotic and abiotic features (Pan et al., 2013). The size-dependent distribution of trees represents the population structure of a forest. The present study revealed that the density and total live biomass varied due to differences in community structure, size of girth classes, site conditions, and degree of disturbance (size of stem cuttings and rate of mortality, etc.) that can be temporary or permanent (Gairola et al., 2011; Sharma et al., 2011; Zhao et al., 2014).

The total tree biomass was recorded to be 292.03 ±139.25 mg/ha, with aboveground 217.82 ±103.97 mg/ha and belowground 74.21 ±35.28 mg/ha which lies closer to or within the range of other forests of the Himalayan region. Ahmad et al. (2014) while estimating the biomass of coniferous forest reported the biomass values (284.81 mg/ha) in mixed conifer. Dar

226 Climate Change and Microbial Diversity

and Sundarapandian (2015) observed the variation of carbon (C) pools with forest type in temperate forests of Anantnag District of Kashmir. It was found that the total ecosystem carbon stocks varied from 112.5 to 205.7 mg C/ha across all the forest types. Also, the vegetation type, stand structure, management history, and altitude influenced the variability of Carbon (C) pools. Recently, the biomass and carbon (C) stock in five temperate forests of Northern Kashmir was assessed by Dar and Sahu (2018). It was revealed that the total carbon (C) stock varies from 143.63 ±93.87 to 228.47 ±128 mgC/ha across all forest types. Singh et al. (2012) assessed the aboveground phytomass (AGP) in the temperate forest of Kashmir Valley and observed that AGP varied from 94.14 to 229.68 mg/ ha in *Cedrus deodara* forests.

In the present study, the AGBD and BGBD to the total biomass density in the temperate forests of Kashmir Himalaya were recorded as 75% and 25%, respectively. Our results are within the range reported by (Chhabra et al., 2002; Sharma et al., 2018) for Indian Himalayan Forest.

During the present study, variability in biomass and carbon (C) sequestration potential of tree species was not only dependent on tree density but also on the size of the tree, rate of productivity, human disturbances, and type of forest management. The species-level differences in biomass stock were also quite revealing, and there was a disproportionate contribution of individual tree species to the total biomass stock in the different forest types of the study area. On comparison, coniferous tree species *Cedrus deodara*, *Pinus wallichiana* contributed maximum biomass stocking potential. Similar results have also been reported from other neighboring Himalayan regions (Negi et al., 2003; Sharma et al., 2010) The slow-growing, long-lived tree species, and mature natural forest stands, having large girth class and with dense wood, density possesses a large amount of biomass stock (Sharma et al., 2010; Gairola et al., 2011). This can be a reason for the accumulation of biomass and carbon in these coniferous tree species. The current study findings are further supported by other similar studies in the Himalayan region, such as studies of Sharma et al. (2016) in the forests of Garhwal Himalaya; Dar et al. (2017) in the forests of Western Himalaya, India; Banday et al. (2018) in subtropical forests of Northwestern Himalaya; Dar and Sahu (2018) in temperate forests of Northern Kashmir Himalaya, India; Mannan et al. (2019) in the Himalayan forests of Northern Pakistan which reported the conifer tree species to have maximum biomass potential as compared witho broad-leaved tree species.

In the present study, we have recorded a maximum (424.33 mg/ha) total biomass density (TBD) on the NE aspect and minimum (154.73 mg/ha) on the SW aspect. Our results are in accordance with the results of Sharma et al. (2011) which attributed this variation to the presence of the moisture and conducive environment on the northern sides. As the north-facing slopes receive less sunlight, they are relatively cooler than the south-facing slopes which are warmer and drier due to higher insolation periods during the day, and because of this there are better growing conditions on the northern aspects than on the southern aspects.

The tree-size distribution of forest and the time since the last disturbance are essential attributes of the forest structure. The forest age is related to many deterministic processes during forest structural development, such as GSVD, mortality, biomass, and litter accumulation (Spies, 1998). Korner (2017) reported that the carbon storage potential for the forest ecosystems is controlled by the tree longevity rather than the growth rate. Size distributions are shifted toward trees with larger girth classes by high carbon stocks (Brienen et al., 2015). In the present study, the low density of mature tree individuals was recorded in lower and middle girth classes 18% and contributed only 2% of the biomass. Low biomass in the lower to middle girth class is likely due to complete loss of middle girth class in the forest structure, and also carbon (C) storage potential is quite low at younger tree ages (Khol et al., 2017).

The forest structure is dependent on the regeneration status of species, species diversity, and the ecological characteristics of sites. In forest ecosystems, tree species maintain the overall physiognomy (Jones et al., 1994), and thus fundamentally explain the structural complexity and environmental heterogeneity. In a climax forest ecosystem, trees are the fundamental components that influence the resources and habitats of almost all other forest organisms (Rawat et al., 2018). Thus, for a comprehendible structure of a community as well as for developing and enacting the conservation strategies, it is essential to have quantitative information regarding the distribution and abundance of tree species. The study shows that DBH and height directly influence carbon (C) accumulation (Yam and Tripathi, 2016). In the present study, DBH value ranges from 5.5 to 175.43 cm with an average value of 78.65 cm which is lower than the value 123 cm, recorded in Terai and Mahabharat Foothills of the region of Nepal (Baral et al., 2009). DBH value of 200 cm was recorded in Central Himalayas (Nautiyal and Singh, 2013) and the value of 250 cm was recorded in US forests (Jenkins et al., 2003) and higher value than average DBH of 39 cm was recorded in the Evergreen forest of Lidder valley, Kashmir Himalaya, India (Rashid et al., 2017), 87.27 cm in subtropical forest Himalayas (Shaheen et al., 2016). Similarly, the average height, 16.58 m, recorded in this study is also lower than the reported value of 26.85–30.05 m in the forests of Nepal (Shrestha et al., 2013) and 18.8–35.1 m as recorded in Indian subtropical forest (Mishra et al., 2009), 13.3 m as in subtropical forest Himalayas (Shaheen et al., 2016), and almost similar to 10.2–25.2 m as recorded in subtropical forests of Northwestern Himalaya (Banday et al., 2017).

The average stem density recorded was recorded as 190trees /ha in this study. The previous studies have shown that in Southern Kashmir, the stem density values ranged between 110 and 530 N/ha (Wani et al., 2014), and it ranged between 103 and 1201 N/hain the Himalayan region (Dar and Sundarapandian, 2015), between 483 and 417 trees/ha in subtropical forests of Northwestern Himalaya (Banday et al., 2018), 492 trees/ha in a subtropical forest type of Himalayas (Shaheen et al., 2016). In lesser Himalayas, it was recorded as 534–620 trees/ha (Ahmed et al., 2006), 530–940 trees/ha in Kumaun Himalayas (Kharkwal, 2009), 380–626 trees/ha (Sharma et al., 2010) and 295–850 trees/ha (Gairola et al., 2010) in Garhwal Himalaya; 1158 trees/ha in the western Himalayas (Sundriyal et al., 1994); 420 and 1300 N/ha in Kumaun Himalaya (Saxena and Singh, 1982). 149.99 trees/ha was reported by Akash and Bhandari (2019) from Garhwal Himalaya, India. One of the reasons for the lower tree density in the current study could be attributed to the extraction of the fuelwood and firewood, unmanaged grazing, and use of NTFPs (non-timber forest products) (Czegledi and Radacsi, 2005; Shaheen et al., 2016). The low stem density reflected the horrible deforestation intensity and immense unchecked anthropogenic pressure (road constructions, fire, logging, and fuelwood practices) on the local forest reserve.

The health of the soil is very important from the ecological point of view. As per different authors (Lemanowicz et al., 2018; Blonska et al., 2017), enzymatic tests are important sources of natural and human impacts on the functioning of the ecosystem. These tests include dehydrogenase, protease, acid phosphatase, and alkaline phosphatase. Soil dehydrogenase is the very important indicator of soil health (Maurya et al., 2011), giving the idea that soil inhabits the good microbiology. The present study revealed that the selected study area "keran" with four selected plots

(plots 1, 2, 3, 4) has notable dehydrogenase soil activity. Increased soil dehydrogenase activity can be attributed to the increased organic carbon (C) and other nutrients which are because of the litter-fall. The working of a group of extracellular enzymes that are present in the soil is reflected by the soil dehydrogenase activity (Maurya et al., 2011). A soil sample can show fluctuation in the dehydrogenase activity when it is air-dried depending upon the nature and horizon of the soil (Roa et al., 2003). The soil dehydrogenase activity is also dependent upon the high organic matter present in the soil surfaces (Nannipieri et al., 2012). The results from Table.9.4 reveal that acid and alkaline phosphatase activities at all selected plots (Plot. 1, Plot. 2, Plot. 3, and Plot. 4) were low as compared with dehydrogenase activity. The availability of low phosphorous in the soil remains the main constraint in worldwide agricultural production especially in places where inorganic fertilizers are unavailable (Maurya et al., 2011). Acid phosphatase enzymes are contributed by the plant roots and micro-organisms in the soil (Masato et al., 2008). Various factors are affecting the activity of enzymes, these include the presence of organic matter, availability of nitrogen interaction between enzyme and the soil (Lemanowiz et al., 2018). When compared with acid phosphatase activity, alkaline phosphatase activity was recorded to be high at selected plots. It has been observed that the phosphatase activity depends upon factors like soil properties, plan cover, leachate inputs, and the presence of inhibitors and activators (Speir and Ross, 1978), the pH of soil also plays a vital role in the enzymatic activity (Saha et al., 2008). Proteases are produced by a wide range of bacteria, actinomycetes, and fungi (Kumar and Takagi, 1999). In the present study, protease activity was not seen much better than dehydrogenase activity but was good enough than phosphatase (acid, and alkaline) activity. When there is an insufficient level of carbon (C), nitrogen (N), or sulfur (S) protease production may be increased (Geisseler et al., 2008) so vice-versa. Certain studies confer that high levels of NH4+ reduce the production of proteases (Beg et al., 2002; Bascaran et al., 1990).

9.5 CONCLUSION

The findings of the present study helped in filling the knowledge gaps in scientific studies on the forest ecosystems in this eco-fragile part of the

Himalayan region and hopefully would help in diminishing the regional climate change by minimizing the anthropogenic impacts within these forest ecosystems in the region, results also highlighted that there is an urgent need for scientific measures to be taken especially by local government to improve the forests so that they can be conserved for sustainable utilization.

ACKNOWLEDGMENT

The authors are great full to the Department of the Botany University of Kashmir for their kind support in providing laboratory facilities to carry out the enzymatic experiments.

KEYWORDS

- carbon stock
- biomass
- tree structural attributes
- enzyme activity
- Kashmir Himalaya

REFERENCES

Agrawal, A.; Nepstad, D.; Chhatre, A. Reducing Emissions from Deforestation and Forest Degradation. *Annu. Rev. Environ. Resour.* **2011,** *36,* 373–396.

Ahmad, A.; Mirza, S. N.; Nizami, S. M. Assessment of Biomass and Carbon Stocks in Coniferous Forest of Dir Kohistan, KPK. *Pak. J. Agric. Sci.* **2014,** *51* (2).

Ahmed, M.; Husain, T.; Sheikh, A. H.; Hussain, S. S.; Siddiqui, M. F. Phytosociology and Structure of Himalayan Forests from Different Climatic Zones of Pakistan. *Pak. J. Bot.* **2006,** *38* (2), 361.

Akash, N.; Bhandari, B. S. A Community Analysis of Woody Species in a Tropical Forest of Rajaji Tiger Reserve. *Environ. Ecol.* **2019,** *37* (1), 48–55.

Banday, M.; Bhardwaj, D. R.; Pala, N. A. Variation of Stem Density and Vegetation Carbon Pool in Subtropical Forests of Northwestern Himalaya. *J. Sustain. Forest.* **2018,** *37* (4), 389–402.

Baral, S. K.; Malla, R.; Ranabhat, S. Above-Ground Carbon Stock Assessment in Different Forest Types of Nepal. *BankoJanakari* **2009**, *19* (2), 10–14.

Barbosa, R. I.; Fearnside, P. M. Incêndiosna Amazônia brasileira: Estimativa da emissão de gases do efeitoestufa pela queima de diferentesecossistemas de Roraima napassagem do evento El Niño (1997/98). *Acta Amaz.* **1999**, *29* (4), 513–534.

Bascarán, V.; Hardisson, C.; Braña, A. F. Regulation of Extracellular Protease Production in Streptomyces Clavuligerus. *Appl. Microbiol. Biotechnol.* **1990**, *34* (2), 208–213.

Beer, C.; Reichstein, M.; Tomelleri, E.; Ciais, P.; Jung, M.; Carvalhais, N.; … Bondeau, A. Terrestrial Gross Carbon Dioxide Uptake: Global Distribution and Covariation with Climate. *Science* **2010**, *329* (5993), 834–838.

Beg, Q. K.; Saxena, R. K.; Gupta, R. De-repression and Subsequent Induction of Protease Synthesis by Bacillus Mojavensis under Fed-Batch Operations. *Process Biochem.* **2002**, *37* (10), 1103–1109.

Bellassen, V.; Luyssaert, S. Carbon Sequestration: Managing Forests in Uncertain Times. *Nature* **2014**, *506* (7487), 153–155.

Błońska, E.; Kacprzyk, M.; Spolnik, A. Effect of Deadwood of Different Tree Species in Various Stages of Decomposition on Biochemical Soil Properties and Carbon Storage. *Ecol. Res.* **2017**, *32* (2), 193–203.

Brienen, R. J.; Phillips, O. L.; Feldpausch, T. R.; Gloor, E.; Baker, T. R.; Lloyd, J.; … Martinez, R. V. Long-Term Decline of the Amazon Carbon Sink. *Nature* **2015**, *519* (7543), 344–348.

Brown, S. L.; Schroeder, P.; Kern, J. S. Spatial Distribution of Biomass in Forests of the Eastern USA. *Forest Ecol. Manage.* **1999**, *123* (1), 81–90.

Bryan, J.; Shearman, P.; Ash, J.; Kirkpatrick, J. B. Estimating Rainforest Biomass Stocks and Carbon Loss from Deforestation and Degradation in Papua New Guinea 1972–2002: Best Estimates, Uncertainties and Research Needs. *J. Environ. Manage.* **2010**, *91* (4), 995–1001.

Buizer, M.; Humphreys, D.; de Jong, W. Climate Change and Deforestation: The Evolution of an Intersecting Policy Domain.

Canadell, J. G.; Le Quéré, C.; Raupach, M. R.; Field, C. B.; Buitenhuis, E. T.; Ciais, P.; … Marland, G. Contributions to Accelerating Atmospheric CO2 Growth from Economic Activity, Carbon Intensity, and Efficiency of Natural Sinks. *Proc. Natl. Acad. Sci.* **2007**, *104* (47), 18866–18870.

Chen, D.; Zhang, C.; Wu, J.; Zhou, L.; Lin, Y.; Fu, S. Subtropical Plantations Are Large Carbon Sinks: Evidence from Two Monoculture Plantations in South China. *Agric. Forest Meteorol.* **2011**, *151* (9), 1214–1225.

Chhabra, A.; Palria, S.; Dadhwal, V. K. Growing Stock-Based Forest Biomass Estimate for India. *Biomass Bioenergy* **2002**, *22* (3), 187–194.

Coulston, J. W.; Wear, D. N.; Vose, J. M. Complex Forest Dynamics Indicate Potential for Slowing Carbon Accumulation in the Southeastern United States. *Sci. Rep.* **2015**, *5*, 8002.

Curtis, J. T.; McIntosh, R. P. An Upland Forest Continuum in the Prairie-Forest Border Region of Wisconsin. *Ecology* **1951**, *32* (3), 476–496.

Dale, V. H.; Pearson, S. M. Quantifying Habitat Fragmentation Due to Land-Use Change in Amazonia. In *Tropical Forest Remnants: Ecology, Management, and Conservation of Fragmented Communities*; Laurance, W. F.; Bierregaard, R. O., Eds.; University of Chicago Press: Chicago, IL, 1997.

Dar, D. A. Assessment of Biomass and Carbon Stock in Temperate Forests of Northern Kashmir Himalaya, India. *Proc. Int. Acad. Ecol. Environ. Sci.* **2018,** *8* (2), 139.

Dar, J. A.; Sundarapandian, S. Variation of Biomass and Carbon Pools with Forest Type in Temperate Forests of Kashmir Himalaya, India. *Environ. Monitor. Assess.* **2015,** *187* (2), 55.

Dar, J. A.; Rather, M. Y.; Subashree, K.; Sundarapandian, S.; Khan, M. L. Distribution Patterns of Tree, Understorey, and Detritus Biomass in Coniferous and Broad-Leaved Forests of Western Himalaya, India. *J. Sustain. Forest.* **2017,** *36* (8), 787–805.

Dick, R. P.; Kandeler, E. Enzymes in Soils/Reference Module in Earth Systems and Environmental Sciences, from Encyclopedia of Soils in the Environment, 2005.

Duke, G. A Participatory Approach to Conservation Safeguarding the Himalayan Forests of the Palas Valley, District Kohistan. *The Destruction of the Forests and Wooden Architecture of Eastern Afghanistan and Northern Pakistan: Nuristan to Baltistan. Islamabad, Pakistan: Asian Study Group,* 1994; pp 40–48.

Eichler, B.; Köppen, D.; Caus, M.; Schnug, E. Soil acid and Alkaline Phosphatase Activities in Regulation to Crop Species and Fungal Treatment. *LandbauforschungVolkenrode* **2004,** *54* (1), 1–5.

Espírito-Santo, F. D.; Gloor, M.; Keller, M.; Malhi, Y.; Saatchi, S.; Nelson, B.; … Palace, M. Size and Frequency of Natural Forest Disturbances and the Amazon Forest Carbon Balance. *Nat. Commun.* **2014,** *5* (1), 1–6.

Franklin, J. F.; Spies, T. A.; Van Pelt, R.; Carey, A. B.; Thornburgh, D. A.; Berg, D. R.; … Bible, K. Disturbances and Structural Development of Natural Forest Ecosystems with Silvicultural Implications, Using Douglas-Fir Forests as an Example. *Forest Ecol. Manage.* **2002,** *155* (1–3), 399–423.

Gairola, S.; Sharma, C. M.; Ghildiyal, S. K.; Suyal, S. Live Tree Biomass and Carbon Variation along an Altitudinal Gradient in Moist Temperate Valley Slopes of the Garhwal Himalaya (India). *Curr. Sci.* **2011,** 1862–1870.

Gairola, S.; Sharma, C. M.; Rana, C. S.; Ghildiyal, S. K.; Suyal, S. Phytodiversity (Angiosperms and Gymnosperms) in Mandal-Chopta Forest of Garhwal Himalaya, Uttarakhand, India. *Nat. Sci.* **2010,** *8* (1), 1–17.

Geisseler, D.; Horwath, W. R. Regulation of Extracellular Protease Activity in Soil in Response to Different Sources and Concentrations of Nitrogen and Carbon. *Soil Biol. Biochem.* **2008,** *40* (12), 3040–3048.

Gupta, N.; Sharma, R. C.; Tripathi, A. K. Study of Bio-Physico-Chemical Parameters of Mothronwala Swamp, Dehradun (Uttarakhand). *J. Environ. Biol.* **2008,** *29* (3), 381.

Januszek, K.; Długa, J.; Socha, J. Dehydrogenase Activity of Forest Soils Depends on the Assay Used. *Int. Agrophys.* **2015,** *29* (1), 47–59.

Jenkins, J. C.; Chojnacky, D. C.; Heath, L. S.; Birdsey, R. A. National-Scale Biomass Estimators for United States Tree Species. *Forest Sci.* **2003,** *49* (1), 12–35.

Johnson, M. G.; Kern, J. S. Quantifying the Organic Carbon Held in Forested Soils of the United States and Puerto Rico. In *The Potential of US Forest Soils to Sequester Carbon and Mitigate the Greenhouse Effect*; CRC Press, 2002; pp 56–81.

Jones, R. H.; Sharitz, R. R.; Dixon, P. M.; Segal, D. S.; Schneider, R. L. Woody Plant Regeneration in Four Floodplain Forests. *Ecol. Monogr.* **1994,** *64* (3), 345–367.

Kharkwal, G. Qualitative Analysis of Tree Species in Evergreen Forests of Kumaun Himalaya, Uttarakhand, India. *Afr. J. Plant Sci.* **2009,** *3* (3), 049–052.

Köhl, M.; Neupane, P. R.; Lotfiomran, N. The Impact of Tree Age on Biomass Growth and Carbon Accumulation Capacity: A Retrospective Analysis Using Tree Ring Data of Three Tropical Tree Species Grown in Natural Forests of Suriname. *PloS one* **2017,** *12* (8), e0181187.

Körner, C. A Matter of Tree Longevity. *Science* **2017,** *355* (6321), 130–131.

Kumar, C. G.; Takagi, H. Microbial Alkaline Proteases: From a Bioindustrial Viewpoint. *Biotechnol. Adv.* **1999,** *17* (7), 561–594.

Ladd, J. N.; Butler, J. H. A. Short-Term Assays of Soil Proteolytic Enzyme Activities Using Proteins and Dipeptide Derivatives as Substrates. *Soil Biol. Biochem.* **1972,** *4* (1), 19–30.

Lemanowicz, J. Dynamics of Phosphorus Content and the Activity of Phosphatase in Forest Soil in the Sustained Nitrogen Compounds Emissions Zone. *Environ. Sci. Pollut. Res.* **2018,** *25* (33), 33773–33782.

Luyssaert, S.; Schulze, E. D.; Börner, A.; Knohl, A.; Hessenmöller, D.; Law, B. E.; … Grace, J. Old-Growth Forests as Global Carbon Sinks. *Nature* **2008,** *455* (7210), 213–215.

Mannan, A.; Zhongke, F.; Khan, T. U.; Saeed, S.; Amir, M.; Khan, M. A.; Badshah, M. T. Variation in Tree Biomass and Carbon Stocks with Respect to Altitudinal Gradient in the Himalayan Forests of Northern Pakistan. *J. Pure Appl. Agric.* **2019,** *4* (1), 21–28.

Masto, R. E.; Chhonkar, P. K.; Purakayastha, T. J.; Patra, A. K.; Singh, D. Soil Quality Indices for Evaluation of Long-Term Land Use and Soil Management Practices in SemiAarid SubTtropical India. *Land Degrad. Dev.* **2008,** *19* (5), 516–529.

Maurya, B. R.; Singh, V.; Dhyani, P. P. Enzymatic Activities and Microbial Population in Agric-Soils of Almora District of Central Himalaya as Influenced by Altitudes. *Int. J. Soil Sci.* **2011,** *6,* 238–248.

Melkania, N. P. Carbon Sequestration in Indian Natural and Planted Forests. *Indian Forester* **2009,** *135* (3), 380.

Mishra, A.; Nautiyal, S.; Nautiyal, D. P. Growth Characteristics of Some Indigenous Fuelwood and Fodder Tree Species of Sub-Tropical Garhwal Himalayas. *Indian Forester,* *135* (3), 373.

Naidu, M. T.; Kumar, O. A. Tree Diversity, Stand Structure, and Community Composition of Tropical Forests in Eastern Ghats of Andhra Pradesh, India. *J. Asia-Pacific Biodiversity* **2016,** *9* (3), 328–334.

Nannipieri, P.; Giagnoni, L.; Renella, G.; Puglisi, E.; Ceccanti, B.; Masciandaro, G.; … Marinari, S. A. R. A. Soil Enzymology: Classical and Molecular Approaches. *Biol. Fertility Soils* **2012,** *48* (7), 743–762.

Nautiyal, N.; Singh, V. Carbon Stock Potential of Oak and Pine Forests in Garhwal Region in Indian Central Himalayas. *J. Pharmacognosy Phytochem.* **2013,** *2* (1).

Negi, J. D. S.; Manhas, R. K.; Chauhan, P. S. Carbon Allocation in Different Components of Some Tree Species of India: A New Approach for Carbon Estimation. *Curr. Sci.* **2003,** *85* (11), 1528–1531.

Nepstad, D. C.; Verssimo, A.; Alencar, A.; Nobre, C.; Lima, E.; Lefebvre, P.; … Cochrane, M. Large-Scale Impoverishment of Amazonian Forests by Logging and Fire. *Nature* **1999,** *398* (6727), 505–508.

Pan, Y.; Birdsey, R. A.; Fang, J.; Houghton, R.; Kauppi, P. E.; Kurz, W. A.; … Ciais, P. A Large and Persistent Carbon Sink in the World's Forests. *Science* 2011, *333* (6045), 988–993.

Pan, Y.; Birdsey, R. A.; Phillips, O. L.; Jackson, R. B. The Structure, Distribution, and Biomass of the World's Forests. *Annu. Rev. Ecol. Evol.Syst.* **2013,** *44,* 593–622.

Peichl, M.; Arain, M. A. Above-and Belowground Ecosystem Biomass and Carbon Pools in an Age-Sequence of Temperate Pine Plantation Forests. *Agric. Forest Meteorol.* **2006,** *140* (1–4), 51–63.

Pielou, E. Association Tests Versus Homogeneity Tests: Their Use in Subdividing Quadrats into Groups. *Vegetatio* **1969,** *18* (1–6), 4–18.

Putz, F. E.; Blate, G. M.; Redford, K. H.; Fimbel, R.; Robinson, J. Tropical Forest Management and Conservation of Biodiversity: An Overview. *Conserv. Biol.* **2001,** *15* (1), 7–20.

Radácsi, L. C. A. Overutilization of Pastures by Livestock. *Acta pascuorum (Grassland Studies) 2005, 3,* 29–36.

Rao, M. A.; Sannino, F.; Nocerino, G.; Puglisi, E.; Gianfreda, L. Effect of Air-Drying Treatment on Enzymatic Activities of Soils Affected by Anthropogenic Activities. *Biol. Fertility Soils* **2003,** *38* (5), 327–332.

Rashid, I.; Bhat, M. A.; Romshoo, S. A. Assessing Changes in the Above Ground Biomass and Carbon Stocks of Lidder Valley, Kashmir Himalaya, India. *Geocarto Int.* **2017,** *32* (7), 717–734.

Rawat, D. S.; Tiwari, J. K.; Tiwari, P.; Nautiyal, M.; Praveen, M.; Singh, N. Tree Species Richness, Dominance and Regeneration Status in Western Ramganga Valley, Uttarakhand Himalaya, India. *Indian Forester* **2018,** *144* (7), 595–603.

Rosenfield, M. F.; Souza, A. F. Biomass and Carbon in Subtropical Forests: Overview of Determinants, Quantification Methods and Estimates. *Neotrop. Biol. Conserv.* **2013,** *8* (2), 103–110.

Saha, S.; Mina, B. L.; Gopinath, K. A.; Kundu, S.; Gupta, H. S. Organic Amendments Affect Biochemical Properties of a Subtemperate Soil of the Indian Himalayas. *Nutr. Cycl. Agroecosyst.* **2008,** *80* (3), 233–242.

Saxena, A. K.; Singh, J. S. A Phytosociological Analysis of Woody Species in Forest Communities of a Part of Kumaun Himalaya. *Vegetatio* **1982,** *50* (1), 3–22.

Schickhoff, U. Himalayan Forest-Cover Changes in Historical Perspective: A Case Study in the Kaghan Valley, Northern Pakistan. *Mountain Res. Dev.* **1995,** 3–18.

Shaheen, H.; Qureshi, R. A.; Ullah, Z.; Ahmad, T. Anthropogenic Pressure on the Western Himalayan Moist Temperate Forests of Bagh, Azad Jammu & Kashmir. *Pak. J. Bot.* **2011,** *43* (1), 695–703.

Shaheen, H.; Khan, R. W. A.; Hussain, K.; Ullah, T. S.; Nasir, M.; Mehmood, A. Carbon Stocks Assessment in Subtropical Forest Types of Kashmir Himalayas. *Pak. J. Bot.* **2016,** *48* (6), 2351–2357.

Sharma, C. M.; Baduni, N. P.; Gairola, S.; Ghildiyal, S. K.; Suyal, S. Tree Diversity and Carbon Stocks of Some Major Forest Types of Garhwal Himalaya, India. *Forest Ecol. Manage.* **2010,** *260* (12), 2170–2179.

Sharma, C. M.; Mishra, A. K.; Krishan, R.; Tiwari, O. P.; Rana, Y. S. Variation in Vegetation Composition, Biomass Production, and Carbon Storage in Ridge Top Forests of High Mountains of Garhwal Himalaya. *J. Sustain. Forest.* **2016,** *35* (2), 119–132.

Sharma, C. M.; Tiwari, O. P.; Rana, Y. S.; Krishan, R.; Mishra, A. K. Elevational Behaviour on Dominance–Diversity, Regeneration, Biomass and Carbon Storage in Ridge Forests of Garhwal Himalaya, India. *Forest Ecol. Manage.* **2018,** *424,* 105–120.

Shrestha, K. B.; Mâren, I. E.; Arneberg, E.; Sah, J. P.; Vetaas, O. R. Effect of Anthropogenic Disturbance on Plant Species Diversity in Oak Forests in Nepal, Central Himalaya. *Int. J. Biodiversity Sci. Ecosyst. Serv. Manage.* **2013,** *9* (1), 21–29.

Simpson, E. H. Measurement of Diversity. *Nature* **1949,** *163* (4148), 688–688.

Singh, S.; Patil, P.; Dadhwal, V. K.; Banday, J. R.; Pant, D. N. Assessment of Above-Ground Phytomass in Temperate Forests of Kashmir Valley, J&K, India. *Int. J. Ecol. Environ. Sci.* **2012,** *38* (2–3), 47–58.

Son, Y.; Hwang, J. W.; Kim, Z. S.; Lee, W. K.; Kim, J. S. Allometry and Biomass of Korean Pine (Pinus Koraiensis) in Central Korea. *Bioresour. Technol.* **2001,** *78* (3), 251–255.

Speir, T. W.; Ross, D. J. Soil Phosphatase and Sulphatase. *Soil Enzymes* **1978,** *203,* 197–250.

Spies, T. A. Forest Structure: A Key to the Ecosystem. *Northwest Sci.* **1998,** *72,* 34–36.

Sundriyal, R. C.; Sharma, E.; Rai, L. K.; Rai, S. C. Tree Structure, Regeneration and Woody Biomass Removal in a Sub-Tropical Forest of Mamlay Watershed in the Sikkim Himalaya. *Vegetatio* **1994,** *113* (1), 53–63.

Swanson, M. E.; Franklin, J. F.; Beschta, R. L.; Crisafulli, C. M.; DellaSala, D. A.; Hutto, R. L.; … Swanson, F. J. The Forgotten Stage of Forest Succession: Early-Successional Ecosystems on Forest Sites. *Front. Ecol. Environ.* **2011,** *9* (2), 117–125.

Upadhyay, T. P.; Sankhayan, P. L.; Solberg, B. A Review of Carbon Sequestration Dynamics in the Himalayan Region as a Function of Land-Use Change and Forest/Soil Degradation with Special Reference to Nepal. *Agric. Ecosyst. Environ.* **2005,** *105* (3), 449–465.

Wei, Y.; Li, M.; Chen, H.; Lewis, B. J.; Yu, D.; Zhou, L.; … Dai, L. Variation in Carbon Storage and Its Distribution by Stand Age and Forest Type in Boreal and Temperate Forests in Northeastern China. *PloS one* **2013,** *8* (8).

Yam, G.; Tripathi, O. P. Tree Diversity and Community Characteristics in Talle Wildlife Sanctuary, Arunachal Pradesh, Eastern Himalaya, India. *J. Asia-Pacific Biodiversity* **2016,** *9* (2), 160–165.

Yanai, A. M.; Nogueira, E. M.; de AlencastroGraça, P. M. L.; Fearnside, P. M. Deforestation and Carbon Stock Loss in Brazil's Amazonian Settlements. *Environ. Manage.* **2017,** *59* (3), 393–409.

Zhang, Y.; Gu, F.; Liu, S.; Liu, Y.; Li, C. Variations of Carbon Stock with Forest Types in Subalpine Region of Southwestern China. *Forest Ecol. Manage.* **2013,** *300,* 88–95.

Zhao, J.; Kang, F.; Wang, L.; Yu, X.; Zhao, W.; Song, X.; … Han, H. Patterns of Biomass and Carbon Distribution across a Chronosequence of Chinese Pine (Pinus Tabulaeformis) Forests. *PLoS one* **2014,** *9* (4).

CHAPTER 10

Rapidly Changing Environment and Role of Microbiome in Restoring and Creating Sustainable Approaches

MANISHANKAR CHAKRABORTY[1], UDAYA KUMAR VANDANA[1], DEBAYAN NANDI[1], LAKKAKULA SATISH[2], and P. B. MAZUMDER[1*]

[1]*Department of Biotechnology, Assam University Silchar, India*

[2]*Department of Biotechnology Engineering, Ben-Gurion University of the Negev, Beer Sheva, Israel*

Corresponding author. E-mail: pbmmbl@gmail.com

ABSTRACT

Environmental degradation is on a rise with increase in deforestation, industrialization, pollution, population explosion, and rapid advancement of technology. Biomes including plants, animals, and microbes present in diverse environments across the globe like soil, ocean, plants, human beings are threatened. These habitats are found to have pivotal role in maintaining and restoring natural environment. Human interventions are impacting the function and diversity of these microbiome that play a pivotal role in maintaining and restoring natural environment, with serious implications. But surprisingly their capability to adapt to different environments and to evolve rapidly had helped them survive and contribute to the environment. This chapter discusses the scientific research on the diverse roles (such as carbon sequestering, bioremediation, alternative source of energy, rhizoremediation, and phytoremediation) played different microbiomes and microbial strains.

10.1 INTRODUCTION

Climate change is one of the intensely discussed topics of the twenty-first century. Global warming, as well as certain, unbalances in various biogeo-chemical cycles are a few of the reasons for these global environmental changes. With the advancements in technology, global energy require-ments have rapidly increased. To satisfy the energy needs of the world, a large number of petroleum-derived products are used. However, the combustion of such fossil fuels acts as a potent emitter of toxic substances and greenhouse gases that have many adverse effects in the environment such as pollution, greenhouse gases effect, and the global environmental changes (Ho et al., 2013). In addition, heavy metal accumulation in the environment is also a major problem in the present time. Although some of them act as co-factor for various metabolic activities in living beings. All metals can reveal adverse impact at increasing concentrations, and the injuriousness of respective metal depends upon the quantity provided to the organisms, the intake of dosage, mode of absorption, and the period of exposure time (Ojuederie and Babalola, 2017). This calls for an urgent need for the removal of such products from the environment.

Bioremediation is one such cost-effective process that can reduce the number of pollutants from the environment by means of microorganisms (Shanahan, 2004). The studies of the microbial world are of extreme importance in this aspect due to their participation in absorption and release of various greenhouse gases, their ability to recycle and transform various essential components, for instance, carbon and nitrogen, and their capability of degrading hydrocarbons and metallic elements in the environment (Das and Chandran, 2011; Dutta and Dutta, 2016; Hansda and Kumar, 2016).

Plants have also been found effective in removing or extracting contaminants in the soil as well as in water through phytoremediation (Adam, 2001). However, an integrated approach of combining phytoremediation with microbial remediation together constitutes rhizoremediation which can exhibit effective degradation of contaminants in soils in a more efficient way (Tang et al., 2010). Apart from below-ground surfaces, microbes also live in close association with plants in above-ground surfaces and are referred to as phyllosphere region. Like rhizosphere microbes, phyllosphere microorganisms also take part in activities such as nutrient cycling, pollutant degradation, etc. (Bao et al., 2019; Dharmasiri et al., 2019).

Microbes play a significant role in stabilizing the atmospheric concentrations of various greenhouse gases such as carbon dioxide (CO_2), methane (CH_4), and nitrous oxide (N_2O) (Singh et al., 2010). Carbon sequestration by microorganisms is found to be an effective tool in reducing the abundant CO_2 in the atmosphere and thus helps in mitigating the problems of global warming and climate change (Lal, 2008). Moreover, the use of biomass-derived biofuels such as biodiesel and bioethanol also serves as a potential source in reducing the risk of global warming by providing a significant reduction of emission of particulate matter, carbon monoxide (CO), and unburned hydrocarbons (Kecebas and Alkan, 2009). The third-generation algae-based biofuel is among the most preferred biofuel for its efficiency over food crops and non-food biomass-derived biofuel (Brennan and Owende, 2010; Noraini et al., 2014).

Microbes are abundantly present in nature and help in the exchange of essential elements, thus maintaining the cycles between land, ocean, and atmosphere. The proper knowledge regarding interactions of microbes with plants, animals, and other microbial communities can open a gateway for the development of new remediation strategies to get rid of the various sources of global climate change.

This chapter focuses on various approaches for mitigating the problem of climate change such as bioremediation techniques including rhizoremediation, phytoremediation, mycoremediation, and bacterial remediation in diminishing the number of petroleum contaminants and heavy metals in the environment, carbon sequestration by microorganisms, and production of algae-based eco-friendly biodiesel and bioethanol.

10.2 RHIZOREMEDIATION OF PETROLEUM CONTAMINANTS

Although petroleum is one of the important resources for global economic development, the contamination caused by petroleum hydrocarbons, however, has become a serious problem worldwide (Chen and Zhong, 2019). Rhizoremediation can be an effective strategy in such cases as it involves the degradation of organic pollutants in the soil rhizosphere through the shared association of plants and microbes (Correa-García et al., 2018; Ubogu et al., 2019). Here, plants through their exudates help in increasing microbial activity for petroleum hydrocarbon (PHC) degradation as well as stabilizing heavy metals (De Cárcer et al., 2007). More than

79 bacterial genera have been found that are capable of degrading PHCs (Tremblay et al., 2017). Efforts are made for increased degradation by combining strains of bacteria with different degrading capabilities in the form of consortium (Samarghandi et al., 2018).

10.2.1   INTERACTION BETWEEN PLANTS AND MICROBES

Interaction of plants with soil microorganisms can take place through various chemical signals which are mediated by plants in the form of certain biologically active compounds or root exudates (Bais et al., 2006). This capability of establishing a relationship with the soil microbes enables plants to be more tolerant to various biotic and abiotic stresses (Saravanan et al., 2020). Plant root exudates help in making pollutants more available for microbial degradation by increasing their solubility (Read and Perez-Moreno, 2003). Also, some components of root exudates can act as co-metabolites in the degradation process of contaminants (Ubogu et al., 2019). It has been found that it is the host plant which releases specific root exudates resulting in particular responses to pollutants (Hussain et al., 2018; Berendsen et al., 2012). Apart from plants, fungi in root rhizosphere also facilitate the bioavailability of hydrophobic substrates to bacteria through bacterial–fungal interactions by uptaking and translocating pollutants using fungal hyphae (Banitz et al., 2013). Examples of few plant–microbe relationships reported to be involved in PHC degradation include an interaction between bacterium *Rhodococcus sp* with Ryegrass (*Lolium perenne L.*) (Kukla et al., 2014) and strains of proteobacteria *Burkholderia fungorum* DBT1 with hybrid poplar plant (willow family) (Andreolli et al., 2013).

10.2.2   MECHANISM OF DEGRADATION OF PETROLEUM HYDROCARBON BY MICROORGANISMS

The degradation of petroleum hydrocarbon can be carried out by enzyme-catalyzed biodegradation under aerobic or anaerobic condition. In aerobic condition, the degradation is generally activated by enzyme oxygenase. Microbial degradation of hydrocarbons involves various approaches that include attachment of microbial cells to the substrates followed by biosurfactants production, organic acids production, and biofilm formation (Ossai et al., 2020; Das and Chandran, 2011). Biosurfactants are bacterial-secreted surface-active compounds having both hydrophilic and hydrophobic nature and are capable of emulsifying complex organic compounds into simpler forms followed by their removal. Bacteria such as *Pseudomonas, Bacillus subtilis*, and *Microbacterium* are reported to be able to degrade hydrocarbon-contaminated soils through the production of biosurfactants (Pathak and Keharia, 2014). Through biofilm, formation microorganisms attach themselves in aggregate over the surface of the matrix secreted by them. This enables them to degrade PHCs with proper protection (Saravanan et al., 2020).

10.2.3 STRATEGIES FOR IMPROVEMENT IN RHIZOREMEDIATION TECHNIQUE

Soil rhizosphere is a huge reservoir of various microbial communities which usually ranges between 1 and 2 mm from the soil surface to the depth. Root exudates are mainly responsible for enriching the soil rhizosphere that promotes increased growth and metabolism in microbes (Hinsinger et al., 2005). For rhizoremediation, tolerances against specific contaminants as well as their extensive root systems are two important prerequisites for plants (Adam and Duncan, 2002). Because of extensive roots and high surface, plants with fibrous root system can offer for more contact between degrading microbes and pollutants certain plant growth-promoting rhizobacteria (PGPR) are abundantly present in close association with root in the rhizosphere region. These PGPR with their intrinsic metabolic capabilities are able to mitigate abiotic stresses in plants. Earlier PGPR were used only for the protection against various plant diseases and uptake of nutrients while the present bioremediation technologies enable them to assist in PHC degradation in collaboration with other pollutant degrading microbes (Hussain et al., 2018). *Acinetobacter, Ralstonia Alcaligenes, Arthrobacter, Burkholderia, Mycobacterium, Micrococcus, Nocardioides, Pseudomonas, Rhodococcus, Sphingomonas, Flavobacterium, and Stenotrophomonas* species are some of the PHC-degrading bacterial strains (Ite and Ibok, 2019). The introduction of such bacteria can take place either in the form of a single bacterial strain or in the form of a bacterial consortium (Samarghandi et al., 2018) produced microbial consortium with bacterial strains of *Stenotrophomonas sp., Staphylococcus sp., Pseudomonas sp., Brevibacillus sp.,* and *Achromobacter sp* isolated from PHC-contaminated soil and found out that the microbial consortium was able to degrade 6% of the pollutant depicting the effectiveness of such consortium in pollutant degradation (Landa-Acuña et al., 2020).

Rhizoremediation of contaminants has also been carried out using certain genetically modified microorganisms by De Cárcer et al., (2007) where Salix sp. plants were treated with two different genetically modified *Pseudomonas fluorescens* strains. Another approach for improved remediation of PHC-contaminated soil was the introduction of multiprocess phytoremediation system by Huang et al., (2004) that comprises of three steps mainly soil aeration, PGPR inoculation, and plant-mediated phytoremediation.

10.2.4   FACTORS INFLUENCING RHIZOREMEDIATION PROCESS

Some of the factors found to be influencing the process of rhizoremediation are temperature, plant root exudates, microbial communities, site condition, type and number of contaminants, and availability of nutrients. Optimum pH and temperature favor microorganisms for more organic acids and enzyme secretion. Similarly, the availability of an optimum amount of nutrients in the soil is essential for healthy microbial growth (Saravanan et al., 2020).

Rhizoremediation also comprises certain limitations such as it is a slow and time-consuming approach and is not effective in dense or clayey soil or in circumstances where there is too much contamination or large distance between the root and contaminated zone (Correa-García et al., 2018). Despite such limitations, this type of bioremediation method is considered effective due to its sustainability, economical, and eco-friendly nature.

10.3   CARBON SEQUESTRATION

Carbon sequestration is the process by which atmospheric inorganic carbon in the form of CO_2 and CO is converted into useful or less harmful organic or inorganic forms by plants and some microbes. The rise of CO_2 emission being a problem in the current time carbon sequestration has proven to be very effective.

The following groups of organisms have been found very efficient:

1. Archaea
2. Clostridium
3. Proteobacteria
4. Algae
5. Cyanobacteria
6. Fungi

10.3.1 ARCHAEA

It is one of the three domains of the living kingdom that consists of single-celled prokaryotes living in extreme environments. Their adaptability to survive in extreme environments has given them extraordinary features of producing organic compounds from the toxic or inhabitable atmosphere such as high salinity, acidic environments, anaerobic environment, and extreme temperatures. Using CO_2 and hydrogen, they produce methane. For instance, *Methanothermobacter marburgensis, M. Thermautotrophicus*, and *Methanosarcina barkeri*, etc.

The most significant application of using methanogenic bacteria is their competence to produce methane as fuel from different substrates. Aceticlastic and hydrogenotrophic are two important groups of methanogens which are efficient in CO_2 sequestration and bioremediation simultaneously.

In sewage treatment where greenhouse gases like CO_2 are used as a substrate, some bacteria like *Methanobacteriaceae, Methanosarcinaceae*, and *Methanospirillaceae* spp. might be enhanced aerobically to produce methane as fuel (Yasin et al., 2015).

The CO_2 sequestration is also successfully accomplished by *Sulfolobus, Metallosphaera, Cenarchaeum*, and *Archaeoglobus* can associate CO_2 through the 3-hydroxypropionate-4-hydroxybutyrate chain. Usually, the enzyme acetyl-CoA or propionyl-CoA carboxylase is utilized in the reaction (Mistry et al., 2019).

The carbonic anhydrase enzyme can variably catalyze by hydrating the CO_2 very quickly (Jo et al., 2013). Thermophiles produce carbonic anhydrase inadequate amount necessary for industrial CO_2 capture.

The six pathways that can assimilate CO_2 into useful compounds are:

- 3-hydroxypropionate bicycle
- Calvin cycle

- Reductive citric acid cycle
- Dicarboxylate–hydroxybutyrate cycle
- Crenarchaeota group I
- Reductive acetyl-CoA pathway.

10.3.2   CLOSTRIDIUM

Clostridia are a genus of gram-positive bacteria. These obligately anaerobic bacteria play an important function in acidogenesis, anaerobic deterioration of organic supplements, carbon cycle, and polysaccharides, etc.

Clostridum spp. possesses the capacity to fix CO_2 and CO, for example, *Clostridium autoethanogenum*, CO_2, and CO fixes into central metabolite acetyl-CoA in the existence of H_2 through the Wood–Ljungdahl pathway (Liew et al., 2016). Apart from these, *Clostridium* spp. also produces commercially important products in particular acetate, acetone, butanol, caproate, caprylate, ethanol, lactate, and valproate through various metabolic pathways and additionally its tolerance to lethal metabolites. Many strains of the *Clostridium* genus can capture CO_2 or CO, as well use other carbon compounds viz., methanol and formate ($HCOO^-$), as the main source of carbon (Mistry et al., 2019).

Conversion of Wastes into Medium-Chain Fatty Acids

The medium-chain fatty acids are the substances that are used as a precursor molecule for biofuel production.

Conversion of Volatile Fatty Acids into Biogas

Biogas is produced in conventional anaerobic digestion process from volatile fatty acids (VFAs) in the presence of H_2 and CO_2 where they are converted into acetic acid by *Moorella thermoacetica* and *C. aceticum.*

Conversion of Weed Ipomoea sp. into Volatile Fatty Acids

Weed Ipomoea (Ipomoea carnea) was converted into VFAs in easy operating reactors. Some acidogenic and cellulolytic microorganisms found in cow dung were applied in the inoculant. VFAs production started immediately after blending of the reactant and reached peak by 10th and 11th day, converting above 10% of the total biomass toward VFAs. These VFAs are indirectly utilized as nourish in several anaerobic digesters for obtaining energy as methane fuel. This increases the scope of utilization of Ipomoea as an essential energy resource.

10.3.3 PROTEOBACTERIA

Proteobacteria is a gram-negative bacterium. They can sequester CO_2 using various metabolic pathways. Certain α-proteobacteria like *Oligotropha carboxidovorans, Xanthobacter flavus, Rhodobacter sphaeroides*, and *R. capsulatus* can sequester CO_2 via Calvin cycle. Both β- and γ-proteobacteria are identically capable of fixing CO_2 through the Calvin cycle. For example, few important β-proteobacteria are *R. eutropha* and *Herbaspirillum autotrophicum* and where γ-proteobacteria are *Acidithiobacillus thiooxidans, Hydrogenovibrio marinus*, and *A. ferrooxidans*. Some commercially useful chemicals are produced from CO_2 by *R. eutropha*, a β-proteobacteria. It can fix the carbon in cell cytoplasm in the form of polyhydroxyalkanates (PHAs), aka bioplastics. These deposited PHAs usually consist of compact chains, that is, poly-3-hydroxybutyrate-co-3-hydroxyvalerate, and poly-3-hydroxybutyrate (Albuquerque et al., 2011). A great advantage of this conversation is that PHAs are almost biodegraded and thus have lesser harm to the environment. Furthermore, *R. eutropha* has medicinal utilization as well (Zeng et al., 2010; Mistry et al., 2019).

Desulfobacter hydrogenophilus, α-proteobacteria can fix CO_2 through TCA cycle that begins by the splitting of citrate molecule into acetyl-CoA and oxaloacetate with the help of an enzyme ATP citrate lyase. Here, organic carbon compound is produced from water and CO_2 (Hügler et al., 2005).

β-proteobacteria is almost found in all places. Although salt marshes are β-proteobacteria and actinomycetes, C- and E-proteobacteria have strong carbon sequestration capability. This indicates toward the fact that saline wetlands may contain potent mineralization ability, whereas less salinity soil marshes favor elevated primary productivity. Likewise, relatively more abundances of c-proteobacteria are observed in low tidelands, implying that growth in inundation may also uphold carbon mineralization (Hu et al., 2016).

10.3.4 ALGAE

One of the most efficient microbes to carry out carbon sequestration is algae. Its adaptation and availability across the globe in different marine and freshwater bodies add to their role in carbon sequestration and photo-synthesis. They exist in varying sizes and mircroalgae to macroalgae.

10.3.4.1   MICROALGAE

Photosynthetic microalgae be a part of one of the eldest forms of life on globe. It includes organisms like diatoms, cyanobacteria, dinoflagellates, red-, brown-, and yellow-green algae, euglenoids and coccolithophores, and a myriad of unknown organisms.

The tremendous carbon sequestration capability is seen in microalgae. It is experimentally found that ~1 kg of cultured microalgae is able to secure ~2 kg of CO_2. Certain algal species like *Anabaena*, *Chlorella vulgaris*, *Nannochloropsis oculate*, and *Scenedesmus obliquus* use Calvin–Benson cycle and produce various biofuels and proteins (Mistry et al., 2019).

Extremophile microalgae are gaining an increasing interest due to their ability for growing under harsh conditions, therefore, letting outdoor cultivation with very little contamination risks and allowing exploit the economic advantages associated with the utilization of open raceways. (Malavasi et al., 2020).

10.3.4.2   MACROALGAE

Macroalgae and microalgae are alike in their photosynthetic capability. However, macroalgae is rather better than microalgae in terms of its wider adaptability varying from ponds to bioreactors and the intended research of several strains used for feeding fish (Aresta et al., 2005). The diverse functioning of macroalgae has been extended to energy production which has encouraged large scale production. They grow rapidly and have produce by-products rich in nutrients. Macroalgae-derived extracts provoke a large number of responses in agriculture specifically, increased plant biomass, greater chlorophyll content in leaves, improved nutrient uptake (nitrogen, potassium, and phosphorus), augmented flower/fruit set prominent to high yields, etc. (Satish et al., 2015, 2016).

The *Gracilaria cornea* strain is mass produced for animal feed. Flue gases are used as CO_2 source and found to be a cheap source at the same time increasing biomass yield. Macroalgae is a great cause for using in carbon reduction at the power plant will decrease the requirement to use commercially synthesized CO_2 in turn delivering cheaper biomass production (Farrelly et al., 2013). Increased levels of CO_2 enhance growth of macroalgal species namely *Gracilaria sp.*, *G. chilensis*, *Hizikia fusiforme*, and *Porphyra yezoensis*. Marine macroalgae also possess numerous essential demands in the alginate industry, biomedicine, and agriculture, etc. Optimum environmental conditions for maximum yield from macroalgae works wonder in carbon mitigation since it has high growth rates, etc. It is also used in human and animal consumption.

10.3.5 CYANOBACTERIA

They are photoautotrophic bacteria utilizing CO_2 to for photosynthesis, thus essential for carbon mitigation. Their optimum temperature for growth is 50–75°C in an anaerobic condition, accompanied by light and absence of N_2 for growth producing H_2 as a by-product (Farrelly et al., 2013).

Cyanobacteria have high energy densities compatibility to present full processing plants and application in food industries, alcohols are mostly produced and studied chemical produced by the strains *Synechococcus elongatus* 7942 and *Synechocystis spp.* 6803. Ethanol has been produced in high titters and improved productivity at 212 mg/L/day and 5.5 g/L in *Synechocystis* 6803.

Lactic acid is a widely used compound in pharmaceutical, food, and polymer industries was optimized for production in *Synechocystis* 6803 (Zhang et al., 2017).

10.3.6 FUNGI

Glomeral fungi in symbiotic relationship with rhizosphere produce a glycoprotein viz., Glomalin which is the most abundant found in soil. It is found to enhance the biological and physiochemical characteristics of the soil. The carbon captured by plants is transferred to the mycorrhizal fungus, converted into glomalin in hyphal and spores' cell wall. This phenomenon holds promising results in the view of rise in global carbon emission (Malyan et al., 2019).

Another significant role was observed by Fitter et al., (2000). Arbuscular mycorrhizal symbioses are essential for the carbon sequestration in various ways. In one of the many ways, it encourages phosphorus uptake of plant, thus acting as carbon sink for plant, reducing carbon in the atmosphere (Fitter et al., 2000).

10.4 ROLE OF PHYLLOSPHERE BACTERIA AND FUNGI IN MAINTAINING PLANT HEALTH

Plants reside in a close communication with the microbes equally above- and below-ground surface. Phyllosphere invokes to the over-ground levels or shoot parts of the plant which is the habitat for a variety of bacteria, fungi, and various other organisms (Wei et al., 2017; Chaudhary et al., 2017). Phyllosphere can be taken into consideration as temporary environment since most of the evergreen plants shed leaves throughout the year (Vorholt, 2012). Approximately 4×10^8 km² on the earth is covered by the phyllosphere which is a home for around 10^{26} bacteria (Wei et al., 2017; Kembel et al., 2014). Leaves are considered to be more influential aerial plant structure and are mostly focused on the study of microbiology of phyllosphere as compared to that of other parts in the phyllosphere.

10.4.1 DIVERSITY OF MICROORGANISMS IN THE PHYLLOSPHERE AND THEIR PROFILING

A diverse population of microorganisms consisting of different genera of bacteria, fungi, yeast, and other microbes resides in the phyllosphere and can establish a relationship with the host plant that can take place in the form of mutualism, commensalism, or parasitism (Rastogi et al., 2013). Among the wide variety of microbial populations, bacteria are the most abundant member inhabiting plant leaves, usually found in numbers up to 10^7 cells/cm² of the leaf (Dulla, 2005). Some of the most available bacterial communities in the phyllosphere include α-proteobacteria, such as *Sphingomonas* and *Methylobacterium*. γ-proteobacteria such as *Pseudomonas* may also have a large frequency of population (Kembel et al., 2014). Phyllosphere also constitutes cyanobacteria like *Scytonema*, *Nostoc*, and *Stigonema* (Vacher et al., 2016).

In the case of fungi, both epiphytic (fungi that are isolated from leaf washings and are in contact with the outer environment) and endophytic fungi (fungi that are in close contact with internal leaf tissue) (Yao et al.,

2019) are found in distinct microenvironments of phyllosphere. The fungal population is dominated by genera of Ascomycota that includes *Cladosporium*, *Aureobasidium*, and *Taphrina* (Kembel and Mueller, 2014). In addition, yeasts associated to the genera of *Cryptococcus* and *Sporobolomyces* also abide in the phyllosphere area (Cordier et al., 2012).

In the case of profiling of microbial communities in the phyllosphere, which is used to be done through traditional culture-dependent techniques and that may likely to exhibit inaccurate data in terms of microbial diversity culture-independent techniques are mostly preferred technique in recent times for better understanding of organization of microbial populations in phyllosphere which includes 16S and 18S rRNA gene amplifications through PCR phospholipid fatty acid analysis based on the constant percentage of phospholipids in microbial cell biomass, amplicon pyrosequencing, and many other techniques (Whipps et al., 2008; Chaudhary, Kumar et al., 2017). However, even with such advancements, there is still no proper knowledge regarding interactions of microbes among the microbial communities and this makes the fact inexplicit how competition and mutualism affect the establishment of various microbial communities.

10.4.2 MICROBIAL COLONIZATION IN THE PHYLLOSPHERE

Microorganisms such as fungi, yeasts, and bacteria may introduce in new plants or leaves through various mediums such as air, insect, seed, crop debris, or animal-borne sources. Microbial colonization mostly takes place at junctions of the wall of epidermal cells, stomata, cuticle, and at the basis of trichomes (Whipps et al., 2008). However, the colonization pattern usually differs among the microorganisms. Figure 10.1 illustrates the phyllosphere environment in a plant. An experiment shows that on inoculating leaves of bean plants with yeasts and bacteria in a growth chamber, yeasts were mostly concentrated at anticlinal walls over the lamina, while the bacteria were found to colonize at anticlinal walls along the veins and stomatal pores. Besides leaves attached to plants, phyllosphere fungi communities also colonize in fallen leaves and are involved in leaf litter decomposition which plays a crucial role in the assessment of nitrogen and phosphorus, as well as in regulating the global carbon cycle (Osono, 2006).

FIGURE 10.1   A figure showing phyllosphere environment.

The middle panel shows an aerial portion of the plant. The left panel illustrates an exaggerated surface of the leaf with bacteria. The right panel shows electron microscopy (a) demonstrates the Arabidopsis thaliana plant leaf edge with visible trichomes, and (b) epiphytic bacteria located in the grooves of epidermal cells on leaf surface. The Image sourced and modified from (Vorholt, 2012; Wei et al., 2017).

10.4.3   IMPORTANCE OF PHYLLOSPHERE BACTERIA AND FUNGI IN PLANT HEALTH MAINTENANCE

Diverse bacterial and fungal communities show a crucial role in plant growth and acclimation to unfavorable environmental conditions, host plant protection against various plant pathogens, etc. (Bao et al., 2019). However, as compared to bacterial communities, very little information is gathered regarding structure and function of fungal communities on leaf surfaces or impact of fungi on the composition of the bacterial community (Rastogi et al., 2013).

10.4.3.1 ANTAGONISTIC ACTIVITY AND PLANT GROWTH HORMONE PRODUCTION

Plant-associated microbial communities have a significant role in maintaining the health of the host plant and can be supported by various evidence such as on leaves of *Arabidopsis thaliana*, strains of bacteria *Sphingomonas* were found to suppress both the growth of causative microbe and symptoms of the disease caused by pathogenic *Pseudomonas syringae* (Innerebner et al., 2011). Also, few *Pantoea* species were found to produce growth hormone (indole-3-acetic acid) for plants as well as reduces growth of pathogens by establishing quorum sensing systems on leaves (Pusey et al., 2011). *Methylobacterium* spp., was reported for the secretion of another growth hormone cytokinin, plays an important function in plant cell division (Holland, 2002). Certain bacteria are also able to show antagonistic activity against pathogenic fungus, for example, *P. synxantha* is reported to exhibit antifungal activity under in vitro conditions against pathogenic *Monilinia laxa*, a causative agent of brown rot of stone fruits (Janakiev et al., 2019). Besides bacteria, certain fungi also exhibit antagonistic activity against other fungi, for example, an endophyte of *Pinus*, *Lophodermium pinastri* is found to reduce the growth of other saprobic fungi in vitro through the production of certain antibiotics. Fungus *Gibberella* mediates plant growth promotion through the production of large amounts of Gibberellins (Bao et al., 2019).

10.4.3.2 DEGRADATION OF AIR POLLUTANTS AND NITROGEN FIXATION

Phyllosphere microbes can degrade air pollutants such as polyaromatic (naphthalene, phenanthrene) and monoaromatic hydrocarbons (toluene, xylene, etc,), etc (Dharmasiri et al., 2019). According to the results of Daane et al., (2001) and Ma et al., (2006), various bacterial species of *Pseudomonas* and *Mycobacterium* were capable in degrading polyaromatic hydrocarbons while cultures of fungi *Cladophialophora* species (Prenafeta-Boldú et al., 2002) are found to be capable of degrading BTX (mixture of isomers, i.e., benzene, toluene, and xylene) in phyllosphere. Apart from pollutant degradation phyllosphere microbes are also involved in nitrogen fixation and phosphate solubilization. Nitrogen fixation usually occurs in the inner part of the leaf in temperate environments while in the tropical environment, the fixation of nitrogen generally takes place on leaf surfaces through epiphytic microbes. This nitrogen-fixation by phyllosphere microorganisms in leaves can provide nitrogen for growth and development in plants in stressed ecosystems (Chowdhury et al., 2007).

10.4.4 FACTORS AFFECTING THE MICROBIAL COMMUNITY IN PHYLLOSPHERE

As compared to the rhizosphere with favorable temperature, moisture, and optimum availability of nutrients, the habitat of phyllosphere is considered to be extreme microorganisms residing in phyllosphere are subjected to multiple stresses that may arise due to various environmental factors and characteristics of the host plant species. The microbes living in such habitats usually develop various strategies for their survival (Leveau, 2009).

Effects mediated by host plant species

- Different plant species could provide different microenvironmental conditions for the microbial community in particular the accessibility of water and nutrients sources, etc.
- The characteristics of host tree-like tree height, branch height, etc., are reported to shape microbial communities in the phyllosphere (Laforest-Lapointe et al., 2016).
- Plant-derived metabolites among which alcohols, amino acids, amines, sugars, as well as plant water, and salts that serve as a source of nutrients are also not directly available for epiphytic microorganisms.
- The presence of waxy cuticles on leaf surfaces limits the water availability for phyllosphere microbes (Bringel and Couée, 2015).

Effects mediated by environmental factors

- The phyllosphere deals with plant photosynthesis and light intensities during the day and this can put microbial colonizers in increased risk to be exposed to reactive oxygen species that can cause a major damage to nucleic acids, lipids, and proteins (Vorholt, 2012).
- Drastic fluctuation in erratic periods also influences microbial community composition (Leveau, 2009).

10.5 PRODUCTION OF BIODIESEL AND BIOETHANOL FROM ALGAE

Biofuel such as biodiesel and bioethanol derived from the macro and microalgae is referred to as third-generation biofuel and it has been found to overcome the major limitations of first- (derived from food crops) and second-generation (derived from lignocellulosic materials) biofuel (Noraini et al., 2014). Moreover, a lower quantity of hemicelluloses and lesser or absence of lignin quantity in the algal biomass makes it suitable for increased hydrolysis and fermentation. Biodiesel and bioethanol production could be developed through advancement in the production technology.

10.5.1 ALGAE AS A BETTER RAW MATERIAL FOR BIOFUEL PRODUCTION

Algae are the diverse population of prokaryotic and eukaryotic organisms ranging in size from microalgae to macroalgae (Li et al., 2014). Algae are abundantly present and are widely available in nature such as in freshwaters, salty waters, deserts, marginal fields, etc., and can grow in every season throughout the year. For their cultivation, ponds are generally used as open systems and photobioreactors as closed systems (Özçimen et al., 2015). Biomass production in the case of algae is found to be 5–10 times higher than that of land-based plants because of more efficiency in the photosynthetic process (Chen et al., 2013). Microalgae could be the most economical source of biodiesel because of its high oil content that may exceed up to 80% by weight of dry biomass while 20–50% oil levels are quite common in most microalgae (Noraini et al., 2014; Chisti, 2007). The microalgal species namely *Chlamydocapsa bacillus*, *Kirchneriella lunaris*, *Ankistrodesmus fusiformis*, and *Ankistrodesmus falcatus* are generally preferred for the production of biodiesel (Behera et al., 2015). Some other algal strains such as *S. Bijugatus*, *S. Abundans, and S. bijuca* are reported to make lipid yields 3–63 mg/L/day. Biodiesel produced by using the alga *S. Bijugatus* suits the ASTM standard and can be used as a suitable substitute of diesel fuel (Ashokkumar et al., 2015). Bioethanol is generally obtained on fermenting sugars from algae by yeast. Microalgae such as *Porphyridium*, *Chlorella*, *Dunaliella*, *Chlamydomonas*, *Scenedesmus*, and *Spirulina* are found to contain a large amount (>40% of the dry weight) of carbohydrates, which is a raw material for ethanol production (Ho et al., 2013). The microalgae *C. Vulgaris*, has been considered as a promising feedstock for the bioethanol production because it holds 37–55% of carbohydrate content of its dry weight (Nguyen, 2012). In the case of macroalgae, 25–50% carbohydrate content is found in the green algae, 30–60% in the red algae, and 30–50% in the brown algae. Macroalgae species with high polysaccharide content are *Ascophyllum* (42–70%), *Porphyra* (40–76%), and *Palmaria* (38–74%). Apart from polysaccharides such as cellulose and hemicellulose, alginate, mannitol, glucan, and laminarin are some of the polymers abundantly present in macroalgal structure (Özçimen et al., 2015, Wargacki et al., 2012).

10.5.2 PROCESS OF BIODIESEL AND BIOETHANOL PRODUCTION FROM ALGAE

The pathways for liquid biofuel production share many common features irrespective of the type of biomass that is to be used. Almost all the pathways include a cultivation step, a collection or harvest step, and a processing step. The production process usually starts with the selection of suitable isolates of algae. The cultivation of selected isolates can be carried out in laboratory (using different growth media in vessels such as beaker, pot, flasks, and tubs) or in ponds (that may be natural or artificial) or in photobioreactors. An efficient low-cost harvesting for the cultivated algae is to be done next for the bioethanol production. Various techniques such as filtration, floatation, flocculation, ultrasonic separation, electrolytic method, etc. (Bibi et al., 2017; Uduman et al., 2010), can be used for the effective harvesting process. The harvested algal biomass is further processed for removing water content from the harvested algae. Some of the efficient drying approaches like spray-drying, drum-drying, freeze-drying, or lyophilization, and sun-drying have widely been applied on microalgal biomass (Richmond, 2008).

10.5.2.1 PRODUCTION OF BIODIESEL

For biodiesel production, the lipids and fatty acid content present in the cell walls of microalgae are to be extracted. In this process, oil extraction is accomplished by using different oil extraction methods such as solvent and mechanical extraction method (Li et al., 2014). Wet processes which use lipids, extracted from disrupted cell membranes in the aqueous medium are likely to consume considerably lower energy than dry processes (Beal et al., 2012). The extracted oil is then further subjected to transesterification. Figure 10.2 illustrates about various steps involved in biodiesel synthesis.

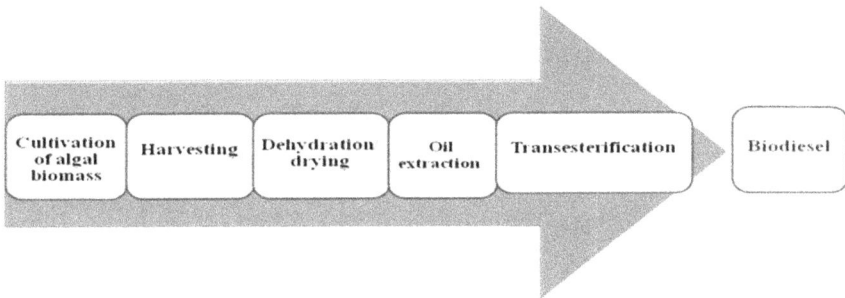

FIGURE 10.2   Steps in biodiesel production

10.5.2.1.1   *Transesterification*

The microalgal-based biodiesel is produced after the transesterification of total extracted lipid content from the microalgae. The microalgal lipids are generally accumulated in the form of triglycerides that are converted over fatty acid alkyl esters through the transesterification reaction. It is a multistep reaction where triglycerides are responsive in methanol in the existence of a catalyst (Mondal et al., 2017). Lipases are mostly used in enzyme catalyzed transesterifications and help in excluding the by-products. Transesterification process is important in reducing the viscosity and increasing the fluidity of algal oil in order to be mixed with petroleum diesel (Adeniyi et al., 2018). Alkaline catalysts in particular KOH and NaOH are very frequently utilized in biodiesel synthesis. These specific chemicals are able to give larger yields of biodiesel in a limited period of time by catalyzing the reaction upon less temperature at atmospheric pressure (Ashokkumar et al., 2015).

10.5.2.2   *PRODUCTION OF BIOETHANOL*

The process of bioethanol production differs subject to the biomass nature. The dehydrated biomass is subjected to pre-treatment phase and scarification which enables the algae cells to release the sugars and poly-mers from their cell wall. Acid pre-treatment and alkaline pre-treatment are extensively utilized for the transition of polymers existing in the cell wall into simple forms (Özçimen et al., 2015). Several enzymes (such as cellulase) and radiations (such as gamma rays) are also utilized to hydro-lyze the algae cells (Schneider et al., 2013). Figure 10.3 illustrates various steps involved in bioethanol synthesis.

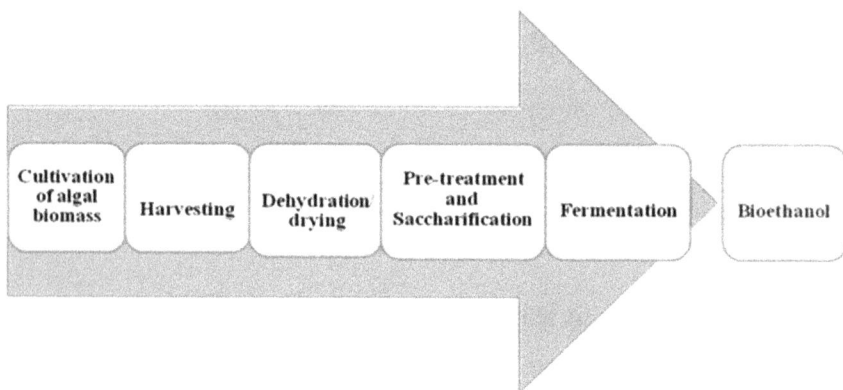

FIGURE 10.3   Steps in bioethanol production

10.5.2.2.1   Fermentation and Purification

For fermentation, various groups of microorganisms such as yeast, bacteria, and fungi can be exploited for the fermentation of pre-treated and saccharified algae biomass through anaerobic condition (Nguyen, 2012; Behera, et al., 2015). *S. cerevisiae* and *Z. mobilis* are most commonly used fermenting microorganisms. Yeast *Pichia angophorae* is found to utilize both mannitol and laminaran (Spolaore et al., 2006) as substrates for ethanol production. The final step involves in the synthesis procedure is the distillation and purification of the final product in order to remove water and other filths to make the bioethanol more concentrated. Various agents for separation such as acetone, benzene, cyclohexane, diethyl ether, hexane, and pentane can also be used in this process (Kumar et al., 2010). Bioethanol is generally concentrated in rectifying column to make the concentration near 95% (Mesa et al., 2011).

10.5.3 CHALLENGES IN BIOFUEL PRODUCTION

There are various challenges faced in the production of bioethanol and biodiesel that includes the risk of contamination in the algal culture during cultivation and harvesting and thus it has to carry out mostly in a controlled environment (Wi et al., 2009). The selection of potent algal strain is an another challenge which is time consuming and tedious job. Some other challenges include high energy consumption in downstream processing and maintenance of the costly equipments. The energy cost of extraction of algal oil is reported to be ten times greater than that of the energy cost of extraction of the soybean seed-derived oil (Savage, 2012).

In the biofuel industry, along with algae some strains of fungi, whole organism has the potential for fermentation for bioethanol production. Some bacteria can produce ethanol from marine biomass alginate (Takeda et al., 2011).

With rapid urbanization and industrialization, there has been rise in the use of heavy metals. Heavy metal toxicity has affected every aspect of the biosphere. Whole microorganisms are used which are effective in controlling pollution. However, more effective results have been obtained using isolated enzymes. Studies reveal that the catalytic activities, self-life, immovability, and stress conditions of the enzymes can be improved by enzyme engineering. The different enzyme engineering techniques are the exclusion of allosteric regulation, improved the substrate-specific features, enhanced thermostability, variation in optimum temperature and pH, etc. (Sharma et al., 2018).

The dramatic rise in CO_2 in the recent past has threatened various lives. Sequestration of CO_2 is performed by both plants and microbes. Algae stand as the strongest contender for its dual role of both removing CO_2 in environment-friendly way and also producing biomass that in turn yields biofuels, bioethanol, etc. (Mistry et al., 2019). Another advantage of the growing algal biomass is that it helps in converting wastewater into freshwater resource. Thus, the wastewater treatment combining with biofuel production can be a significant step toward bioremediation (Cabanelas et al., 2013).

In the direction to fight against the damage caused to the environment and maintaining a sustainable environment. It is unambiguous that we need to rely more on microorganisms. More isolation and characterization along with parallel progression by technological techniques or advancement in background biotechnology can help to speedup the process. The knowledge of metagenomics, meta-transcriptomics, and meta-proteomics, etc., has boosted the process. The intricate signaling network of microbes should be studied in addition to their adaptive behavior in a diverse environment.

Recently, some cost-effective methodologies in the synthesis of nanomaterials and nanoparticles-based resources are fascinating wide interest for their unique properties and immense application potential in diverse areas (Sharma et al., 2018).

The goal of the science fraternity should be on bringing efficient, and environment-friendly approaches centered on microbes with the application of evolved technology.

10.6 BIOREMEDIATION

Bioremediation literally nothing but the use of microorganisms to restore or destroy or to immobilize the pollutants from the environment. The process of a regular bioremediation system has been utilized by peoples for wastewater management. However, premeditated use for mitigation of hazardous waste is a novel and modern development. The incorporation of genetic engineering techniques holds promise in accelerating the process of bioremediation.

Bioremediation is classified into two different systems viz., in situ and ex situ bioremediation. The former involves treatment of material at the site of contamination, whereas the latter involves complete removal of contaminated material from one site and its transfer to another site, where it has been treated using biological agents.

Overtime various anthropogenic and industrial activities have resulted in increased contaminated sites due to the lack of strict guidelines against damage caused due to manmade activities. This method of bioremediation uses many agents such as bacteria, algae, fungi, yeast, and higher plants as major tools in treating heavy metals and oil spills.

The concept of microbiome extends from prokaryotic bacteria to eukaryotic yeast and protista. Microbes have been a field of extensive study due to their wider adaptability and comparatively shorter genome. From inhabiting the human gut to hot spring to toxic wasteland, they have played pivotal roles in contributing to maintaining a sustainable environment. The dynamic changes in the microbial genome make them diametrically variable thus suitable for survival in every environment. Due to their small genome, presence of unique outer membrane proteins, they can adapt to the changing and harsh environment (Lin et al., 2002). A substantial amount of data is available on various approaches such as bioremediation, rhizoremediation, phylloremediation, biofuel production, carbon sequestration, etc. One of the many methods of bioremediation is biosorption which is very effective as it uses dead or inactive biomaterial. Biofertilizers and biopesticides with fungal base are available in the market which reduce the use of chemical fertilizers. The phyllosphere is vulnerable to environmental factors such as high temperature, UV rays, fluctuating climatic conditions, and limited availability of nutrients. Recent studies have revealed that certain bacteria and fungi can alleviate the stresses and also promote the production of growth hormones pollutant degradation acquisition of nutrient and antimicrobial activities against pathogens.

10.6.1 BIOREMEDIATION OF ORGANIC DETRITUS

Pollutants from industrialization and marine transport comprise dissolved and suspended organic material, following carbon chains that can be exploited by algae and microbes (Dasand Chandran, 2011). Microbes that belong to the genus *Pseudomonas* are supposed to be capable of degrading the organic detritus and efficaciously cleanup carbonaceous wastes from waterbodies. As for instance, *B. licheniformis*, *B. coagulans*, and *B. subtilis*, and bacteria of genus *Phenibacillus*, such as *P. polymyxa* are suitable for bioremediation of organic wastes. With the introduction of some *Bacillus* strains into the water creates competition for the naturally available microbes for the organic matter in water, such as leftover feed and shrimp faeces (Jacques et al., 2008). These *Bacillus* strains are released into the waterbodies after mixing with sand or clay (Obed Ntwampe, 2014). In some cases, *Lactobacillus* has also been utilized along with *Bacillus* spp. to accelerate the breakdown of organic materials. This led to the breakdown of different enzymes like proteins and starch into their smaller subunits, which in turn are used as a major source of energy by other microorganisms present in that environment.

10.6.2   *BIOREMEDIATION OF THE NITROGENOUS COMPOUNDS*

The presence of nitrogenous compounds (e.g., ammonia and nitrite) in excess amount in the river water leads to deterioration of water quality and causes toxicity in aquatic life. A huge quantity of ammonia is excreted by fish excretion and other aquatic organisms.

The process of nitrification is as follows:

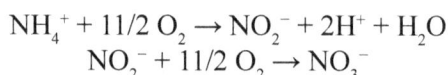

$$NH_4^+ + 11/2\ O_2 \rightarrow NO_2^- + 2H^+ + H_2O$$
$$NO_2^- + 11/2\ O_2 \rightarrow NO_3^-$$

The role of symbiotic nitrogen-fixing bacteria is well known. It is also effective in keeping in check on pollution. Some of the genus that oxidizes ammonia are *Nitrosovibrio*, *Nitrolobus*, *Nitrosomonas*, *Nitrosococcus*, and *Nitrospira* and the three genus known for oxidizing nitrite are *Nitrospira*, *Nitrococcus*, and *Nitrobacter* (De Oliveira et al., 2016). In some cases, heterotrophic nitrifies are found to produce a small amount of nitrate and nitrite with the utilization of organic sources of nitrogen. The nitrification process increases the acidity to promote the bioavailability of various soluble substances together with nitrate production (Hlihor et al., 2017). Moreover, the regions rich in nitrates are ideal for the growth of anaerobic bacteria. These microbes produce nitrogen gas reducing nitrate.

The process of nitrification is as follows:

$$NO_3 \rightarrow NO_2 \rightarrow NO \rightarrow N_2O \rightarrow N_2$$

So far, around 14 geneses of bacteria are found which can reduce nitrate. Among them *Alcaligenes*, *Bacillus*, and *Pseudomonas* are the most efficient among all the other species. Bioremediation of water through a combination of *Nitrobacteria* and grass plant species *Lotium perenne* improves wastewater quality (Pan et al., 2007).

10.6.3   *BIOREMEDIATION OF HYDROGEN SUFHIDE*

Sulfur is an essential element of amino acids. Some bacteria use sulfate for metabolism in place of O_2 under anaerobic phase and lead to the synthesis of H_2S gas through numerous microbes advanced reductions (Szabõ, 2007).

$$SO_4^{2-} + 4H_2 + 2H^+ \rightarrow H_2S + 4H_2O$$

Organic materials in water initiate H_2S production and cause harm to benthic fauna, damage to gills and other physiological issues in fishes. Photosynthetic bacteria utilize H_2S, which is pivotal for maintaining a satisfactory river environment. They decompose organic matters such as H_2S, NO_2, and other potentially hazardous wastes from the river through decreasing electrons from H_2S with less energy than H_2O excruciating photoautotrophs, and therefore, reducing the need for photosynthesis. Some of the prominent photosynthetic bacteria are *Chlorobium, Clathrochloris, Amoebobacter, Chloropseudomonas, Pelodictyon, Thiosarcina, Ectothiorhodospira, Lamprocystis, Thiopedia, Thiodictyon, Thiocystis, Thiocapsa, Prosthecochloris, Rhodopseudomonas, Rhodospirillum, Rhodomicrobium, Chromatium*, and *Thiospirillum*, etc. (Shishi, et al., 2019).

10.6.4 BIOREMEDIATION OF PETROLEUM

Oil spill causes damage to aquatic life as it contains less molecular mass toxic aromatic compounds viz., toluene, benzene, and xylene were found in petroleum. However, they are successfully naturally degraded by a number of freshwater and marine microorganisms. Biodegradation of the petroleum in the river water is performed by *Pseudomonas* spp that commonly found in the seawater. These *Pseudomonas* spp. together with algae, plants, and living beings use hydrocarbons and some other substrates, that is, proteins and carbohydrates in river for ameliorating the water quality. Pollution, particularly with petroleum hydrocarbons triggers rapid bacterial division. A few of the potent petroleum humiliating cyanobacteria are *Aphanocapsa* spp., *Plectonema terebrans,* and *Oscillatoria salina* (Vijayakumar, 2012; Zinicovscaia and Cepoi, 2016).

10.6.5 BIOREMEDIATION OF HEAVY METAL CONTAMINATIONS

Heavy metal pollution stands as a threat to the biosphere. Both pedogenetic processes and anthropogenic processes are causes of heavy metal pollution. Fortunately, a myriad of microorganisms ranging from bacteria to fungi has been found very effective against toxic heavy metals which are immobilized and mineralized by them using different biochemical pathways. These microbial communities form covalent or electrovalent (ionic) bonds through metal ions for absorbing them (Kang et al., 2016). Very few sulfate-reducing bacteria produce H_2S by minimizing the heavy metals including CdS, ZnS, and CuS as they have very poor water solubility for depositing and controlling contamination of the heavy metals.

10.6.6 BIOREMEDIATION OF EUTROPHICATION

Eutrophication depends mostly on the occurrence of carbon, nitrogen, and phosphorus in water, which are very critical factors for the growth of algae in waterbodies which have stagnant water and become deposit of toxic compounds.

Elimination of eutrophication is based on the principle of removing phosphorus, nitrogen, and other organic carbon molecule origins. Technical and mechanical withdrawal is also used in some studies. Besides, water draining and bioremediation are found to be potent over eutrophication as well (Shan et al., 2009).

Several types of bacteria are used to endure the essential compounds that cause eutrophication. Bacterial species present in the same area are much better options since they have high acclimatization with less environmental risk factors. In a recent study, some protozoans, cyanobacteria, and chlorophyta including *Pediastrum asterionella,* etc., were used for eutrophication. Contrastive analysis disclosed that the various microbial preparations can upsurge the water volume of self-purification, decline the turbidness, prevents the algae growth, and that improves water quality progressively at a considerably lower cost. Thus, the problem of stream eutrophication can be resolved radically by bioremediation.

Klebsiella oxytoca was found to be productive for the elimination of N_2 from the water bodies (Shawabkeh et al., 2007). Furthermore, *Phormidium bohneri* is a photoautotrophic microorganism that removes both phosphorus and nitrogen at specific conditions accompanied by solar power systems (Chevalier et al., 2000). Microbial treatments need more acceptance for their high yield and economic values.

10.6.7 BIOREMEDIATION OF PESTICIDE

With the start of industrial revolution (despite the rise in productivity), the deposition of pesticides in land and water has threatened the life. While they are frequently used to enhance the process of cultivation, their bioaccumulation has led to some fatal diseases across the trophic levels with man being the most affected. Different strains of bacteria and fungi have been documented to have been very effective against this.

10.6.7.1 BACTERIAL DEGRADATION OF PESTICIDES

The most efficient group of bacterial species that are involved in the degradation of pesticides belongs to *Azotobacter, Flavobacterium, Burkholderia, Arthobacter, Raoultella,* and *Pseudomonas.* During the process of oxidation, the parent compound is oxidized to yield CO_2 and water, which give them energy. It is advisable to add pesticide degrading bacteria from an external source in the sites where the local microbial population is incapable of managing pesticides. Environmental factors like pH, temperature, etc., are necessary for the efficient activity of microbial enzymes. While some pesticides get easily degraded, some are difficult to degrade due to the existence of anionic species within the compound. *Pseudomonas* species are found to be efficient in degrading organophosphorus as well as the neonicotinoid substances.

10.6.7.2 FUNGAL BIODEGRADATION OF PESTICIDES

The insignificant structural modifications that fungi do to reduce/degrade the pesticides and condense them into non-toxic materials and release them into soils where it is liable to furthermore degradation. The various fungi which have shown the ability to degrade pesticides are given in the table below.

Species of fungi	Potential for degrading pesticide	Reference
Hypholoma fasciculare, Coriolus versicolor, Agrocybe semiorbicularis, Flammulina velutipes, Auricularia auricula, Dichomitus squalens, Avatha discolor. and *Stereum hirsutum*	Phenylurea, triazine, dicarboximide	Bending et al., 2002
White-rot fungi	Chlordane mirex, Heptachlor atrazine, lindane, metalaxyl, DDT, gamma-hexachlorocyclohexane (g-HCH), dieldrin, terbuthylazine, diuron, aldrin, etc.	

The presence of napropamide was found to promote the growth of pesticide degrading bacteria. Similarly, the role of vinclozolin in soil also proliferates and triggers pseudomonas of pesticide-degrading bacteria. The pesticide degrading genes that are present in plasmids are transferred through horizontal gene transfer. A bacterial strain was found possessing manifold plasmid genes for pesticide degradation (Uqab et al., 2016).

10.6.8   ENZYMES USED IN BIOREMEDIATION

Although enzymes secreted by microbes are efficient in degrading enzymes but the natural course of reaction that takes place in situ is very time consuming which reduced the feasibility of the process. Table 10.1 illustrates the list of enzymes that are involved in bioremediation process. In the recent past, to overcome the problem steps have been taken for the isolation of the enzyme from their source. Enzymes are biocatalysts, which act on numerous biochemical reactions that are involved in degradation pathways of various pollutants (Table 10.2). The following tables provide details on their source organisms, pollutants they act on and mechanism pathway (Sharma et al., 2018).

TABLE 10.1   List of Enzymes That Are Involved in Bioremediation.

Enzymes	Role
Laccases	Molecular O_2 in water is reduced and produced free radicals
Oxygenases	Catalyze oxidation in the compounds such as aliphatic olefins, chlorinated biphenyls, etc. By incorporation of either one or two O_2 molecules, making them liable to additional transformation and to promote mineralization
Peroxidases	Catalyze reduction process in the presence of peroxides, including H_2O_2 and produce free radicals' nest to oxidation of organic molecules
Haloalkane dehydrogenases	Applied for biodegradation of the 1,2,3-trichloropropane, a halogenated aliphatic compound
Lipases	Breakdown of triglycerol into fatty acid and glycerol
Cellulases	Complex cellulosic materials are broken down into simple sugars
Carboxylesterases	The hydrolysis of carboxyl ester-bond shows in pesticides, for example, organophosphate in presence of H_2O, is catalyzed by it
Phosphotriesterases	Catalyzes hydrolysis of phosphotriester bonds, the important components of organophosphorous compounds. Mostly present in pesticides, cause poisoning and death.

TABLE 10.2 List of Enzymes That Are Associated in Degradation of Pollutants.

Enzymes	Producing microbes	Pollutant	Mechanism	Reference
Peroxidases	Phanerochaete chrysosporium	Tnt (2,4,6-trinitrotoluene) nitroaromatic compounds	4-amino-2,6-dinitrotoluene (4amdnt) and 2-amino-4,6-dinitrotoluene (2amdnt) are the initial and intermediate products of TNT biodegradation, which is further mineralized into simpler compounds	Cameron et al., (2000)
Atrazine dechlorinase, triazine hydrolase	Rhodococus sp., Nocardioides sp., Pseudomonas sp.	Triazine herbicides	Metabolizes atrazine into ammonia, chloride, and carbon dioxide in an enzyme-mediated reaction	Scott et al., (2010)
Laccase	Pleurotusostreatus, Trametes versicolor	Pcbs (polychlorinated biphenyls)	Pcbs metabolizes into its constituent elements and laccase dechlorinate chloro-phenols by oligomerization of the substrate	Mayer and Staples (2002)
Chromium reductase	Pseudomonas, bacillus, enterobacter, deinococcus, shewanella, agrobacterium, escherichia, thermus	Chromium	Converts Cr (VI) into Cr (III) in the presence of O_2 where, initially, Cr (VI) accepts one electron from one molecule of NADH to generate Cr (V) and then Cr (v) accepts two electrons to form Cr (III)	Thatoi et al., (2014)
Horseradish peroxidase (hrp)	Horseradish (armoracia rusticana)	Chlorophenol, phenol	Both meta and ortho pathway are used to degrade the phenolic compounds. Phenol is converted into aldehyde and pyruvate after degradation by meta pathway	Flock et al., (1999)
Carboxylesterases	Pseudomonas aeruginosa pa1	Malathion and parathion	It transforms malathion into malathion monocarboxylic acid and di-carboxylic acid	Qiao et al., (2003); Singh et al., (2012)
Nitrilases	*Rhodococcus* sp. *Nocardia* sp.,	Nitrile compounds	Catalyzes hydrolysis of nitriles into acids and ammonia	Rao et al., (2010)

TABLE 10.2 *(Continued)*

Enzymes	Producing microbes	Pollutant	Mechanism	Reference
Cr^{6+} reductases	*Escherichia coli,* *Pseudomonas putida* MK1	Heavy metal Cr^{6+} (hexavalent species of chromium)	It can reduce Cr^{6+} to Cr^{3+} in aerobic and anaerobic conditions	Cheung et al., (2007)
Lipases	*Candida rugosa*	Oil spills	Catalyze the hydrolysis of triacylglycerols to glycerol and free-fatty acids by breaking the triglyceride ester bond.	Karigar et al., (2003)

10.7 HEAVY METAL

One of the most important environmental issues of the present time is heavy metal pollution. Many of the heavy metal ions are detrimental to living beings. These metal ions are non-degradable in nature and are persistent in environment.

They are mostly released from anthropogenic actions of factories and industries. They are taken up by the plants and various animals from the environment. Plants while absorbing necessary nutrient elements take some toxic heavy metals as well and they enter the biosphere. Although some of them act as co-factor for various metabolic activities in living beings but all metals can have harmful impact depending upon their concentrations and exposure of an organism to the particular metal. Moreover, the duration of exposure and the route also matters (Ojuederie et al., 2017). The heavy metals that are detrimental across all living kingdoms including plants, animals, and microorganisms are arsenic, cadmium, chromium, lead, mercury, aluminium, iron, etc.

Various microorganisms can utilize heavy metals but all of them cannot do all the metals. So, according to their capability, their inoculation should be used.

10.7.1 PATHWAYS AND MECHANISMS

Decontamination can be performed by physicochemical methods but recently biological processes namely biosorption, metal–microbe interaction, bioaccumulation, bioleaching, biomineralization, and biotransformation have gained popularity for their effectiveness.

Various parts of the cell are involved in metal sequestration using different processes: precipitation, complexation, coordination, chelation, and reduction. Among them, biosorption is a significant one. It is able to sequester dissolved metals from dilute complex solutions very actively. The most commonly applied microbes are *Staphylococcus, Aspergillus, Pseudomonas,* and *Bacillus* spp. After the treatment, *Bacillus* and *Pseudomonas* spp. are reduced the Cu 56 and 68% and Ni 48 and 65%, respectively. *A. niger* reduced the Cd 50% and Zn 58% whereas *Staphylococcus* spp. reduced Cr 45%, Cu 42%, and Pb 93% (Hansda et al., 2016). The major research area are biotreatability, cometabolism, kinetics, biotransformation, and modeling of biogeochemical processes (Hansda et al., 2016).

10.7.2 PLANT–MICROBE INTERACTION

Plants and microorganism combination are found to be more effective in sequestering heavy metals, for example, *Pseudomonas fluorescens and P. tolaasii.* The improvements in Cd-phytoextraction were due to only microbial plant growth promotion (PGP), as Cd concentrations in the plant tissues remained constant.

Microbial PGP can be attained by the secretion of PGP compounds such as indole-3-acetic acid, gibberellins or cytokinins, or by improving plant nutrition via the solubilization of inorganic phosphates and ferric iron (Fe [III]) or nitrogen fixation. It is worth noting that a number of PGP processes can also mobilize heavy metals in the soil: Microbial sidero-phores promote plant growth by solubilizing ferric iron and increasing plant Fe (III) acquisition, but can also chelate and mobilize heavy metals (Wood et al., 2016).

A typical strategy for using microorganisms to improve phytoextraction is to "mine" PGP or metal-solubilizing microorganisms from contaminated soils and rhizospheres.

10.7.3 CADMIUM RESISTANCE BY PSEUDOMONAS AERUGINOSA

Pseudomonas aeruginosa expresses high efficiency in Cd resistance. The mechanisms used are bioaccumulation, biosorption, antibiotic resistances, heavy metal, adaptive and cross-resistance, biosurfactant and extra poly-saccharides production, bioinoculant, PGPR activity, and phytoextraction procedure. Microbial cadmium resistance was found in around six different methods such as altered accumulation of the toxic compound, confession of the toxic metal ions in the cell wall, and changes in the cell wall plasma membrane complex. Bacterial cells can uptake Cd through divalent cation uptake systems viz. Zn_2^+ and Mn_2^+ or gene amplification, active Cd efflux, etc. (Mishra et al., 2017).

10.7.4 HEAVY METAL TOLERANT–PLANT GROWTH PROMOTING MICROORGANISMS

Agricultural soils are being contaminated with heavy metal causing threat to the soil ecosystem and biosphere at large. Application of plant–microbe associations to renew contaminated soils is an optimistic method that is even continued today. Heavy metal tolerant–plant growth promoting (HMT–PGP) microorganisms are immensely beneficial for their expertise in improving soil quality, enhancing plant growth, detoxifying and removing heavy metals from the soil. Some recent microcosm-scale phytoextraction experiments found that amalgamation of HMT–PGP microbes with thiosulfate increased mobilization and interest of Hg and As in *L. albus* and *B. juncea* grew in polluted soils containing the aforesaid metals. Some microorganisms producing enzyme 1-aminocyclopropane-1-carboxylate deaminase showed enhanced growth in heavy metals-contaminated regions (Mishra et al., 2017).

10.7.5 FUNGI IN BIOSORPTION

An experiment by Hameed *et.al* found several fungal species which were obtained from rock phosphate and phosphate fertilizer grow in the presence of Cu^{++} and Zn^{++} and 18 isolates showed positive response in the presence of the cations of Co, Cr, Pb, and Cd.

When grown in liquid medium containing different heavy metals in very high concentration some showed rapid growth. After morphological analysis, they were found to be Aspergillus spp. (Hameed et al., 2015).

Molds and yeasts are a great choice for their biosorption capacity, easy cultivation, genetic manipulation, and high yield. They are used in large scale as they provide essential compounds such as kojic acid, ferrichrome, gallic acid, amylases, lipases, pectinases, glucose isomerase, and glucanases. They are excellent biosorbents as they can remove toxic metals from polluted waterbodies with outstanding capabilities for metal absorption and recovery. For neutralizing toxicity from heavy metals, they have developed a complex defense mechanism (Ojuederie et al., 2017).

10.7.6 CURRENT STATUS OF BIOLOGICAL APPROACHES

Biological approaches show more specificity in comparison to physical and chemical approaches, its compatibility to in situ methods, and the scope of improvement by different techniques of genetic engineering. The structure of biomass with exact metal-binding qualities, metal chelating peptides and proteins, metal precipitation methods, and the prolusion of metal transformation actions in stressful environmental types. However, extensive use of biological methods is even seldom in use (Valls et al., 2002).

10.7.7   BACTERIA AND BIOREMEDIATION

A wide range of bacterial spp. has been documented to be capable of converting inorganic metals into some organic forms. The role of the plasmid is found in this case. ATP-binding cassette transporter proteins and few microbes harboring plasmids possessing tolerance specific genes.

Several transmembrane proteins have reported alongside P-type ATPases (hydrolyse ATP) ATP-binding cassette transporters for zinc. Such proteins are found to be active in the process of detoxification. Bacterial change is due to the collective role of transposon, chromosomal, and plasmid-mediated resistance system. The concentration of zinc influences soil fertility and ecology because of its varying concentrations (Joshi et al., 2013).

10.7.8   ALGAE AND ROLE IN BIOSORPTION

Marine algae are identified as a potential biosorbent, due to their high uptake capacities, cheaper price, availability, renewability across the globe. Out of all, most in-depth studies have been conducted on brown algae. Results are available on the study conducted for the disposal of cadmium, chromium, copper, gold, lead, nickel, uranium, and zinc. The brown algae remove the highly harmful metal ions, that is, chromium and lead. The non-living, inactive biomass removes metal ions based on the principle of affinities between biomass and the metal ions. The basic biochemical composition of the marine algae supported by cell wall components, viz., alginate and fucoidan that are mainly liable for heavy metal repossession. The most common skeletal material of the cell wall is cellulose. The surrounding matrix is sulfated galactans for red algae, and where the alginate and sulfated polysaccharide (fucoidan) are for brown algae. The main active and functional groups sufficiently present in both green and brown algae, that is, amine, carboxyl, hydroxyl, phosphate, and sulfate groups that play a prominent role in the metal binding (He et al., 2014).

10.8 CONCLUSION

In the present study, rhizoremediation of petroleum contaminants, carbon sequestration, the role of phyllosphere bacteria, and fungi in maintaining plant health, production of biodiesel and bioethanol from marine algae, bioremediation, plant–microbe interactions, and heavy metal degradation was described. Considering the importance of bacterial communities in the environment that enhancing the bioremediation of organic detritus, heavy metals, fuel contaminants, etc., that are stubborn in the nature, thorough studies desirable to be performed for increasing the rate of the microbe's existence after cultured in the contaminated ecological niches. More determinations are still essential in terms of biological and technological methods used for planning and implementing the efficient bioremediation policy. The combinatorial effect of bacteria with fungi, marine algae/cyanobacteria, or plants is not well known and further importance must be given in this extent.

KEYWORDS

- **rhizoremediation**
- **bacteria**
- **fungi**
- **carbon sequestration**
- **mycoremediation**

REFERENCES

Adam, G. A Study into the Potential of Phytoremediation for Diesel Fuel Contaminated Soil; Doctoral dissertation, University of Glasgow, 2001.

Adam, G.; Duncan, H. Influence of Diesel Fuel on Seed Germination. *Environ. Pollut.* **2002,** *120* (2), 363–370.

Adeniyi, O. M.; Azimov, U.; Burluka, A. Algae Biofuel: Current Status and Future Applications. *Renew. Sustain. Energy Rev.* **2018,** *90*, 316–335.

Albuquerque, M. G. E.; Martino, V.; Pollet, E.; Avérous, L.; Reis, M. A. M. Mixed Culture Polyhydroxyalkanoate (PHA) Production from Volatile Fatty Acid (VFA)-Rich Streams:

Effect of Substrate Composition and Feeding Regime on PHA Productivity, Composition and Properties. *J. Biotechnol.* **2011,** *151* (1), 66–76.

Andreolli, M.; Lampis, S.; Poli, M.; Gullner, G.; Biró, B.; Vallini, G. Endophytic Burkholderia fungorum DBT1 can improve phytoremediation efficiency of polycyclic aromatic hydrocarbons. *Chemosphere* **2013,** *92* (6), 688–694.

Antczak, M. S.; Kubiak, A.; Antczak, T.; Bielecki, S. Enzymatic Biodiesel Synthesis–Key Factors Affecting Efficiency of the Process. *Renew. Energy* **2009,** *34* (5), 1185–1194.

Aresta, M.; Dibenedetto, A.; Barberio, G. Utilization of Macro-Algae for Enhanced CO2 Fixation and Biofuels Production: Development of a Computing Software for an LCA Study. *Fuel Process. Technol.* **2005,** *86* (14–15), 1679–1693.

Ashokkumar, V.; Salam, Z.; Tiwari, O. N.; Chinnasamy, S.; Mohammed, S.; Ani, F. N. An Integrated Approach for Biodiesel and Bioethanol Production from Scenedesmus Bijugatus Cultivated in a Vertical Tubular Photobioreactor. *Energy Convers. Manage.* **2015,** *101*, 778–786.

Bais, H. P.; Weir, T. L.; Perry, L. G.; Gilroy, S.; Vivanco, J. M. The Role of Root Exudates in Rhizosphere Interactions with Plants and Other Organisms. *Annu. Rev. Plant Biol.* **2006,** *57*, 233–266.

Banitz, T.; Johst, K.; Wick, L. Y.; Schamfuß, S.; Harms, H.; Frank, K. Highways Versus Pipelines: Contributions of Two Fungal Transport Mechanisms to Efficient Bioremediation. *Environ. Microbiol. Rep.* **2013,** *5* (2), 211–218.

Bao, L.; Cai, W.; Zhang, X.; Liu, J.; Chen, H.; Wei, Y.; ... Bai, Z. Distinct Microbial Community of Phyllosphere Associated with Five Tropical Plants on Yongxing Island, South China Sea. *Microorganisms* **2019,** *7* (11), 525.

Beal, C. M.; Hebner, R. E.; Webber, M. E.; Ruoff, R. S.; Seibert, A. F. The Energy Return on Investment for Algal Biocrude: Results for a Research Production Facility. *BioEnergy Res.* **2012,** *5* (2), 341–362.

Behera, S.; Singh, R.; Arora, R.; Sharma, N. K.; Shukla, M.; Kumar, S. Scope of Algae as Third Generation Biofuels. *Front. Bioeng. Biotechnol.* **2015,** *2*, 90.

Bending, G. D.; Frioux, M.; Walker, A. Degradation of Contrasting Pesticides by White Not Fungi and Its Relationship with Ligninolytic Potential. *FEMS Microbiol. Lett.* **2002,** *212* (1), 59–63.

Berendsen, R. L.; Pieterse, C. M.; Bakker, P. A. The Rhizosphere Microbiome and Plant Health. *Trends Plant Sci.* **2012,** *17* (8), 478–486.

Bhatnagar, S.; Kumari, R. Bioremediation: A Sustainable Tool for Environmental Management–A Review. *Annu. Res. Rev. Biol.* **2013,** 974–993.

Bibi, R.; Ahmad, Z.; Imran, M.; Hussain, S.; Ditta, A.; Mahmood, S.; Khalid, A. Algal Bioethanol Production Technology: A Trend Towards Sustainable Development. *Renew. Sustain. Energy Rev.* **2017,** *71*, 976–985.

Bringel, F.; Couée, I. Pivotal Roles of Phyllosphere Microorganisms at the Interface between Plant Functioning and Atmospheric Trace Gas Dynamics. *Front. Microbiol.* **2015,** *6*, 486.

Chaudhary, D.; Kumar, R.; Sihag, K.; Kumari, A. Phyllospheric Microflora and Its Impact on Plant Growth: A Review. *Agric. Rev.* **2017,** *38* (1), 51–59.

Chen, C. Y.; Zhao, X. Q.; Yen, H. W.; Ho, S. H.; Cheng, C. L.; Lee, D. J.; ... Chang, J. S. Microalgae-Based Carbohydrates for Biofuel Production. *Biochem. Eng. J.* **2013,** *78*, 1–10.

Chen, S.; Zhong, M. Bioremediation of Petroleum-Contaminated Soil. In *Environmental Chemistry and Recent Pollution Control Approaches*. IntechOpen, 2019.

Chisti, Y. Biodiesel from Microalgae. *Biotechnol. Adv.* **2007,** *25* (3), 294–306.

Chowdhury, S. P.; Schmid, M.; Hartmann, A.; Tripathi, A. K. Identification of Diazotrophs in the Culturable Bacterial Community Associated with Roots of Lasiurus sindicus, a Perennial Grass of Thar Desert, India. *Microbial Ecol.* **2007,** *54* (1), 82–90.

Cordier, T.; Robin, C.; Capdevielle, X.; Fabreguettes, O.; Desprez-Loustau, M. L.; Vacher, C. The Composition of Phyllosphere Fungal Assemblages of European Beech (F agus Sylvatica) Varies Significantly Along an Elevation Gradient. *New Phytol.* **2012,** *196* (2), 510–519.

Correa-García, S.; Pande, P.; Séguin, A.; St-Arnaud, M.; Yergeau, E. Rhizoremediation of Petroleum Hydrocarbons: A Model System for Plant Microbiome Manipulation. *Microbial Biotechnol.* **2018,** *11* (5), 819–832.

Daane, L. L.; Harjono, I.; Zylstra, G. J.; Häggblom, M. M. Isolation and Characterization of Polycyclic Aromatic Hydrocarbon-Degrading Bacteria Associated with the Rhizosphere of Salt Marsh Plants. *Appl. Environ. Microbiol.* **2001,** *67* (6), 2683–2691.

Das, N.; Chandran, P. Microbial Degradation of Petroleum Hydrocarbon Contaminants: An Overview. *Biotechnol. Res. Int.* **2011,** *2011*.

De Cárcer, D. A.; Martín, M.; Mackova, M.; Macek, T.; Karlson, U.; Rivilla, R. The Introduction of Genetically Modified Microorganisms Designed for Rhizoremediation Induces Changes on Native Bacteria in the Rhizosphere But Not in the Surrounding Soil. *ISME J* **2007,** *1* (3), 215–223.

de Oliveira, V. P.; Martins, N. T.; Guedes, P. D. S.; Pollery, R. C. G.; Enrich-Prast, A. Bioremediation of Nitrogenous Compounds from Oilfield Wastewater by Ulva Lactuca (Chlorophyta). *Bioremediation J.* **2016,** *20* (1), 1–9.

Dharmasiri, R. B. N.; Nilmini, A. H. L.; Undugoda, L. J. S.; Nugara, N. N. R. N.; Udayanga, D.; Manage, P. M. Phenanthrene Degradation Ability of Phyllosphere Bacteria Inhabiting the Urban Areas in Sri Lanka, 2019. Available at SSRN 3497475.

Dulla, G. A Closer Look at Pseudomonas Syringae as a Leaf Colonist. *Asm News* **2005,** *71,* 469–475.

Dutta, H.; Dutta, A. The Microbial Aspect of Climate Change. *Energy Ecolo. Environ.* **2016,** *1* (4), 209–232.

Farrelly, D. J.; Everard, C. D.; Fagan, C. C.; McDonnell, K. P. Carbon Sequestration and the Role of Biological Carbon Mitigation: A Review. *Renew. Sustain. Energy Rev.* **2013,** *21,* 712–727. DOI:10.1016/j.rser.2012.12.038

Fitter, A. H.; Heinemeyer, A.; Staddon, P. L. The Impact of Elevated CO2 and Global Climate Change on Arbuscular Mycorrhizas: A Concentric Approach. *New Phytol.* **2000,** *147* (1), 179–187.

Hameed, A. H. A. E.; Ewada, W. E.; Abou-Taleb, K. A. A. Biosorption of Uranium and Heavy Metals Using Some Local Fungi Isolated from Phosphatic Fertilizer. *Ann. Agric. Sci.* **2015,** *60* (2), 345–351.

Hansda, A.; Kumar, V. A Comparative Review of the Potential of Microbial Cells for Heavy Metal Removal with an Emphasis on Biosorption and Bioaccumulation. *World J. Microbiol. Biotechnol.* **2016,** *32* (10), 170.

He, J.; Chen, J. P. A Comprehensive Review on Biosorption of Heavy Metals by Algal Biomass: Materials, Performances, Chemistry, and Modelling Simulation Tools. *Bioresour. Technol.* **2014,** *160,* 67–78.

Hinsinger, P.; Gobran, G. R.; Gregory, P. J.; Wenzel, W. W. Rhizosphere Geometry and Heterogeneity Arising from Root-Mediated Physical and Chemical Processes. *New Phytol.* **2005,** *168* (2), 293–303.

Hlihor, R. M.; Gavrilescu, M.; Tavares, T.; Favier, L.; Olivieri, G. Bioremediation: An Overview of Current Practices, Advances, and New Perspectives on Environmental Pollution Treatment, 2017.

Ho, S. H.; Huang, S. W.; Chen, C. Y.; Hasunuma, T.; Kondo, A.; Chang, J. S. Bioethanol Production Using Carbohydrate-Rich Microalgae Biomass as Feedstock. *Bioresour. Technol.* **2013,** *135,* 191–198.

Holland, M. A. Methylobacterium spp.: Phylloplane Bacteria Involved in Cross-Talk with the Plant Host? *Phyllosphere Microbiol.* **2002.**

Huang, X. D.; El-Alawi, Y.; Penrose, D. M.; Glick, B. R.; Greenberg, B. M. A Multi-Process Phytoremediation System for Removal of Polycyclic Aromatic Hydrocarbons from Contaminated Soils. *Environ. Pollut.* **2004,** *130* (3), 465–476.

Hügler, M.; Wirsen, C. O.; Fuchs, G.; Taylor, C. D.; Sievert, S. M. Evidence for Autotrophic CO2 Fixation via the Reductive Tricarboxylic Acid Cycle by Members of the ε Subdivision of Proteobacteria. *J. Bacteriol.* **2005,** *187* (9), 3020–3027.

Hussain, I.; Puschenreiter, M.; Gerhard, S.; Schöftner, P.; Yousaf, S.; Wang, A.; … Reichenauer, T. G. Rhizoremediation of Petroleum Hydrocarbon-Contaminated Soils: Improvement Opportunities and Field Applications. *Environ. Exp. Bot.* **2018,** *147,* 202–219.

Innerebner, G.; Knief, C.; Vorholt, J. A. Protection of Arabidopsis Thaliana against Leaf-Pathogenic Pseudomonas Syringae by Sphingomonas Strains in a Controlled Model System. *Appl. Environ. Microbiol.* **2011,** *77* (10), 3202–3210.

Ite, A. E.; Ibok, U. J. Role of Plants and Microbes in Bioremediation of Petroleum Hydrocarbons Contaminated Soils. *Int. J.* **2019,** *7* (1), 1–19.

Janakiev, T.; Dimkić, I. Z.; Unković, N.; Ljaljević Grbić, M.; Opsenica, D. M.; Gašić, U. M.; … Berić, T. Phyllosphere Fungal Communities of Plum and Antifungal Activity of Indigenous Phenazine-Producing Pseudomonas Synxantha against Monilinia Laxa. *Front. Microbiol.* **2019,** *10,* 2287.

Jasmin, M. Y.; Syukri, F.; Kamarudin, M. S.; Karim, M. Potential of Bioremediation in Treating Aquaculture Sludge. *Aquaculture* **2020,** *519,* 734905.

Joshi, A.; Jaiswal, P. Microorganisms Living in Zinc Contaminated Soil-a Review. *IOSR J. Pharm. Biol. Sci.* **2013,** *6,* 67–72.

Kang, C. H.; Kwon, Y. J.; So, J. S. Bioremediation of Heavy Metals by Using Bacterial Mixtures. *Ecol. Eng.* **2016,** *89,* 64–69.

Kecebas, A.; Alkan, M. A. Educational and Consciousness-Raising Movements for Renewable Energy in Turkey. *Energy Educ. Sci. Technol. B-Social Educ. Stud.* **2009,** *1* (3–4), 157–170.

Kembel, S. W.; Mueller, R. C. Plant Traits and Taxonomy Drive Host Associations in Tropical Phyllosphere Fungal Communities. *Botany* **2014,** *92* (4), 303–311.

Kembel, S. W.; O'Connor, T. K.; Arnold, H. K.; Hubbell, S. P.; Wright, S. J.; Green, J. L. Relationships between Phyllosphere Bacterial Communities and Plant Functional Traits in a Neotropical Forest. *Proc. Natl. Acad. Sci.* **2014,** *111* (38), 13715–13720.

Kukla, M.; Płociniczak, T.; Piotrowska-Seget, Z. Diversity of Endophytic Bacteria in Lolium Perenne and Their Potential to Degrade Petroleum Hydrocarbons and Promote Plant Growth. *Chemosphere* **2014,** *117,* 40–46.

Kumar, M. R.; Tauseef, S. M.; Abbasi, T.; Abbasi, S. A. Control of Amphibious Weed Ipomoea (Ipomoea Carnea) by Utilizing It for the Extraction of Volatile Fatty Acids as Energy Precursors. *J. Adv. Res.* **2015,** *6* (1), 73–78.

Kumar S, Singh N, Prasad R. Anhydrous Ethanol: A Renewable Source of Energy. *Renew. Sustain. Energy Rev.* **2010,** *14* (7), 1830–1844.

Laforest-Lapointe, I.; Messier, C.; Kembel, S. W. Host Species Identity, Site and Time Drive Temperate Tree Phyllosphere Bacterial Community Structure. *Microbiome* **2016,** *4* (1), 27.

Lal, R. Carbon Sequestration. *Philos. Trans. R. Soc. B: Biol. Sci.* **2008,** *363* (1492), 815–830.

Landa-Acuña, D.; Acosta, R. A. S.; Cutipa, E. H.; de la Cruz, C. V.; Alaya, B. L. Bioremediation: A Low-Cost and Clean-Green Technology for Environmental Management. In *Microbial Bioremediation & Biodegradation*; Springer: Singapore, 2020; pp 153–171.

Leveau, J. Life on Leaves. *Nature* **2009,** *461* (7265), 741–742.

Li, Y.; Naghdi, F. G.; Garg, S.; Adarme-Vega, T. C.; Thurecht, K. J.; Ghafor, W. A.; … Schenk, P. M. A Comparative Study: The Impact of Different Lipid Extraction Methods on Current Microalgal Lipid Research. *Microb. Cell Factor.* **2014,** *13* (1), 14.

Liew, F.; Henstra, A. M.; Winzer, K.; Köpke, M.; Simpson, S. D.; Minton, N. P. Insights into CO2 Fixation Pathway of Clostridium Autoethanogenum by Targeted Mutagenesis. *MBio* **2016,** *7* (3).

Lin, J.; Huang, S.; Zhang, Q. Outer Membrane Proteins: Key Players for Bacterial Adaptation in Host Niches. *Microbes Infect.* **2002,** *4* (3), 325–331.

Ma, Y.; Wang, L.; Shao, Z. Pseudomonas, the Dominant Polycyclic Aromatic Hydrocarbon-Degrading Bacteria Isolated from Antarctic Soils and the Role of Large Plasmids in Horizontal Gene Transfer. *Environ. Microbiol.* **2006,** *8* (3), 455–465.

Malavasi, V.; Soru, S.; Cao, G. Extremophile Microalgae: The Potential for Biotechnological Application. *Journal of Phycology* **2020**.

Malyan, S. K.; Kumar, A.; Baram, S.; Kumar, J.; Singh, S.; Kumar, S. S.; Yadav, A. N. Role of Fungi in Climate Change Abatement through Carbon Sequestration. In *Recent Advancement in White Biotechnology through Fungi*. Springer, Cham, 2019; pp 283–295.

Mesa, L.; González, E.; Cara, C.; González, M.; Castro, E.; Mussatto, S. I. The Effect of Organosolv Pretreatment Variables on Enzymatic Hydrolysis of Sugarcane Bagasse. *Chem. Eng. J.* **2011,** *168* (3), 1157–1162.

Mingjun, S. H. A. N.; Yanqiu, W. A. N. G.; Xue, S. H. E. N. Study on Bioremediation of Eutrophic Lake. *J. Environ. Sci.* **2009,** *21,* S16–S18.

Mishra, J.; Singh, R.; Arora, N. K. Alleviation of Heavy Metal Stress in Plants and Remediation of Soil by Rhizosphere Microorganisms. *Front. Microbiol.* **2017,** *8,* 1706.

Mistry, A. N.; Ganta, U.; Chakrabarty, J.; Dutta, S. A Review on Biological Systems for CO_2 Sequestration: Organisms and Their Pathways. *Environ. Progress Sustain. Energy* **2019,** *38* (1), 127–136.

Mondal, M.; Goswami, S.; Ghosh, A.; Oinam, G.; Tiwari, O. N.; Das, P.; … Halder, G. N. Production of Biodiesel from Microalgae through Biological Carbon Capture: A Review. *3 Biotech* **2017,** *7* (2), 99.

Nguyen, T. H. M. Bioethanol Production from Marine Algae Biomass: Prospect and Troubles. *J. Vietnamese Environ.* **2012,** *3* (1), 25–29.

Noraini, M. Y.; Ong, H. C.; Badrul, M. J.; Chong, W. T. A Review on Potential Enzymatic Reaction for Biofuel Production from Algae. *Renew. Sustain. Energy Rev.* **2014,** *39,* 24–34.

Ojuederie, O. B.; Babalola, O. O. Microbial and Plant-Assisted Bioremediation of Heavy Metal Polluted Environments: A Review. *Int. J. Environ. Res. Public Health* **2017,** *14* (12), 1504.

Osono, T. Role of Phyllosphere Fungi of Forest Trees in the Development of Decomposer Fungal Communities and Decomposition Processes of Leaf Litter. *Can. J. Microbiol.* **2006,** *52* (8), 701–716.

Ossai, I. C.; Ahmed, A.; Hassan, A.; Hamid, F. S. Remediation of Soil and Water Contaminated with Petroleum Hydrocarbon: A Review. *Environ. Technol. Innov.* **2020,** *17,* 100526.

Özçimen, D.; İnan, B.; Biernat, K. An Overview of Bioethanol Production from Algae. In *Biofuels—Status and Perspective*. InTech, 2015.

Pathak, K. V.; Keharia, H. Application of Extracellular Lipopeptide Biosurfactant Produced by Endophytic Bacillus Subtilis K1 Isolated from Aerial Roots of Banyan (Ficus Benghalensis) in Microbially Enhanced Oil Recovery (MEOR). *3 Biotech* **2014,** *4* (1), 41–48.

Prenafeta-Boldú, F. X.; Vervoort, J.; Grotenhuis, J. T. C.; Van Groenestijn, J. W. Substrate Interactions during the Biodegradation of Benzene, Toluene, Ethylbenzene, and Xylene (BTEX) Hydrocarbons by the Fungus Cladophialophora sp. Strain T1. *Appl. Environ. Microbiol.* **2002,** *68* (6), 2660–2665.

Pusey, P. L.; Stockwell, V. O.; Reardon, C. L.; Smits, T. H. M.; Duffy, B. Antibiosis Activity of Pantoea Agglomerans Biocontrol Strain E325 against Erwinia Amylovora on Apple Flower Stigmas. *Phytopathology* **2011,** *101* (10), 1234–1241.

Rastogi, G.; Coaker, G. L.; Leveau, J. H. New Insights into the Structure and Function of Phyllosphere Microbiota through High-Throughput Molecular Approaches. *FEMS Microbiol. Lett.* **2013,** *348* (1), 1–10.

Read, D. J.; Perez-Moreno, J. Mycorrhizas and Nutrient Cycling in Ecosystems–A Journey towards Relevance? *New Phytol.* **2003,** *157* (3), 475–492.

Richmond, A., Ed. *Handbook of Microalgal Culture: Biotechnology and Applied Phycology;* John Wiley & Sons, 2008.

Ross, A. B.; Jones, J. M.; Kubacki, M. L.; Bridgeman, T. Classification of Macroalgae as Fuel and Its Thermochemical Behaviour. *Bioresour. Technol.* **2008,** *99* (14), 6494–6504.

Samarghandi, M. R.; Arabestani, M. R.; Zafari, D.; Rahmani, A. R.; Afkhami, A.; Godini, K. Bioremediation of Actual Soil Samples with High Levels of Crude Oil Using a Bacterial Consortium Isolated from Two Polluted Sites: An Investigation of the Survival of the Bacteria. *Global Nest J.* **2018,** *20* (2), 432–438.

Saravanan, A.; Jeevanantham, S.; Narayanan, V. A.; Kumar, P. S.; Yaashikaa, P. R.; Muthu, C. M. Rhizoremediation–A Promising Tool for the Removal of Soil Contaminants: A Review. *J. Environ. Chem. Eng.* **2020,** *8* (2), 103543.

Satish, L.; Rameshkumar, R.; Rathinapriya, P.; Pandian, S.; Rency, A.S.; Sunitha, T.; Ramesh, M. Effect of Seaweed Liquid Extracts and Plant Growth Regulators on in Vitro Mass Propagation of Brinjal (*Solanum Melongena* L.) through Hypocotyl and Leaf Disc Explants. *J. Appl. Phycol.* **2015,** *27* (2), 993–1002.

Satish, L.; Rathinapriya, P.; Rency, A. S.; Ceasar, S. A.; Pandian, S.; Rameshkumar, R.; Ramesh, M. Somatic Embryogenesis and Regeneration Using *Gracilaria Edulis* and *Padina Boergesenii* Seaweed Liquid Extracts and Genetic Fidelity in Finger Millet (*Eleusine Coracana*). *J. Appl. Phycol.* **2016,** *28* (3), 2083–2098.

Savage, P. E. Algae under Pressure and in Hot Water. *Science* **2012,** *338* (6110), 1039–1040.

Schneider, R. C.; Bjerk, T. R.; Gressler, P. D.; Souza, M. P.; Corbellini, V. A.; Lobo, E. A. Potential Production of Biofuel from Microalgae Biomass Produced in Wastewater. *Biodiesel–Feedstocks, Prod. App.* **2013.**

Shanahan, P. *Bioremediation. Waste Containment and Remediation Technology*; Spring, Massachusetts Institute of Technology, 2004.

Sharma, B.; Dangi, A. K.; Shukla, P. Contemporary Enzyme-Based Technologies for Bioremediation: A Review. *J. Environ. Manage.* **2018,** *210*, 10–22.

Shawabkeh, R.; Khleifat, K. M.; Al-Majali, I.; Tarawneh, K. Rate of Biodegradation of Phenol by Klebsiella Oxytoca in Minimal Medium and Nutrient Broth Conditions. *Bioremediation J.* **2007,** *11* (1), 13–19.

Shin, D. Y.; Cho, H. U.; Utomo, J. C.; Choi, Y. N.; Xu, X.; Park, J. M. Biodiesel Production from Scenedesmus Bijuga Grown in Anaerobically Digested Food Wastewater Effluent. *Bioresour. Technol.* **2015,** *184*, 215–221.

Shishir, T. A.; Mahbub, N. Review on Bioremediation: A Tool to Resurrect the Polluted Rivers. *Pollution* **2019,** *5* (3), 555–568.

Singh, B. K.; Bardgett, R. D.; Smith, P.; Reay, D. S. Microorganisms and Climate Change: Terrestrial Feedbacks and Mitigation Options. *Nat. Rev. Microbiol.* **2010,** *8* (11), 779–790.

Spolaore, P.; Joannis-Cassan, C.; Duran, E.; Isambert, A. Commercial Applications of Microalgae. *J. Biosci. Bioeng.* **2006,** *101* (2), 87–96.

Szabó, C.; Ischiropoulos, H.; Radi, R. Peroxynitrite: Biochemistry, Pathophysiology and Development of Therapeutics. *Nat. Rev. Drug Disc.* **2007,** *6* (8), 662–680.

Takeda, H.; Yoneyama, F.; Kawai, S.; Hashimoto, W.; Murata, K. Bioethanol Production from Marine Biomass Alginate by Metabolically Engineered Bacteria. *Energy Environ. Sci.* **2011,** *4* (7), 2575–2581.

Tang, J. C.; Wang, R. G.; Niu, X. W.; Wang, M.; Chu, H. R.; Zhou, Q. X. Characterisation of the Rhizoremediation of Petroleum-Contaminated Soil: Effect of Different Influencing Factors. *Biogeosciences* **2010,** *7* (12), 3961.

Tremblay, J.; Yergeau, E.; Fortin, N.; Cobanli, S.; Elias, M.; King, T. L.; … Greer, C. W. Chemical Dispersants Enhance the Activity of Oil-and-Gas Condensate-Degrading Marine Bacteria. *ISME J.* **2017,** *11* (12), 2793–2808.

Ubogu, M.; Odokuma, L. O.; Akponah, E. Enhanced Rhizoremediation of Crude Oil–Contaminated Mangrove Swamp Soil Using Two Wetland Plants (Phragmites Australis and Eichhornia Crassipes). *Braz. J. Microbiol.* **2019,** *50* (3), 715–728.

Uduman, N.; Qi, Y.; Danquah, M. K.; Forde, G. M.; Hoadley, A. Dewatering of Microalgal Cultures: A Major Bottleneck to Algae-Based Fuels. *J. Renew. Sustain. Energy* **2010,** *2* (1), 012701.

Uqab, B.; Mudasir, S.; Nazir, R. Review on Bioremediation of Pesticides. *J. Bioremediat. Biodegrad.* **2016,** *7* (343), 2.

Vacher, C.; Hampe, A.; Porté, A. J.; Sauer, U.; Compant, S.; Morris, C. E. The Phyllosphere: Microbial Jungle at the Plant–Climate Interface. *Annu. Rev. Ecol. Evol. Systemat.* **2016,** *47*, 1–24.

Valls, M.; De Lorenzo, V. Exploiting the Genetic and Biochemical Capacities of Bacteria for the Remediation of Heavy Metal Pollution. *FEMS Microbiol. Rev.* **2002,** *26* (4), 327–338.

Vorholt, J. A. Microbial Life in the Phyllosphere. *Nat. Rev. Microbiol.* **2012,** *10* (12), 828–840.

Wargacki, A. J.; Leonard, E.; Win, M. N.; Regitsky, D. D.; Santos, C. N. S.; Kim, P. B.; … Lakshmanaswamy, A. An Engineered Microbial Platform for Direct Biofuel Production from Brown Macroalgae. *Science* **2012**, *335* (6066), 308–313.

Wei, X.; Lyu, S.; Yu, Y.; Wang, Z.; Liu, H.; Pan, D.; Chen, J. Phylloremediation of Air Pollutants: Exploiting the Potential of Plant Leaves and Leaf-Associated Microbes. *Front. Plant Sci.* **2017**, *8*, 1318.

Whipps, J.; Hand, P.; Pink, D.; Bending, G. D. Phyllosphere Microbiology with Special Reference to Diversity and Plant Genotype. *J. Appl. Microbiol.* **2008**, *105* (6), 1744–1755.

Wi, S. G.; Kim, H. J.; Mahadevan, S. A.; Yang, D. J.; Bae, H. J. The Potential Value of the Seaweed Ceylon Moss (Gelidium Mansion) as an Alternative Bioenergy Resource. *Bioresour. Technol.* **2009**, *100* (24), 6658–6660.

Wood, J. L.; Liu, W.; Tang, C.; Franks, A. E. Microorganisms in Heavy Metal Bioremediation: Strategies for Applying Microbial-Community Engineering to Remediate Soils. *AIMS Bioeng.* **2016**, *3* (2), 211.

Yao, H.; Sun, X.; He, C.; Maitra, P.; Li, X. C.; Guo, L. D. Phyllosphere Epiphytic and Endophytic Fungal Community and Network Structures Differ in a Tropical Mangrove Ecosystem. *Microbiome* **2019**, *7* (1), 57.

Yasin, N. H. M.; Maeda, T.; Hu, A.; Yu, C. P.; Wood, T. K. CO2 Sequestration by Methanogens in Activated Sludge for Methane Production. *Appl. Energy* **2015**, *142*, 426–434.

Zeng, Q.; Holder, J. W.; Mahan, A. E.; Brigham, C. J.; Budde, C. F.; Rha, C.; Sinskey, A. J. Elucidation of Beta-Oxidation Pathways in RalstoniaEutropha H16 by Examination of Global Gene Expression, 2010.

Zhang, A.; Carroll, A. L.; Atsumi, S. Carbon Recycling by Cyanobacteria: Improving CO2 Fixation through Chemical Production. *FEMS Microbiol. Lett.* **2017**, *364* (16), fnx165.

Zinicovscaia, I.; Cepoi, L.. Eds. *Cyanobacteria for Bioremediation of Wastewaters*; Springer International Publishing, 2016.

Index

C

For Product Safety Concerns and Information please contact our EU
representative GPSR@taylorandfrancis.com
Taylor & Francis Verlag GmbH, Kaufingerstraße 24, 80331 München, Germany

www.ingramcontent.com/pod-product-compliance
Lightning Source LLC
Chambersburg PA
CBHW060340220326
41598CB00023B/2766